# Electronic and Electrical Servicing

# Electronic and Electrical Servicing

## Consumer and commercial electronics

Second Edition

*Ian Sinclair*

and

*John Dunton*

ELSEVIER

AMSTERDAM • BOSTON • HEIDELBERG • LONDON • NEW YORK • OXFORD
PARIS • SAN DIEGO • SAN FRANCISCO • SINGAPORE • SYDNEY • TOKYO
Newnes is an imprint of Elsevier

Newnes

Newnes is an imprint of Elsevier Ltd
Linacre House, Jordan Hill, Oxford OX2 8DP
30 Corporate Road, Burlington, MA 01803

First published 2002
Reprinted 2003
Second edition 2007



**British Library Cataloguing in Publication Data**
A catalogue record for this book is available from the British Library

**Library of Congress Cataloguing in Publication Data**
A catalogue record for this book is available from the Library of Congress

ISBN: 978-0-7506-6988-7

For information on all Newnes publications
visit our web site at www.books.elsevier.com

Typeset by Charon Tec Ltd (A Macmillan Company), Chennai, India
www.charontec.com

Printed and bound in Great Britain

# Contents

# Preface to the second edition

This new edition of *Electronic and Electrical Servicing* reflects the rapid changes that are taking place within the electronics industry. In particular, we have to recognise that much of the equipment that requires servicing will be of older design and construction; by contrast, some modern equipment may require to be replaced under guarantee rather than be serviced. We also need to bear in mind that servicing some older equipment may be totally uneconomical, because it will cost more than replacement. With all this in mind, this new edition still provides information on older techniques, but also indicates how modern digital systems work and to what extent they can be serviced.

This volume is intended to provide a complete and rigorous course of instruction for Level 2 of the City & Guilds Progression Award in Electrical and Electronics Servicing – Consumer/Commercial Electronics (C&G 6958). For those students who wish to progress to Level 3, a further set of chapters covering all of the core units at this level is available as free downloads from the book's companion website or as a print-on-demand book with ISBN 978-0-7506-8732-4.

**Companion website**

Level 3 material available for free download from
http://books.elsevier.com/companions/9780750669887

# Acknowledgements

The development of this series of books has been greatly helped by the City & Guilds of London Institute (CGLI), the Electronics Examination Board (EEB) and the Engineering & Marine Training Authority (EMTA). We are also grateful to the many manufacturers of electronics equipment who have provided information on their websites.

*Ian Sinclair*
*John Dunton*

# Unit 1
# D.c. technology, components and circuits

## Outcomes

1. Demonstrate an understanding of electrical units, primary cells and secondary cells and apply this knowledge in a practical situation
2. Demonstrate an understanding of cables, connectors, lamps and fuses and apply this knowledge in a practical situation
3. Demonstrate an understanding of resistors and potentiometers and apply this knowledge in a practical situation

   Health and Safety. Note: The content of this topic has been placed later, as Chapter 25.

# 1 Direct current technology

Electric current consists of the flow of small particles called electrons in a circuit. Its rate of flow is measured in units called **amperes**, abbreviated either to 'amps' or 'A'. One ampere is the amount of electric charge, in units called coulombs, that passes a given point in a circuit per second. The coulomb has a value of about $6.289 \times 10^{18}$ electrons ($10^{18}$ means a 1 followed by 18 zeros). The measurement of current is done, not by actually counting these millions of millions of millions of electric charges, but by measuring the amount of force that is exerted between a magnet and the wire carrying the current that is being measured. Current can flow in any material that allows electrons to move, but in such materials there is always some resistance to the flow of current (except for materials called *superconductors*). Resistance is measured in units called ohms, symbol $\Omega$ (the Green letter omega).

> Electric current can be direct current (d.c.) or alternating current (a.c.) or a mixture of both

Direct current is a steady flow of current, the type that occurs in a circuit fed by a battery. This type of current is used to operate most types of electronic circuit. Alternating current is not a steady flow; it is a current that rises to a peak in one direction, reverses and reaches a peak in the opposite direction and reverses again. This means that at times the current becomes zero and at other times it can be flowing in either direction.

Electronics makes use of a.c. with much smaller times for one cycle, and we usually prefer to refer to the number of cycles in a second, a quantity called **frequency**, rather than the time of one cycle. The unit of frequency is one complete cycle per second, called 1 hertz, abbreviation Hz. Looked at this way, the mains frequency is 50 Hz. The frequencies used for radio broadcasting are measured in millions of hertz, MHz. Computers typically work with thousands of millions of hertz, gigahertz, abbreviation GHz. In following chapters we'll look at how a.c. behaves in circuits and the differences between a.c. and d.c.

In the same way as a pressure is needed to cause a flow of water through a pipe, so an electrical 'pressure' called **electromotive force** or **voltage** is needed to push a current through a resistance. Electromotive force (emf) is measured in units of volts, symbol V. A voltage is always present when a current is flowing through a resistance, and the three quantities of volts, amps and ohms (the unit of resistance) are related. Like current, voltage can be direct or alternating.

The ampere is a fairly large unit, and for most electronics purposes the smaller units milliamp (one-thousandth of an ampere) and microamp (one-millionth of an ampere) are more generally used. The abbreviations for these qualities are mA and μA, respectively. All electrical units can use the same set of smaller units **(submultiples)** and larger units (**multiples**), and some of the most common are listed in Table 1.1. The abbreviation list is of **SI prefixes**, the standard letters used to indicate the multiple or submultiple.

**Table 1.1** Multiples and submultiples

| *Number* | *Power* | *Written as* | *Abbreviation* |
|---|---|---|---|
| 0.000 000 000 001 | $10^{-12}$ | pico- | p |
| 0.000 000 001 | $10^{-9}$ | nano- | n |
| 0.000 001 | $10^{-6}$ | micro- | μ |
| 0.000 01 | $10^{-3}$ | milli- | m |
| 1000 | $10^{3}$ | kilo- | k |
| 1 000 000 | $10^{6}$ | mega- | M |

Two simple examples will help to show how the system works. A current flow of 0.015 amperes can be more simply written as 15 mA (milliamps), which is $15 \times 10^{-3}$ A. A resistance of 56 000 ohms, which is equal to $56 \times 10^{3}$ ohms, is written as 56 k (**k** for kilohms). The ohm sign Ω is often left out.

Do not use K for kilo, because the K abbreviation is used for temperatures measured in kelvin. You may see the K in some circuit diagrams that were drawn before agreement was reached on how to represent kilohms.

There are two other important electrical units, of **energy** and of **power**. The energy unit is called the **joule**, abbreviation J, and it measures the amount of work that an electrical current can do, such as in an electric motor or a heater. The power unit measures the rate of doing work, which is the amount of work per second, and its unit is the **watt**, symbol W. Both the units can also make use of the multiples and submultiples in Table 1.1.

## Circuits and current

An electric **circuit** is a closed path made from conducting material. When the path is not closed, it is an **open circuit**, and no current can flow. In a circuit that contains a battery and a lamp, for example, the lamp will light when the circuit is closed, and we take the direction of the conventional flow of current as from the positive (+) pole of the battery to the negative (−). This convention was agreed centuries ago, and we now know that the movement of electrons is in the opposite direction, from negative to positive. For most purposes, we stay with the old convention, but for some purposes in electronics we need to know the direction of the electron flow.

An electrical circuit such as the lamp and battery can be shown in two ways. One is to draw the battery and the bulb as they would appear to the eye. The other is to draw the shape of the circuit, representing items such as the battery and the lamp, the components of the circuit, as symbols, and the conductor as a line. We draw these **circuit diagrams** to show the path that the current takes, because this is more important than the appearance of the components. To avoid confusion, there are some rules (conventions) about drawing these circuits.

- A line represents a conductor
- Where lines cross, the conductors are NOT joined
- Where two lines meet in a T junction, with or without a dot, conductors are connected.

Figure 1.1 shows some symbols that are used for common components. Most of these use two connections only, but a few use three or more. These symbols are UK [British Standard (BS)] and European standards, but circuit diagrams from the USA and Japan may use the alternative symbols for resistors and capacitors.

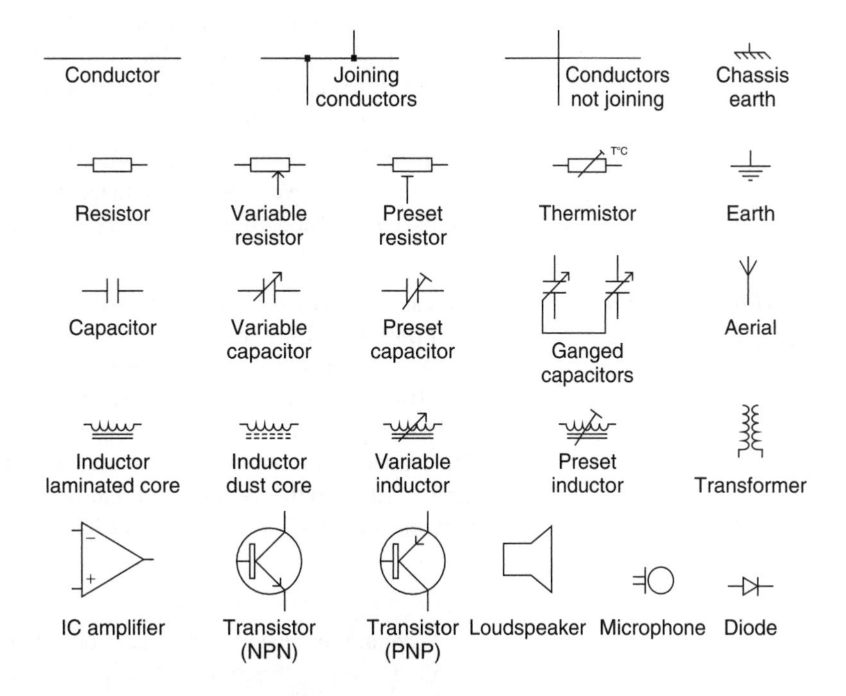


**Figure 1.1** Some symbols used in UK circuit diagrams

Circuit diagrams are important because they are one of the main pieces of information about a circuit, whether it is a circuit for the wiring of a house or the circuit for a television receiver. For servicing purposes you must be able to read a circuit diagram and work out the path of currents.

## Effects of current

Electric current causes three main effects, which have been known for several hundred years.

- **Heating effect**: when a current flows through a conductor, heat is generated so that the temperature of the conductor rises.
- **Magnetic effect**: when a current flows through a conductor it causes the conductor to become a magnet.
- **Chemical effect**: when a current flows through a chemical solution it can cause chemical separation (in addition to heating and magnetism).

All of these effects can be either useful or undesirable. We use the heating effect in electric fires and cookers, but we try to minimize the loss of energy from transmission cables by using high voltages with low current for transmission. The heating effect is the same whether the current is d.c. or a.c. The magnetic effect is used in electric motors, relays and solenoids (meaning magnets that can be switched on and off). Less desirable effects include the unwanted interference that comes from the magnetic fields created around wires. One notable wanted chemical effect is that of chemical energy being converted to electrical energy in a cell or battery. Some other chemical effects are, however, undesirable. A current that is passed through a solution of a salty material dissolved in water will cause a chemical change in the solution which can release corrosive substances. This is the effect that causes electrolytic corrosion, particularly on electrical equipment that is used in ships.

All of these effects have been used at one time or another to measure electric current. We now use the magnetic effect in the older type of instruments, but the modern digital meters work on quite different principles.

## Working with numbers

We can write any denary number as a number between 1 and 10 multiplied by a **power** of 10. For example, the number 100 is a denary number, equal to 10 tens, and we count in multiples (powers) of 10, using 1000 ($10 \times 10 \times 10$), 10 000, 100 000 and so on. The number 89 is a denary number, equal to eight tens plus nine units. The number 0.2 is also a denary number equal to 2/10, a decimal fraction. The number 255 can be written as $2.55 \times 100$ (or $2.55 \times 10^2$) and this is often useful in calculations because it avoids the need to work with numbers that contain a large set of zeros.

> A **denary** number is one that is either greater than unity, such as 2, 50 or 350, or a fraction such as 1/10, 3/40 or 7/120 that is made up of digits 0–9 only.

Denary numbers can be added, subtracted, multiplied and divided digit by digit, starting with the least significant figures (the units of a whole number, or the figure farthest to the right of the decimal point of a fraction), and then working left towards the most significant figures.

**Decimals** are denary numbers that are fractions of 10, so that the number we write as 0.2 means 2/10, and the number we write as 3.414 is 3 + 414/1000. The advantage of using decimals is that we can add, subtract, multiply and divide with them using the same methods as for whole numbers. Even the simplest of calculators can work with decimal numbers.

The numbers 0.047, 47 and 47 000 are all denary numbers. Each of them consists of the two figures 4 and 7, along with a power of 10 which is shown by zeros put in either before or after the decimal point (or where a decimal point would be). The number 0.047 is the fraction 47/1000, and 47 000 is $47 \times 1000$. The figures 4 and 7 are called the significant figures of all these numbers, because the zeros before or after them simply indicate a power of 10. Zero can be a significant figure if it lies between two other significant figures as, for example, in the numbers 407 and 0.407. The zeros in a number are not significant if they follow the significant figures, as in 370 000, or if they lie between the decimal point and the significant figures, as in 0.000 23.

Powers of 10 are always written in this index form, as shown in Table 1.2. A positive index means that the number is greater than one (unity), and a negative index means that the number is less than unity; for instance, the number $1.2 \times 10^3 = 1200$, the number $47 \times 10^{-2} = 0.047$, and so on.

**Table 1.2** Powers of 10 in index form

| *Number* | *Power* | *Written as* |
|---|---|---|
| 1/1 000 000 or 0.000 001 | −6 | $10^{-6}$ |
| 1/100 000 or 0.000 01 | −5 | $10^{-5}$ |
| 1/10 000 or 0.0001 | −4 | $10^{-4}$ |
| 1/1000 or 0.001 | −3 | $10^{-3}$ |
| 1/100 or 0.01 | −2 | $10^{-2}$ |
| 1/10 or 0.1 | −1 | $10^{-1}$ |
| 1 | 0 | $10^{0}$ |
| 10 | 1 | $10^{1}$ |
| 100 | 2 | $10^{2}$ |
| 1000 | 3 | $10^{3}$ |
| 10 000 | 4 | $10^{4}$ |
| 100 000 | 5 | $10^{5}$ |
| 1 000 000 | 6 | $10^{6}$ |

The British Standard (BS) system of marking values of resistance (BS1852/1977) uses the standard prefix letters such as k and M, but with a few changes. The main difference is that the ohm sign (Ω) and the decimal point are **never** used. This avoids making mistakes caused by an unclear decimal point, or by a spot mark mistaken for a decimal point, or the Ω sign mistaken for a zero. This is particularly important for circuit diagrams that are likely to be used in workshop conditions. In this BS system, all values in ohms are indicated by the letter R, all values in kilohms by the letter k, and all values in megohms by M. These letters are then placed where the decimal point would normally be found, and the point is not used. Thus R47 = 0.47

ohms; 5k6 = 5.6 kilohms; 2M2 = 2.2 megohms, and so on. The BS system is illustrated throughout this book. In this system there is no space between the number and the letter. The BS value system is used also for capacitance values and for some voltage values such as the stabilized value of a Zener diode.

## Relationships between units

The electrical units of volts, amps and ohms are related, and the relationship is commonly known (not quite correctly) as Ohm's law, which as an equation is written as $V = R \times I$. In words, it means that the voltage measured across a given resistor (in volts) is equal to the value of the resistance (in ohms) multiplied by the amount of current flowing (in amperes). Any equation like this can be rearranged, using a simple rule:

> An equation is unaltered if the quantities on each side of the equals sign are multiplied or divided by the same amount.

For example, the Ohm's law equation can be rearranged, as illustrated in Figure 1.2, as either $R = V/I$ (resistance equals volts divided by current) or $I = V/R$ (current equals volts divided by resistance). We get the first of these by taking $V = R \times I$ and dividing both sides by $I$ to get $V/I = (R \times I)/I$. Because $I/I$ must be 1, this boils down to $V/I = R$ (the same as $R = V/I$). Now try for yourself the effect on $V = R \times I$ of dividing each side by $R$.

These equations are the most fundamentally important ones you will meet in all your work on electricity and electronics. In electrical circuits the units in which the law has been quoted (volts, amperes, ohms) should normally always be used; but in electronic circuits it is in practice much easier to measure resistance in k and current in mA. Ohm's law can be used in any of its forms when both $R$ and $I$ are expressed in these latter units, but the unit of voltage in these other expressions always remains the volt.

There are, therefore, two different combinations of units with which you can use Ohm's law as it stands: either VOLTS AMPERES OHMS or VOLTS MILLIAMPERES KILOHMS. Never mix the two sets of units. Do not use milliamperes with ohms, or amperes with kilohms. If in doubt, convert your quantities to volts, amps and ohms before using Ohm's law.

**Example:** What is the resistance of a resistor when a current of 0.1 A causes a voltage of 2.5 V to be measured across the resistor?

**Solution:** Express Ohm's law in the form in which the unknown quantity $R$ is isolated: $R = V/I$. Substitute the data in units of volts and amperes.

$$R = 2.5/0\ 1 = 25\ \Omega$$

**Example:** What value of resistance is present when a current of 1.4 mA causes a voltage drop of 7.5 V?

**Solution:** The current is measured in milliamps, so the answer will appear in kilohms. $R = VI = 7.5/1.4 = 5.36$ kilohms, or about 5k4.

**Example:** What current flows when a 6k8 resistor has a voltage of 1.2 V across its terminals?

**Solution:** The data is already in workable units, so substitute in $I = V/R$. Then $I = 1.2/6.8\,\text{A} = 0.176\,\text{A}$, or 176 mA.

**Example:** What current flows when a 4k7 resistor has a voltage of 9 V across its terminals?

**Solution:** With the value of the resistor quoted in kilohms, the answer will appear in milliamps. So substitute in $I = V/R$, and $I = 9/4.7 = 0.001\,915\,\text{A} = 1.915\,\text{mA}$.

The importance of Ohm's law lies in the fact that if only two of the three quantities current, voltage and resistance are known, the third of them can always be calculated by using the formula. The important thing is to remember which way up Ohm's law reads. Draw the triangle illustrated in Figure 1.2. Put $V$ at its Vertex, and $I$ and $R$ down below and you will never forget it. The formula follows from this arrangement automatically using a 'cover-up' procedure. Place a finger over $I$ and $V/R$ is left, thus $I = V/R$. The other ratios can be found in a similar way.

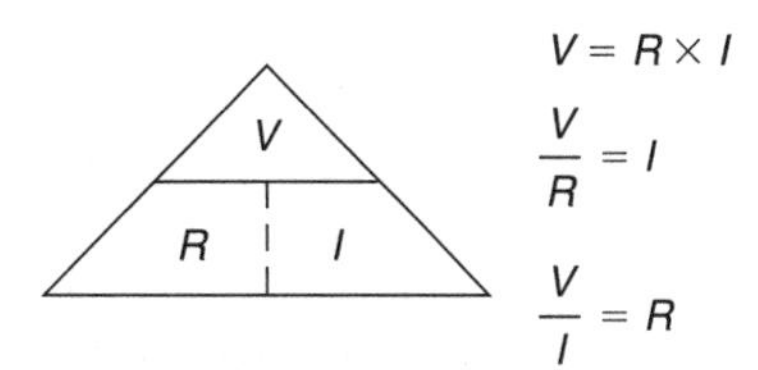


**Figure 1.2** The *V R I* triangle

## Work, power and energy

The related quantities of work, energy and power are often confused. Mechanical **work** is done whenever a force $F$ causes movement in the same direction as the force through a distance $d$. The force is measured in units of newtons (N) and 1 N is the force necessary to accelerate a mass of 1 kilogram by 1 metre per second per second ($1\,\text{m/s}^2$). Work is therefore the product of $F \times d$ (measured in newton metres), and this unit is called the **joule**. Work is also directly related to the **torque** or turning moment applied to a rotating shaft. The joule is also the unit of work that is used in electrical measurements.

Power is the rate at which work is done and is measured in **watts** (which are joules of work per second). Work also generates heat and this is also measured in watts. Electric motors were often specified by their work loading in horse power or brake horse power (HP or BHP) on the rating plate, where 1 HP is equivalent to 746 W. Therefore, a 1/2 HP a.c. motor would draw just over 1.5 A from the nominal 240 V supply mains.

**Energy** is the capacity to do work and, because it is easier to measure power, energy is often calculated as the product of power and time. Thus, 1 J is equal to 1 watt-second (**not** one watt per second but watts multiplied by seconds). This means that 1 kWh = 3600 kilojoules (kJ) or 3.6 megajoules (MJ).

When a current flows through a resistor, electrical energy is converted into heat energy, and this heat is passed on to the air around the resistor, and dissipated, spread around. The rate at which heat is dissipated, which is the rate of working, is **power**, and is measured in units of watts.

The amount of power dissipated can be calculated from any two of the quantities $V$ (in volts), $I$ (in amps) and $R$ (in ohms), as follows:

- Using $V$ and $I$ Power = $V \times I$ watts
- Using $V$ and $R$ Power = $V^2/R$ watts
- Using $I$ and $R$ Power = $I^2R$ watts

Most electronic circuits use small currents measured in mA, and large values of resistance measured in k, and we seldom know both volts and current. The power dissipated by a resistor is therefore often more conveniently measured in milliwatts using volts and k or using milliamps and k. Expressing the units $V$ in volts, $I$ in milliamps and $W$ in milliwatts, the equations to remember become:

The milliwatts dissipated = $V^2/R$ (using volts and k)
= $I^2R$ (using milliamps and k)

**Example:** How much power is dissipated when: (a) 6 V passes a current of 1.4 A, (b) 8 V is placed across 4 ohms, (c) 0.1 A flows through 15 R?

**Solutions:** (a) Using $V \times I$, Power = $6 \times 1.4 = 8.4$ W, (b) using $V^2/R$, Power = $8^2/4 = 64/4 = 16$ W, (c) using $I^2R$, Power = $0.1^2 \times 15 = 0.01 \times 15 = 0.15$ W.

**Example:** How much power is dissipated when (a) 9 V passes a current of 50 mA, (b) 20 V is across a 6k8 resistor, (c) 8 mA flows through a 1k5 resistor?

**Solutions:** (a) Using $V \times I$, Power = $9 \times 50 = 450$ mW, (b) using $V^2/R$, Power = $20^2/6.8 = 400/6.8 = 58.8$ mW, (c) using $I^2R$, Power = $8^2 \times 1.5 = 64 \times 1.5 = 96$ mW.

The amount of energy that is dissipated as heat is measured in joules. The watt is a rate of dissipation equal to the energy loss of one joule per second, so that joules = watts × seconds or watts = joules/second. The energy is found by multiplying the value of power dissipation by the amount of time during which the dissipation continues. The resulting equations are: Energy dissipated $= V \times I \times t$ joules or $V^2t/R$ joules or $I^2Rt$ joules, where $t$ is the time during which power dissipation continues, measured in seconds. In electronics you seldom need to make use of joules except in heating problems, or in calculating the stored energy of a capacitor.

Electrical components and appliances are rated according to the power that they dissipate or convert. A 3 W resistor, for example, will dissipate 3 J of energy per second; a 3 kW motor will convert 3000 J of energy per second into motion (if it is 100% efficient). As a general rule, the greater the power dissipation required, the larger the component needs to be.

## Calculations

At one time tables of values were used to help in solving complicated calculations, or calculations that used numbers containing many **significant figures**. Significant figures are the digits that need to be used in calculations, so that zeros ahead of or following other digits are not significant, but zeros between other digits are. For example, the zeros in 12 000 or 0.0053 are not significant because it is only the other digits, the 1 and 2 or 5 and 3, that we really need to work with. When you multiply 12 000 by 3, you don't need to start by thinking 'three times zero is zero, three times zero is zero' and so on. You simply think 'three times 12 (thousand) is 36 (thousand)'. Zeros in 26 005 **are** significant because they are part of the number.

Nowadays we use electronic calculators in place of tables, but a calculator is useful only if you know how to use it correctly. Calculators can be simple types that can carry out addition, subtraction, multiplication and division only, and these can be useful for most of your calculations.

To solve some of the other types of calculations you will meet in the course of electronics servicing, a scientific calculator is more useful. A good scientific calculator, such as the Casio, need not be expensive and it will be able to cope with any of the calculations that will need to be made throughout this course. You should learn from the manual for your calculator how to carry out calculations involving squares, square roots and powers, angle functions (particularly sines and cosines), and the use of brackets.

The **square** of a number means that number multiplied by itself. For example, 2 squared (written as $2^2$) is $2 \times 2 = 4$. Five squared is 25. It is simple enough for whole numbers, but when it comes to numbers with fractions, like $6.75^2$ (equal to 45.5625), then you need a calculator.

Many of the quantities used in electronics measurements are **ratios**, such as the ratio of the current flowing in the collector circuit of a transistor ($I_c$) to the current flowing in its base circuit ($I_b$). A ratio consists of one number divided by another, and can be expressed in several different ways:

- as a common fraction, such as 2/25
- as a decimal fraction, such as 0.47; this is the most common method
- as a percentage, such as 12% (which is another way of writing the fraction 12/100).

To convert a decimal fraction into a common fraction, first write the figures of the decimal, but not the point. For example, write 0.47 as 47. Now draw a fraction bar under this number (called the **numerator**) and under it write a power of 10 with as many zeros as there are figures above. In this example, you would use 100, with two zeros because there are two digits in 47. This makes the fraction 47/100.

To convert a common fraction into a decimal, do the division using a calculator. For example, the fraction 2/27 uses the 2, division and 27 keys and comes out as 0.074074, which you would round to 0.074.

To convert a decimal ratio into a percentage, shift the decimal point two places to the right, so that 0.47 becomes 47%. If there are empty places, fill them with zeros, so that 0.4 becomes 40%.

To convert a percentage to a decimal ratio, imagine a decimal point where the % sign was, and then shift this point two places to the left, so that 12% becomes 0.12. Once again, empty places are filled with zeros, so that 8% becomes 0.08.

## Averages

The **average value** of a set of numbers is found by adding up all the numbers in the set and then dividing by the number of items in the set. Suppose that a set of resistors has the following values: one 7R, two 8R, three 9R, four 10R, four 11R, three 12R and two 13R. This is a set of 19 values, and the average value of the set is found as follows:

$$\frac{\begin{bmatrix} 7+8+8+9+9+9+10+10+10+10+ \\ 11+11+11+11+12+12+12+13+13 \end{bmatrix}}{19} = \frac{196}{19}$$

This divides out to 10.32 (using two places of decimals), so that the average value of the set is 10.32 ohms or 10R32.

An average value like this is often not 'real', in the sense that there is no actual resistor in the set that has the average value of 10.32R. It is like saying that the average family size in the UK today is 2.2 children. This may be a perfectly truthful average value statement, but you will seldom meet a family containing two children and 0.2 of a third one.

## Chemical cells

Cells convert chemical energy into d.c. electrical energy without any intermediate stage of conversion to heat. Only a few chemical reactions can at present be harnessed in this way, although work on fuel cells has enabled electricity to be generated directly without any fuel having to be burned to provide heat. Cells and batteries, however, although important as a source of electrical energy for electronic devices, represent only a tiny (and expensive) fraction of the total electrical energy that is generated.

A cell converts chemical energy directly into electrical energy. A collection of cells is called a **battery**, but we often refer to a single cell as a 'battery'. Cells may be connected in **series** to increase the voltage available or in **parallel** to increase the current capacity, but parallel connection is usually undesirable because it can lead to the rapid discharge of all cells if one becomes faulty and the others pass current into the faulty cell.

Cells may be either primary or secondary cells. A **primary cell** is one that is ready to operate as soon as the chemicals composing it are put together.

Once the chemical reaction is finished, the cell is exhausted and can only be thrown away. A **secondary cell** generally needs to be charged by connecting it to a voltage higher than the output voltage of the cell before it can be used. Its chemical reaction takes place in one direction during charging, and in the other direction during discharge (use) of the cell. The cell can then be recharged.

Cells are classed according to their open-circuit voltage (usually 1.2–1.6 V, except for lithium cells) and their capacity. **Open circuit** means that nothing is connected to the cell that could allow current to flow. The **capacity** of a cell is its stored energy, measured in mA-hours. In principle, a cell rated at 500 mA-hours could supply 1 mA for 500 hours, 2 mA for 250 hours, 10 mA for 50 hours, and so on. In practice, the figure of energy capacity applies for small discharge currents and is lower when large currents are delivered.

Cells also have **internal resistance**, the resistance of the current-carrying chemicals and conducting metals in the cell. This limits the amount of current that the cell can deliver to a load, because even if the cell is short-circuited the internal resistance will limit the amount of current. Rechargeable cells usually have lower values of internal resistance than the non-rechargeable type.

Most primary cells are of the zinc/carbon (Leclanché) type, of which a cross-section is shown in Figure 1.3. The zinc case is sometimes steel coated to give extra protection. The ammonium chloride paste is an acidic material which gradually dissolves the zinc. This chemical action provides the energy from which the electrical voltage is obtained, with the zinc the negative pole.

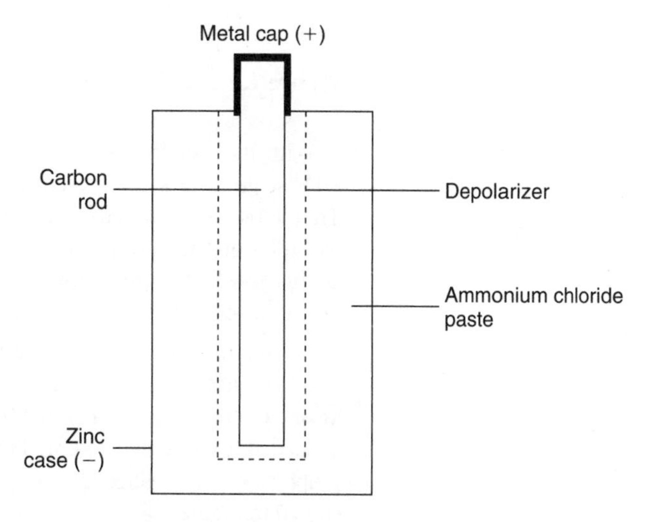


**Figure 1.3** A typical (Leclanché) dry cell construction

The purpose of the manganese dioxide *depolarizer* mixture that surrounds the carbon rod is to absorb hydrogen gas, a by-product of the chemical reaction. The hydrogen would otherwise gather on the carbon, insulating it so that no current could flow. The zinc/carbon cell is suitable for most purposes for which batteries are used, having a reasonable shelf-life and yielding a fairly steady voltage throughout a good working life.

Other types of cell such as alkaline manganese, mercury or silver oxide and lithium types are used in more specialized applications that need high working currents, very steady voltage or very long life at low current drains. However, mercury-based cells are not considered environmentally friendly when discarded unless they can be returned to the manufacturer. The use of a depolarizer is needed only if the chemical action of the cell has generated hydrogen, and some cell types do not.

## Practical 1.1

Connect the circuit of Figure 1.4(a) using a 9 V transistor radio battery. Draw up a table on to which readings of output voltage *V* and current *I* can be entered.

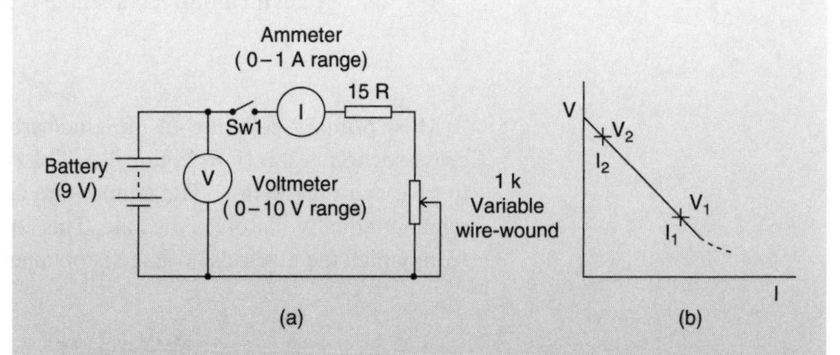


**Figure 1.4** (a) Circuit for Practical 1.1, and (b) graph

With the switch Sw1 open, note the voltmeter reading (using the 10 V scale). Mark the current column 'zero' for this voltage reading. Then close Sw1 and adjust the variable resistor until the current flow recorded on the current meter is 50 mA. Note the voltage reading *V* at this level of current flow, and record both readings on the table. Open switch Sw1 again as soon as the readings have been taken.

Go on to make a series of readings at higher currents (75 mA, 100 mA, etc.) until voltage readings of less than 5 V are being recorded. Take care that for every reading Sw1 remains closed for only as long as is needed to make the reading. Plot the readings you have obtained on a graph of output voltage against current. It should look like the example shown in Figure 1.4(b).

(Continued)

## Practical 1.1 (Continued)

Now pick from the table a pair of voltage readings *V*1 and *V*2, with *V*2 greater than *V*1, together with their corresponding current readings, *I*1 and *I*2, expressed in amperes. Work out the value of the expression shown, left, and you will get the internal resistance of the battery in units of ohms.

Note that false readings can be obtained if a cell passes a large current for more than a fraction of a second. Try to make your readings quickly when you are using currents approaching the maximum, and switch off the current as soon as you have taken a reading.

Most primary cells have an open-circuit voltage (or emf) of around 1.4–1.5 V. The important exception is the lithium cell (see later), which provides around 3.5 V. A lithium cell must **never** be opened, because lithium will burst into flames on exposure to air or water. Lithium cells should not be recharged or put into a fire.

Towards the end of the useful life of a cell or battery, the value of its internal resistance rises. This causes the output voltage at the terminals of the cell or battery to drop below its normal value when current flows through the cell or battery, which is then said to have poor **regulation**. A voltage check with this cell or battery removed from the equipment will show a normal voltage rating, but the cell or battery should nevertheless be replaced.

The only useful check on the state of a cell or battery is a comparison of voltage reading on load (with normal current flowing) with the known on-load voltage of a fresh cell. Simply reading the voltage of a cell that is not connected to a load is pointless.

At one time, the term **secondary cell** meant either the type of lead-acid cell which is familiar as the battery in a car, or the nickel–iron alkaline (NiFe) cell used in such applications as the powering of electric milk floats. In present-day electronics, both types have to some extent been superseded by the nickel–cadmium (NiCd or nicad), nickel–metal-hydride (Ni-MH) and lithium-ion secondary (meaning rechargeable) cells. There is, however, a large difference in the emf of secondary cells. The old lead-acid type has an emf of 2.0 V (2.2 V when fully charged), but the nickel–cadmium and Ni-MH types have a much lower emf of only 1.2 V, and the lithium-ion cell can provide around 3.6 V.

The NiCd cell uses as its active material cadmium (a metal like zinc), in powdered form, that is pressed or sintered into perforated steel plates, which then form the negative pole of the cell. The positive pole is a steel mesh coated with solid nickel hydroxide. The electrolyte is potassium hydroxide (caustic potash), usually in jelly form (Figure 1.5).

Nickel–cadmium (nicad) cells are sealed so that no liquids can be spilled from them, and they have a fairly long working life provided they are correctly used. They can deliver large currents, so they can be used for

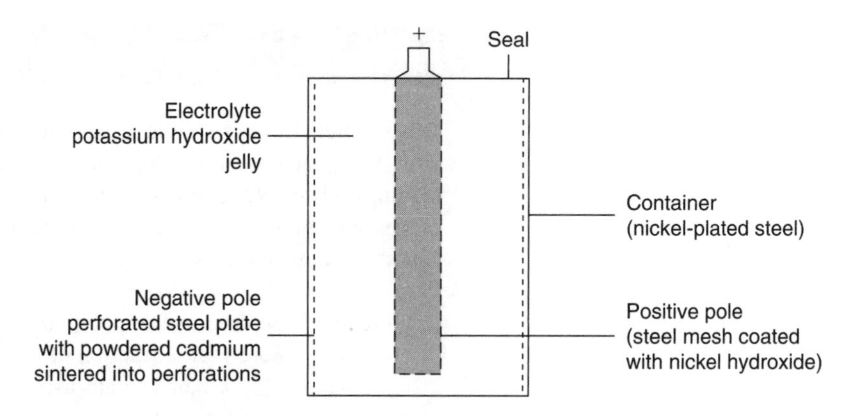


**Figure 1.5** The nickel–cadmium cell

equipment that demands higher power than could be supplied by primary cells. They have much longer life than other cells in applications for which they are rapidly discharged at intervals. Long periods of inactivity can cause the cells to fail, although it is often possible to restore their action by successive cycles of discharging and charging. One major problem is the memory effect, a reduction in capacity caused by recharging before the cell is fully discharged. Because of this, it is common practice to use a discharging cycle before charging a nickel–cadmium cell.

More recently, the sealed Ni-MH battery has been introduced. This type has up to 40% higher capacity than its nickel–cadmium counterpart of the same size, and also offers benefits of faster charge and discharge rates, and longer life. The Ni-MH cell contains no cadmium and is therefore more environmentally acceptable. The operating voltage is about the same as that of the NiCd cell. The memory effect can be greatly reduced if the Ni-MH battery is on occasions completely discharged before recharging. The usual recommendation is that this should be done after three to five normal charge–discharge sequences.

Table 1.3 shows the advantages and disadvantages of using batteries as power sources for electronic equipment, as compared to mains supplies.

**Table 1.3** Battery-operated equipment

| *Advantages* | *Disadvantages* |
|---|---|
| Equipment is portable | Limited energy capacity |
| An ordinary battery is smaller and lighter than is any form of connection to the main supply | Voltage generally low |
| Safer to use, no high voltages being involved | Batteries deteriorate during storage |
| Equipment requires no trailing leads | Batteries for high voltage or high current operation are heavier and more bulky than the equivalent mains equipment |

## Lithium-ion rechargeable cell

The lithium-ion rechargeable (Li-ion) cell avoids the direct use of the metal lithium because it oxidizes too readily. A lithium-ion cell must never be broken open. The carbon anode is formed from a mixture of compounds at about 1100°C and then electrochemically treated with a lithium compound. The cathode is formed from a mixture of the compounds of lithium, cobalt, nickel and manganese. A mixture of chemicals, avoiding the use of water, is used for the electrolyte. This cell is nominally rated at 3.6 V and has a long self-discharge period, typically falling by 30% after 6 months. The recharge period is typically about 3 hours and the cell can withstand at least 1200 charge–recharge cycles.

In addition to its advantages with holding increased energy, the Li-ion cell tends not to suffer from the memory effect, ensuring a longer life even when poorly treated. Its features include high energy density and high output voltage with good storage and cycle life. Lithium-ion cells are used in desktop personal computers (to back up memory), camcorders, cellular phones and also for portable compact disc (CD) players, laptop computers, personal digital assistants (PDAs) and similar devices. Lithium-ion cells have even been used in an electric sports car (the Venturi Fetish).

The cell operates on the principle that both charging and discharging actions cause lithium ions to transfer between the positive and negative electrodes. Unlike the action of other cells, the anode and cathode materials of the lithium-ion cell remain unchanged through its life.

## Charging cells

Lead-acid cells need to be recharged from a constant-voltage supply, so that when the cell is fully charged, its voltage is the same as that of the charger, and no more current passes. By contrast, nickel–cadmium cells must be charged at constant **current**, with the current switched off when the cell voltage reaches its maximum. Constant-current charging is needed so that excessive current cannot pass when the cell voltage is low. Using the wrong charging method can damage cells.

Nickel–metal-hydride cells need a more complicated charger circuit, and one typical method charges at around 10% of the maximum rate, with the charging ended after a set time. A charger suitable for Ni-MH cells can be used also for NiCd, but the opposite is not true. Some types of cell include a temperature sensor that will open-circuit the cell when either charge or discharge currents cause excessive heating. For some applications trickle charging at 0.03% of maximum can be used for an indefinite period.

Lithium-ion cells can be charged at a slow rate using trickle-chargers intended for other cell types, but for rapid charging they require a specialized charger that carries out a cycle of charging according to the manufacturer's instructions.

A completely universal battery charger needs to be microprocessor controlled and is an expensive item, although useful if you use a variety of different cells.

Table 1.4 compares primary and secondary cells.

**Table 1.4** Primary and secondary cells compared

| *Primary cells* | *Secondary cells* |
|---|---|
| Low cost | Expensive |
| Small size | Some fairly large |
| Short life | Comparatively long life |
| Throw away when exhausted | Rechargeable |
| Light weight | Generally heavier than equivalent primary cell |
| Readily available | Specialized products, less easily obtainable |

Capacitors with a value of about 1–5 farads (F), which can be charged to 5 V through a high value of resistance, can support a small (backup) discharge current for many hours. Modern construction provides a device of only about 5 mm high with the same diameter, so that these can be used to provide a backup power supply for circuits that need only a small current to maintain operation throughout short duration power failures.

## Connecting cells

When a set of cells is connected together, the result is a **battery**. The cells that form a battery could be connected in series, in parallel, or in any of the series–parallel arrangements, but in practice the connection is nearly always in series. The effect of both series and parallel connection can be seen in Figure 1.6. When the cells are connected in series, the open-circuit voltages (emfs) add, and so do the internal resistance values, so that the overall voltage is greater, but the current capability is the same as that of a single cell.

When the cells are connected in parallel, the voltage is as for one cell, but the internal resistance is much lower, because it is the result of several internal resistances in parallel. This allows much larger currents to be drawn, but unless the cells each produce *exactly* the same emf value, there is a risk that current will flow between cells, causing local overheating. For this reason, primary cells are never used connected in parallel, and even secondary cells, which are more able to deliver and to take local charging current, are seldom connected in this way.

Higher currents are therefore obtained by making primary cells in a variety of (physical) sizes, with the larger cells being able to provide more current, and having a longer life because of the greater quantity of essential chemicals. The limit to size is portability, because if a primary cell is not portable it has a limited range of applications. Secondary cells have much

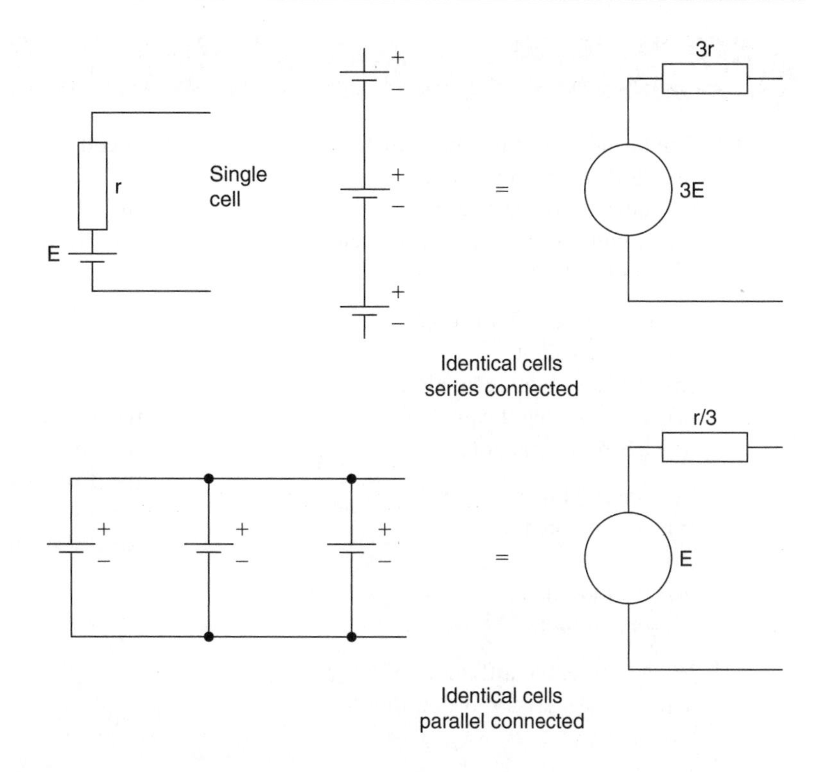


**Figure 1.6** Connecting cells in series and in parallel

lower internal resistance values, so that if high current capability is required along with small volume, a secondary cell is always used in preference to a primary cell. One disadvantage of the usual type of nickel–cadmium secondary cell in this respect, however, is a short *shelf-life*, so that if equipment is likely to stand for a long time between periods of use, secondary cells may not be entirely suitable, because they will always need to be recharged just before use.

The important parameters for any type of cell are its open-circuit voltage (the emf), its 'typical' internal resistance value, its shelf-life, active life and energy content. The **internal resistance** is the resistance of the electrolyte and other conductors in the cell, and its value limits the amount of current that a cell can provide because it causes the output voltage of the cell to drop when current flows. The **shelf-life** indicates how long a cell can be stored, usually at a temperature not exceeding 25°C, before the amount of internal chemical action seriously decreases the useful life. The **active life** is less easy to define, because it depends on the current drain, and it is usual to quote several figures of active life for various average current drain values. The **energy content** is defined as emf × current × active life, and will usually be calculated from the most favourable product of current and time. The energy content is affected more by the type of chemical reaction and the weight of the active materials than by details of design.

## Multiple-choice revision questions

1.1 The quantity 'voltage' measures:
(a) flow of electric current
(b) quantity of electric current
(c) driving force of electric current
(d) stored electric current.

1.2 In the number 537, the digit 7 is:
(a) a binary digit
(b) the least significant digit
(c) the most significant digit
(d) a decimal fraction.

1.3 The prefix M means:
(a) one hundred
(b) one thousand
(c) ten thousand
(d) one million.

1.4 When a potential difference of 6 V exists across a 1k0 resistor, the current flowing will be:
(a) 6 mA
(b) 3 A
(c) 0.16 A
(d) 1/6 mA.

1.5 A cell or a battery converts:
(a) heat energy into electrical energy
(b) electrical energy into light energy
(c) electrical energy into chemical energy
(d) chemical energy into electrical energy.

1.6 A nickel–cadmium (NiCd) cell must be recharged:
(a) from a source of constant current
(b) from a low-impedance source
(c) from a source of constant voltage
(d) from a high-impedance source.

# 2 Conductors, insulators, semiconductors and wiring

## Conductors and insulators

**Conductors** are materials that allow a steady electric current to flow easily through them, and which can therefore form part of a circuit in which a current flows. All metals are good conductors. Gases at low pressure (as in neon tubes) and solutions of salts, acids or alkalis in water will also conduct electric current well.

**Insulators** are materials that do not allow a steady electric current to flow through them and they are therefore used to prevent such a flow. Most of the insulators that we use are solid materials that are not metals. Natural insulators, such as sulphur and pitch, are no longer used, and plastic materials such as polystyrene and polythene have taken their place. Pure water is an insulator, but any trace of impurity will allow water to conduct some current, so that this provides one way of measuring water purity.

**Semiconductors** are materials whose ability to conduct current can be enormously changed by adding microscopic amounts of chemical elements. In a pure state, a semiconductor is an insulator, although light or high temperature will greatly lower the resistance.

A good example of the contrasting uses of insulators and conductors is provided by printed circuit boards. The boards are made of an insulator, typically stiff bonded paper called SRBP (synthetic-resin bonded paper), which is impregnated with a plastic resin; but the conducting tracks on the boards are made of conducting copper or from metallic inks. Boards made from fibreglass are used for more demanding purposes.

Both *insulation and conduction* are relative terms. A conductor that can pass very small currents may not conduct nearly well enough to be used with large currents. An insulator that is sufficient for the low voltage of a torch cell could be dangerously unsafe if it were used for the voltage of the mains (line) supply.

## Resistance

The amount of conduction or insulation of materials is measured by their **resistance**. A very low resistance means that the material is a conductor; a very high resistance means that the material is an insulator. The resistance of any sample of a substance measures the amount of opposition it presents to the flow of an electric current. A long strip of the material has more resistance than a short strip of the same material. A wide sample has less resistance than a narrow sample of the same length and same material. These effects are illustrated in Figure 2.1.

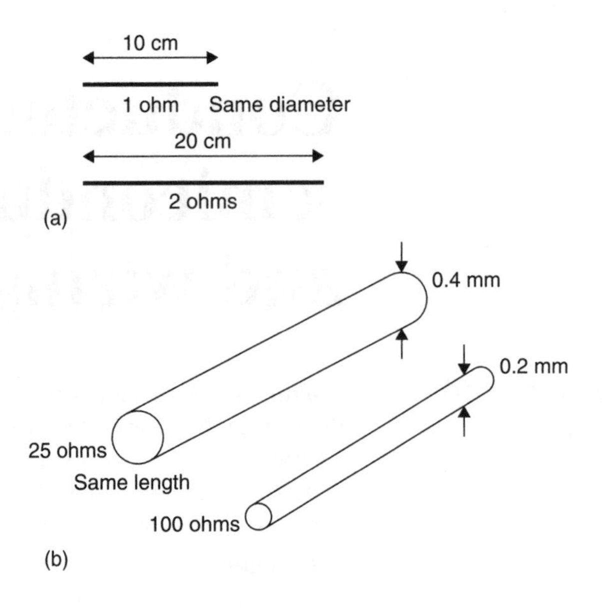


**Figure 2.1** Both wire length (a) and wire diameter (b) affect resistance

The resistance of a sample of a given substance therefore depends both on its dimensions and on the material itself. As a formula, this is

$$R \propto L/A$$

where $L$ represents the length of the sample and $A$ its area of cross-section. The sign $\propto$ means 'is proportional to'. The formula therefore reads in full as: 'Resistance is proportional to sample length divided by the sample's area of cross-section'. This assumes that the area of cross-section is constant, as is normal for a wire. Resistance is measured in units called **ohms** (symbol Ω, the Greek letter omega), which will be defined shortly.

We can use the formula at present by using proportion. To illustrate this, look at two examples.

**Example:** A 2 m long sample of wire has a resistance of 5 ohms. What resistance value would you expect for a length of 1.5 m of the same wire?

**Solution:** If a 2 m sample has a resistance of 5 ohms, the resistance of a 1 m sample must be 5/2 = 2.5 ohms. A 1.5 m length will therefore have a resistance of 2.5 × 1.5 = 3.75 ohms.

**Example:** A sample of wire of radius 0.2 mm has a resistance of 12 ohms. What would be the resistance of a sample of the same material of the same length but with radius 0.3 mm?

**Solution:** The area of a cross-section is proportional to the *square* of the radius. Therefore Resistance $A \times$ (Radius of $A$)$^2$ = Resistance $B \times$ (Radius of $B$)$^2$. Substitute the data, and simplify to get $12 \times 0.2^2 = R = 0.3^2$ so that $R = 5.33$ ohms.

The resistance value of a sample of material does not depend only on its dimensions, however, because every material has a different value of resistance per standard sample. This resistance per standard sample is called the **resistivity** of the material in question. It is measured in units called ohm-metres (ohms multiplied by metres, written as Ωm). The resistance of a sample of any material is then given by the formula:

$$R = \rho/A$$

where $\rho$ (the Greek letter called 'rho') represents the resistivity of the material, $L$ is the length, and $A$ is the area of cross-section.

**Example:** Copper has value of resistivity equal to $1.7 \times 10^{-8}$ Ωm. What is the resistance of 12 m of copper wire of 0.3 mm radius?

**Solution:** Substitute the data in the equation $R = \rho L/A$, putting all lengths into metres and remembering that $A = \pi r^2$, then:

$$R = \frac{1.7 \times 10^{-8} \times 12}{\pi \times (0.0003)^2} = \frac{20.4 \times 10^{-8}}{2.83 \times 10^{-7}} = 0.72 \text{ ohms}$$

The resistivity values of some common materials are listed in Table 2.1. The resistivity figures need to be multiplied by $10^8$ to give values in ohm-metres.

**Table 2.1** Resistivity of some common materials

| *Material* | *Resistivity* $\times$ $10^{-8}$ |
|---|---|
| Copper | 1.7 |
| Gold | 2.3 |
| Aluminium | 2.7 |
| Iron | 10.5 |
| Manganin | 43 |
| Mercury | 96 |
| Bakelite | $10^5$ |
| Glass | $10^{12}$ |
| Quartz | $10^{20}$ |
| PTFE | $10^{20}$ |

The material listed as **manganin** is a copper-based alloy containing manganese and nickel, much used in the construction of wire-wound resistors. PTFE is the high-resistivity plastic material whose full name is polytetrafluoroethylene.

Copper has a very low value of resistivity, so that copper is a good conductor and is used in electrical cables. Of all the metals, only silver has lower resistivity, but silver is too expensive to use for cables, although it is often used for small lengths of conductors. Aluminium is used for high-tension cables because an aluminium cable that has the same resistance as a copper cable is thicker but has less weight. Gold has a lower resistivity than aluminium and is used where the features of low resistivity and softness are particularly useful, such as on contacts.

### Practical 2.1

Calculate the resistance of 10 m of wire, diameter 0.3 $mm^2$ if a sample of the same material 50 cm long with diameter 0.2 $mm^2$ has a resistance of 0.5 ohms.

### Practical 2.2

Use a resistance meter and (if available) a Megger, to measure the resistance of a metre of copper wire, a metre of nichrome wire, and a square of paper.

## Alternating current mains supply

The mains supply to a house is alternating current (a.c.) at around 240 V, 50 Hz. This has been transformed down from the very high voltages (up to 250 000 V) used for distribution, but factories normally use a higher voltage supply of around 415 V a.c.

The wiring in a house or factory uses mains cable. For a house this consists of stranded copper made into three cores, each insulated and with an outer insulated polyvinyl chloride (PVC) cover for protection. House wiring uses a ring construction, so that each wiring socket is connected in a ring, taking current from all of the cables in the circuit. This scheme has been used for some 60 years, and it makes fusing and cabling simpler, as well as requiring less cable. Wiring in a factory is likely to use either three or four conductors in a cable, and it may require cables that are insulated with fireproof minerals and with a metal outer cover. In a house, electrical equipment will be connected through a flexible mains cable to a three-pin plug.

The conventional house sockets use **live**, **neutral** and **earth** connectors. The live connection provides the full 240 V a.c., and the neutral connection is used to complete the circuit. Any equipment that is plugged in will therefore be connected between the live and the neutral. The earth connection is taken to buried metalwork in the house, and is used to ensure that equipment with a metal casing can be used safely, because the casing will be connected through the three-pin plug, to earth. Wiring regulations specify that **protective multiple earthing (PME)** will be used, so that all metal pipes in a house must be connected using large-capacity cable to the same earth.

> The neutral wire is earthed at the source, but there may be several volts of a.c. between earth and neutral in a household. **Never** connect the neutral line to earth.

The mains supply, which in the UK and in most of Europe is at 240 V a.c., is the greatest hazard in most electronic servicing work. Three connections are made at the usual UK domestic supply socket, labelled live (L), neutral (N) and earth (E). The live contact at 240 V can pass current to either of the other two. The neutral connection provides the normal return path for current with the earth connection used as an emergency path for returning current in the event of a fault. Earth leakage circuit breakers (ELCBs) work by detecting any small current through the earth line and using this to operate a relay that will open the live connection, cutting off the supply. Connection to domestic supplies is made by a three-pin plug, either the BS type or the International Electrotechnical Commission (IEC) type of connector shown opened in Figure 2.2. The BS plug is designed so that shutters within the socket are raised only when the plug is correctly inserted.

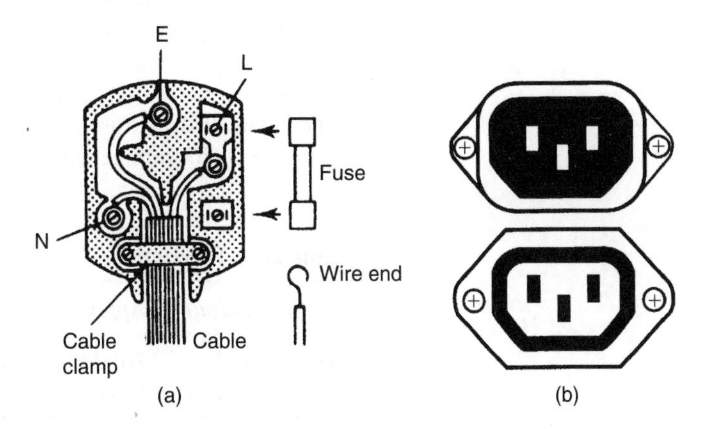


**Figure 2.2** (a) BS, and (b) IEC mains connectors

Table 2.2 lists the colour coding of the wire connections to the plug. Although it is many years since the coding colours were changed, older equipment can still be found bearing the older colours (and any such equipment should be checked carefully to ensure that it conforms to modern requirements for insulation). Note that these are the colour codes for flexible wiring; internal house wiring still uses the older red, black, green scheme.

**Table 2.2** Colour codes for flexible wiring

| | *Live* | *Neutral* | *Earth* |
|---|---|---|---|
| Modern | Brown | Blue | Green/yellow |
| Old colours | Red | Black | Green |

Particular care should be taken in working on cables that have been colour coded to non-UK standards, although all European equipment should now be using the same coding. A plug must be wired so that:

- the cable is firmly held and clamped without damaging the insulation or the conductors
- all connections are tight with no loose strands of wire; the ends of the cable can be coated with solder to prevent loose strands from separating, but the soldered end should not be used for clamping because the wire is more brittle and will loosen off after some time.

The wires should be cut to length so that the live lead will break and pull free before the earth lead if there is excessive force on the cable. The design of some plugs can make this very difficult, and you should select plugs that permit the use of a live lead that will pull out before the earth lead becomes strained.

A **fuse** of the correct rating must be used (Table 2.3). The standard fuse ratings for domestic equipment are 3 A, colour coded red, and 13 A, colour coded brown. Most domestic electronic equipment can use the 3 A rated fuses, and the few items that require a larger fuse should preferably be used with a (non-standard) 7 A fuse rather than the 13 A type, because the appliance cables for such equipment are seldom rated for 13 A.

**Table 2.3** Fuse ratings

| *Value* | *Colour* | *Applications* |
|---|---|---|
| 3 A | Red | All domestic electronic equipment; test gear |
| 13 A | Brown | Heaters and kettles |

**Note** that red and brown are easily confused if you are slightly colour-blind and working at low light levels. Always check wiring using a bright light. Use a torch if room lighting is inadequate.

The cable should be clamped where it enters the equipment, or alternatively a plug and socket of the standard type can be used so that the mains lead consists of a domestic plug at one end and an IEC socket (Eurosocket), *never a plug* at the other.

**Note** that non-domestic equipment often makes use of standard domestic plugs, but where higher power electronic equipment is in use the plugs and socket will generally be of types designed for higher voltages and current, often for three-phase 440 V a.c. In some countries, flat two-pin plugs and sockets are in use, with no earth provision except for cookers and washing machines, although eventually uniform standards should have prevailed in Europe, certainly for new buildings.

## Practical 2.3

Connect a domestic three-pin plug to a cable and check the resistance of the connections. Check the insulation between live, neutral and earth.

The circuit should also include:

- a fuse whose rating matches the consumption of the equipment. This fuse may have blowing characteristics that differ from those of the fuse in the plug. It may, for example, be a fast-blowing type that will blow when submitted to a brief overload, or it may be of the slow-blow type that will withstand a mild overload for a period of several minutes.
- a double-pole switch that breaks both live and neutral lines; the earth line must never be broken by a switch.
- a mains warning light or indicator which is connected between the live and neutral lines.

All these items should be checked as part of any servicing operation, on a routine basis. As far as possible all testing should be done on equipment that is disconnected and switched off. The absence of a pilot light or the fact that a switch is in the OFF position should never be relied on. Mains-powered equipment in particular should be completely isolated by unplugging from the mains. If the equipment is, like most non-domestic equipment, permanently wired then the fuses in the supply line must be removed before the covers are taken from the equipment. Many pieces of industrial electronics equipment have safety switches built into the covers so that the mains supply is switched off at more than one point when the covers are removed.

The UK has used the three-pin plug with rectangular pins for some time, but in the lifetime of most of us these may be replaced by the pattern used on the continent. The continental pattern is that most plugs are two-pin, with three used only for a few items such as electric files and cookers. There is no fuse in the plug, but an ELCB is incorporated into the socket to cut off current if any leakage to earth is detected; similar contact breakers are available in the UK for use with power tools out of doors.

Special regulations apply to electrical tools and other equipment that can be described as **double-insulated**. No earth connection is required for such equipment because the metal parts that are connected to the supply (such as the motor of a power tool) are insulated from the casing, even if the casing is metal. Equipment with plastic casings is normally of the double-insulated type of construction.

The current rating of domestic plugs in the UK is a maximum of 13 A, but for electronic equipment, fusing at 3 A is more common. Despite attempts to standardize only the 13 A and 3 A fuses, you can buy and use intermediate values such as 5 A and 7 A.

Fuses are pieces of thin wire in an insulating container, and the principle is that when excessive current flows the fuse will melt, and so cut off the current. A fuse will not melt immediately, and at the rated current it may take several minutes to blow. A fuse provides protection from a disastrous short-circuit that causes a large current to flow, but it provides no protection to people or to some types of electronic equipment. Contact breakers, which can break the circuit on a precise and very small value of excess current, are to an increasing extent replacing fuses as a method of protection.

## Signal connections

Cables that carry signals use different connectors, most of which are rated for low voltages only. Two very common types are the coaxial plug used for television aerial inputs, and the DIN plug for interconnections of audio equipment. Coaxial plugs are often wired without using solder, but soldering makes a more reliable contact. Only the inner lead of a coaxial cable is soldered, with the outer braid wrapped round and clamped. Great care should be taken not to melt the insulation of the plug by keeping the soldering iron in contact too long. It is often useful to insert the end of the plug into a coaxial socket when soldering so as to keep the centre pin in line when the heat of the soldering iron softens the insulator.

DIN plugs are more difficult to work with because of the number of connections (typically three or five) on one small plug. The body of the plug should be held in a vice, and the wire clamped to the pins before soldering. The soldered joint should be made quickly to avoid overheating.

Required tools for working on cables and terminations are screwdriver, tweezers, soldering iron and pliers. You should be familiar with the safe use of all these tools. For testing circuits you will need a good multimeter; a suitable instrument will have a low-resistance range for checking continuity and a high-resistance range for checking insulation.

### Practical 2.4

Connect (a) a coaxial cable to a coaxial plug, and (b) a five-pin DIN plug to a five-core cable. Carry out continuity and insulation tests on the cable and connectors.

Cables that carry signals are often of the type that uses an earthed outer screen to avoid interference (or to prevent the signals on the cable from interfering with other equipment). In some examples, such as coaxial cable, the outer screening is not earthed by a direct connection but only to a.c. Connector terminations for signal cables are almost all soldered, but screw terminations are more common for mains cables.

## Multiple-choice revision questions

2.1 The feature that is most typical of a semiconductor is:
(a) its resistivity value is between that of a conductor and an insulator
(b) its resistivity can be changed by adding tiny quantities of impurity
(c) it is composed of crystals
(d) it melts easily.

2.2 Two strips of the same metal A and B have the same lengths and thicknesses but A is half the width of B. The resistance ratio A to B is:
(a) 4:1
(b) 2:1
(c) 1:4
(d) 1:2.

2.3 The unit of resistivity is the:
(a) ohm
(b) ohm-metre
(c) ohm per metre
(d) metre.

2.4 Which of the following has the highest resistivity?
(a) aluminium
(b) PTFE
(c) wood
(d) mercury.

2.5 A distribution box uses an RCD. This will:
(a) replace the on/off switch
(b) blow when any appliance draws excessive current
(c) light up to indicate a fault
(d) disconnect when earth leakage occurs.

2.6 The three-pin plug on an audio amplifier lead should have:
(a) the green/yellow lead on the earth terminal and a 3 A fuse
(b) the brown lead on the earth terminal and a 13 A fuse
(c) the blue lead on the earth terminal and a 3 A fuse
(d) the green/yellow lead on the earth terminal and a 13 A fuse.

# 3 Resistors and resistive circuits

## Resistors

Resistors are components that have a stated fixed or maximum value of resistance. Methods of construction include wire-wound, carbon moulded, carbon film and metal film resistors. The two commonly used symbols for a resistor in a circuit are shown in Figure 3.1. The older zigzag symbol is still used, particularly in the USA and Japan, and the rectangular symbol is used in Europe, but both are British Standard symbols. Resistors can be fixed, with a fixed value of resistance, or variable, with a value that can be altered from almost zero to a set maximum value.

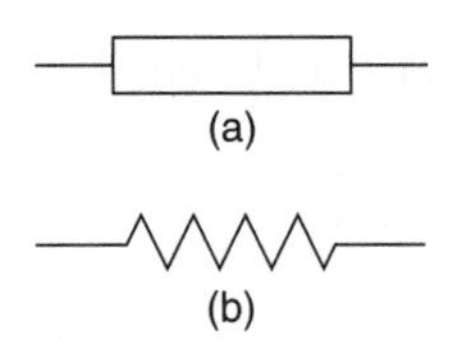


**Figure 3.1** Resistor symbols

Values of resistance are given in ohms (Ω) or multiples of ohms. In circuit diagrams, they are coded R for ohms, k for kilohms (thousands of ohms) and M for megohms, or millions of ohms. The letters R, k or M are placed in the position of the decimal point when a resistance value is written against the resistor symbol, as explained in Chapter 1.

Resistors, whether fixed or variable, are manufactured with certain average values aimed at. These values are called **preferred values**. They are chosen so that no resistor, whatever its actual value of resistance, can possibly lie outside the range of all tolerances. Table 3.1 lists the 20%, 10%, 5% and 1% tolerance preferred values. Within Europe, resistor tolerances are described by the letter E and a digit that indicates the number of resistors per decade, and this is related to the value tolerance as follows: 20%–E6, 10%–E12, 5%–E24 and 1%–E96. To accommodate the greater number of values in the series, E96 devices are coded (see later) by using five or six coloured bands.

**Table 3.1** Preferred values tolerance series

| | | | | | | | | | | | | | | | |
|---|---|---|---|---|---|---|---|---|---|---|---|---|---|---|---|
| *E6 series 20% tolerance* | | | | | | | | | | | | | | | |
| 1.0 | 1.5 | 2.2 | 3.3 | 4.7 | 6.8 | | | | | | | | | | |
| *E12 series 10% tolerance* | | | | | | | | | | | | | | | |
| 1.0 | 1.2 | 1.5 | 1.8 | 2.2 | 2.7 | 3.3 | 3.9 | 4.7 | 5.6 | 6.8 | 8.2 | | | | |
| *E24 series 5% tolerance* | | | | | | | | | | | | | | | |
| 1.0 | 1.1 | 1.2 | 1.3 | 1.5 | 1.6 | 1.8 | 2.0 | 2.2 | 2.4 | 2.7 | 3.0 | 3.3 | 3.6 | 3.9 | 4.3 |
| 4.7 | 5.1 | 5.6 | 6.2 | 6.8 | 7.6 | 8.2 | 9.1 | | | | | | | | |
| *E96 series 1% tolerance* | | | | | | | | | | | | | | | |
| 1.00 | 1.02 | 1.05 | 1.07 | 1.10 | 1.13 | 1.15 | 1.18 | 1.21 | 1.24 | 1.27 | 1.30 | 1.33 | 1.37 | 1.40 | 1.43 |
| 1.47 | 1.50 | 1.54 | 1.58 | 1.62 | 1.65 | 1.69 | 1.74 | 1.78 | 1.82 | 1.87 | 1.91 | 1.96 | 2.00 | 2.05 | 2.10 |
| 2.15 | 2.21 | 2.26 | 2.32 | 2.37 | 2.43 | 2.49 | 2.55 | 2.61 | 2.67 | 2.74 | 2.80 | 2.87 | 2.94 | 3.01 | 3.09 |
| 3.16 | 3.24 | 3.32 | 3.40 | 3.48 | 3.57 | 3.65 | 3.74 | 3.83 | 3.92 | 4.02 | 4.12 | 4.22 | 4.32 | 4.42 | 4.53 |
| 4.64 | 4.75 | 4.87 | 4.99 | 5.11 | 5.23 | 5.36 | 5.49 | 5.62 | 5.76 | 5.90 | 6.04 | 6.19 | 6.34 | 6.49 | 6.65 |
| 6.81 | 6.98 | 7.15 | 7.32 | 7.50 | 7.68 | 7.87 | 8.06 | 8.25 | 8.45 | 8.66 | 8.87 | 9.09 | 9.31 | 9.53 | 9.76 |

These figures are used for all resistor values of a given tolerance range. Values of 2R2, 3K3, 47K, 100K, lM5 could thus lie in either range; but values of 1R2, 180R, 2K7, 39K, 560K could lie only in the 10% (or closer) tolerance range.

Take, for example, a 2k2 resistor in the 20% series. It could have a value of 2k2±20%, that is to say, a range from 2k64 to 1k76. But if such a resistor has a measured value of 2k7 or 1k6, it would not become a reject. The reason is that the next larger preferred value is 3k3 and 3k3–20% is 2k64. So the 2k7 resistor would quite legitimately be reclassified as a 3k3. Similarly, on the low side, the next value down from 2k2 is 1k5, and 1k5+20% is 1k8. A resistor with a measured value of 1k6 would therefore be acceptable as a 1k8.

In other words, the preferred values are so chosen that no resistor that is manufactured can be rejected simply because of its measured value. The preferred value of a resistor is either printed on the resistor or coded on it by reference to the colour code illustrated and explained below.

The basic relationship between values and colours is shown in Table 3.2 and Figure 3.2. One of the many ancient mnemonics designed to aid memorizing this series goes: 'Bye Bye Rosie Off You Go Birmingham Via Great Western'.

**Table 3.2** Colour codes

| *Figure* | *Colour* | *Figure* | *Colour* | *Figure* | *Colour* | *Figure* | *Colour* | *Figure* | *Colour* |
|---|---|---|---|---|---|---|---|---|---|
| 0 | Black | 1 | Brown | 2 | Red | 3 | Orange | 4 | Yellow |
| 5 | Green | 6 | Blue | 7 | Violet | 8 | Grey | 9 | White |
| Multiplier colours can also be: | | | | | | | | | |
| 0.1 | Gold | 0.01 | Silver | | | | | | |
| No tolerance band is used if the resistor has 20% tolerance | | | | | | | | | |
| Tolerance: | | | | | | | | | |
| 10% | Silver | 5% | Gold | 2% | Red | 1% | Brown | | |

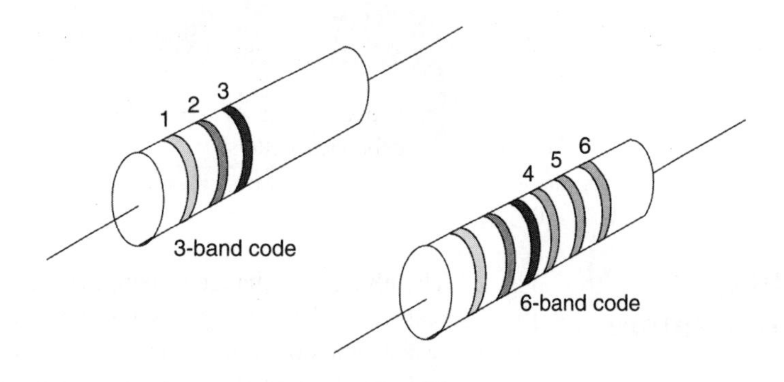


**Figure 3.2** Resistor colour coding using three or six bands

In the three-band code resistor pictured in Figure 3.2, all the colour bands are printed towards one of its ends. The first band (the one nearest the end) represents the first figure of the coded value, the second band represents the second figure, and the third band the number of zeros following this second figure. The three bands painted in the sequence, beginning from the end of the resistor – blue, grey, red – would mean (see Table 3.2) a resistor with a preferred value of 6–8–two zeros, or 6800 ohms. Such a resistor would in practice always be described as a 6k8. This colour coding system can be extended by adding a fourth band to indicate the component tolerance. No fourth band means ±20%, but a gold band indicates ±5% and a silver band indicates ±10%.

Up to six bands may be used for close-tolerance resistors as shown in the six-band example of Figure 3.2. Here, the first three bands are used to represent significant figures, while the fourth band indicates the multiplier or number of following zeros. Bands five and six then indicate component tolerance and temperature coefficient, respectively.

**Moulded carbon** resistors are usually colour coded. Wire-wound and film type resistors generally have the resistance value printed on them, using the R, k, M notation. These resistors are cheap to manufacture but cannot be made to precise values of resistance. Their large tolerances mean that the values of individual resistors have to be measured and suitable ones selected if a precise value of resistance is needed. Moulded resistors are no longer in use in the UK, although they can be found on some imported equipment.

**Wire-wound** resistors are constructed from lengths of insulated wire, often a nickel–chromium alloy, and they are used when comparatively low values of resistance are required. They are particularly useful when the resistor may become very hot in use, or when precise resistance values are needed.

**Carbon film** and **metal film** resistors are made, as the name suggests, from thin films of conducting material. They can be manufactured in quantity batches, but to fairly precise values. They are now the standard type of resistor used in domestic electronic equipment, and are often found in the miniature surface mount (SM) form.

### Practical 3.1

Inspect a set of six colour-coded resistors and write down their resistance values.

## Effect of temperature

The effect of a change of temperature on a resistor is to change its value of resistance. This change is caused by alteration of the resistivity value of the resistor material rather than by the very small change in dimensions (length and cross-sectional area) that also takes place.

This effect of temperature on resistance is measured by the **temperature coefficient of resistance**, which is defined as the fractional change in resistance per degree of temperature change. As a formula this is:

$$\frac{\text{Change in resistance}}{\text{Original resistance} \times \text{Change in temperature}}$$

or it can be written as $\Delta R/R\ \Delta T$, with the $\Delta$ (Greek delta) symbol meaning 'change in'.

Temperature coefficient values may be either positive or negative. A **positive** temperature coefficient means that the resistance value of a resistor at high temperature is greater than its resistance value at low temperatures. A **negative** temperature coefficient means that the resistance value at high temperatures is lower than its resistance value at low temperatures. Coefficients are often quoted as parts per million per degree Celsius.

For example, if a resistor is quoted as having a temperature coefficient of +250 ppm/°C, this means a change of 250 units for each million units of resistance for each degree Celsius of temperature rise.

For a 100 k resistor raised in temperature from 20°C to 140°C (a change of 120°C), this means: (250/1 000 000) × 1 000 000 × 120 = 3000 ohms, or 3k0, making the value a total of 103 k at the higher temperature.

Taking another example, if an insulator has a resistance of 100 M at 20°C and has a temperature coefficient of −500 ppm/°C, then at 200°C (a *rise* of 180°C) its resistance will **drop** by: (500/1 000 000) × 1 000 000 × 180 = 9 000 000 or 9 M, so that the resistance at 200°C is 91 M.

Materials that are widely used for manufacturing resistors normally have positive coefficients of small value. Insulating materials generally have negative temperature coefficients, and insulation becomes less effective at high temperature. There are also materials that have very large coefficients of temperature change, and some that have very low values.

## Power dissipation

When selecting a resistor, you have to consider its **power dissipation** as well as its resistance. A resistor that is rated at 1/4 W runs noticeably hot when it is asked to dissipate power of 1/4 W, and it will be damaged if it is expected to dissipate more power than that. In electronic circuits, most resistors have to dissipate considerably less than 1/4 W, so that 1/4 W or even 1/8 W types can be used. Component lists therefore specify only those few resistors that need higher ratings. The description 2k2 W/W 5W, for example, means 2200 ohm resistance, wire-wound, 5 W dissipation. For calculations of how power is dissipated in watts see later in this chapter.

Some resistors appear on circuit diagrams with **safety-critical** warnings, meaning that if this resistor fails it must be replaced by another that is of exactly the same type. This is important, and a service engineer could be held responsible for any damage caused if an ordinary resistor was used to replace a special component.

When resistors fail, they commonly become open-circuit (o/c), rarely short-circuit (s/c), or they may change their value. This change in value may be either upwards or downwards, and a faulty resistor that has changed value is said to have gone high or gone low. Excessive dissipation is one main cause of resistor failure, and when a resistor has to be replaced, the replacement must be of a suitable power rating, and one must also ensure that the resistor is not surrounded by other components that would make cooling difficult.

## Thermistors

Thermistors have a large temperature coefficient of resistance in the order of $15–50 \times 10^{-3}$. These devices may be in rod, bead, washer or disc form. They are made from carefully controlled mixtures of certain metallic oxides. These are sintered at very high temperatures to produce a ceramic finish. Negative temperature coefficient (NTC) thermistors have resistances that fall with a rise in temperature and are commonly made from mixtures of metal oxides. Positive temperature coefficient (PTC) components whose resistance increases with a rise in temperature can be made from barium titanate with carefully controlled amounts of metals added.

Thermistors are used extensively for temperature measurement and control up to about 400ºC. An obvious use is in temperature sensors, but they are also extensively used along with conventional resistors in compensating for the effects of changing temperature in semiconductor amplifiers. Another application is to provide temperature compensation for the change in the winding resistance of alternators and other generators which affects their performance when the operating temperature rises.

## Resistive circuits

Resistors can be connected either in series or in parallel, or in any combination of series and parallel. When resistors are connected in series (Figure 3.3a), the same current must flow through each resistor in turn. The total resistance encountered by the current is the sum of all their separate values.

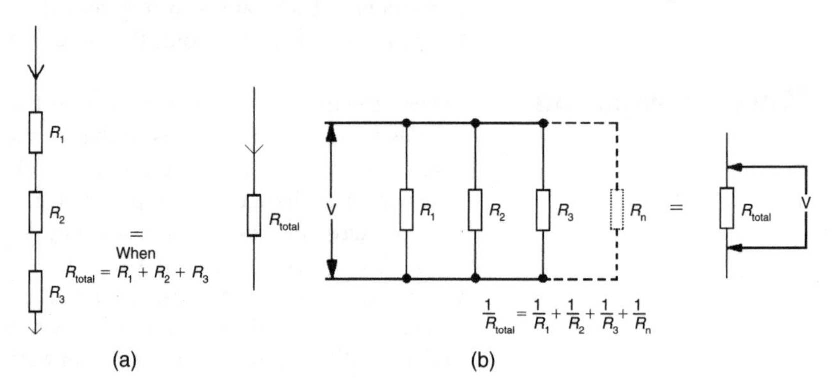


**Figure 3.3** Connecting resistors (a) in series, and (b) in parallel

When resistors are connected in parallel (Figure 3.3b), the current flow is divided among the resistors according to their values, with the most current flowing through the lowest value resistor(s). The same voltage, however, is maintained across all the resistors, and the total current is equal to the sum of the currents in each resistor.

Any combination of resistors connected in series and in parallel is capable of being replaced, as far as the flow of current is concerned, by a single resistance of equivalent value. These single resistances are then combined by using the formulae given in Figure 3.3, until a single resistance equivalent in value to the resistance of the combination as a whole is achieved. This is called the **equivalent resistance**.

The equivalent value of a set of resistors in series is simply the sum of the resistance values in the series. For example, the equivalent value of 6k8 + 3k3 + 2k2 connected in series is 6.8 + 3.3 + 2.2 = 12.3k, or 12k3. The equivalent resistance of the series is thus greater than the value of any of the single resistors.

The equivalent value for resistors connected in parallel is rather more difficult to calculate. The rule is to calculate the inverse ($1/R$) of the value of every resistor connected in parallel, to add all the inverses together, and then to invert the result. As a formula this is

$$1/R_{\text{total}} = 1/R_1 + 1/R_2 + 1/R_3 + \ldots$$

for however many resistors there may be in the parallel combination. For two resistors this can be simplified to product/sum, so that for two resistors $R_1$ and $R_2$ in parallel, the equivalent is $R_1R_2/(R_1 + R_2)$.

**Example:** Find the equivalent resistance of 2k2, 3k3, 6k8 connected in parallel.

**Solution:** Working in kilohms:

$$\frac{1}{R} = \frac{1}{2.2} + \frac{1}{3.3} + \frac{1}{6.8} = 0.4545 + 0.3030 + 0.4170 = 0.9046$$

so that if $1/R$ is 0.9046, then $R = 1/0.9046$, which is 1.1054 k, or 1105.4 ohms. A calculator will invert a number when you press the Inv or $x$–1 key.

When a circuit contains resistors that are connected both in series and in parallel, the calculations must be carried out in sequence. The example of Figure 3.4 shows two circuits whose total equivalent resistance needs to be found. The procedure is set out in the following.

In the example (a), the value of the parallel resistors is calculated first. The 10 k and the 15 k combine as follows

$$\frac{1}{R} = \frac{1}{10} + \frac{1}{15} = 0.166$$

so that $R = 6$ k (units of k have been used throughout). This 6 k is now added to the 6k8 series connected resistor to give a total equivalent resistance of 12 k, which is the total resistance of the circuit. Note that we need

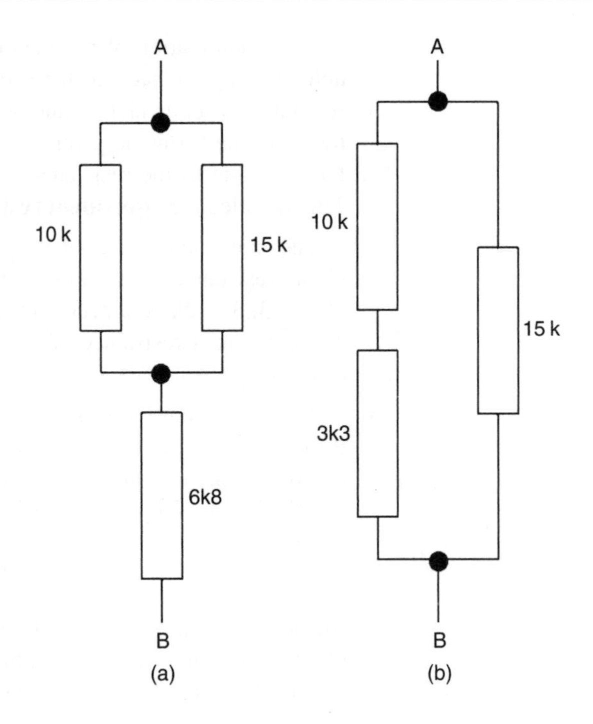


**Figure 3.4** Solving series/parallel circuits

to find the equivalent of the parallel resistors before we can add in the value of the series resistance.

In example (b), the series resistance values are combined first. The 10k plus the 3k3 gives 13k3, and this value is now combined in parallel with 15k.

$$\frac{1}{R} = \frac{1}{13.3} + \frac{1}{15} = 0.1418$$

so that $R = 7.05$ k or 7k05.

Both series and parallel combinations of resistors can produce values of total resistance that are unobtainable within the normal series of preferred values. This can often be useful as a way of obtaining unusual values.

## Potential divider

A highly important circuit using two resistors connected in series is the potential divider circuit illustrated in Figure 3.5. The total resistance is $R_1 + R_2$, which in the example shown is 5k5. By Ohm's law, the current flow must be = 1.8182 mA.

Such a current flowing through $R_2$ requires voltage $V$ which, by Ohm's law again, must be $I \times R_2$. Substitute the known values of current and resistance, and the voltage across $R_2$ works out at 1.8182 mA (2k2), or about 4 V in the example shown.

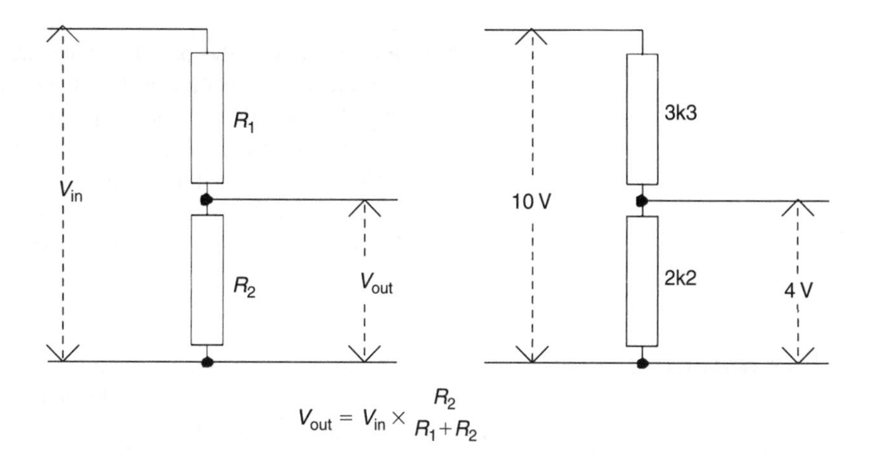


**Figure 3.5** The potential divider circuit

A potential divider (or attenuator) circuit is used to obtain a lower voltage from a source of high voltage. Where $E$ is the higher voltage and $V$ the lower voltage, the lower voltage can be calculated by using the formula:

$$V = \frac{ER_2}{R_1 + R_2}$$

For example, if 10 V is applied across a series circuit of 2 k and 3 k resistors, the voltage across the 3 k resistor is:

$$V = \frac{10 \times 3}{3 + 2} = 6\text{ V}$$

**Example:** Given the circuit shown in Figure 3.6(a), calculate $V_{out}$, $I_1$, $I_2$ and $I_3$.

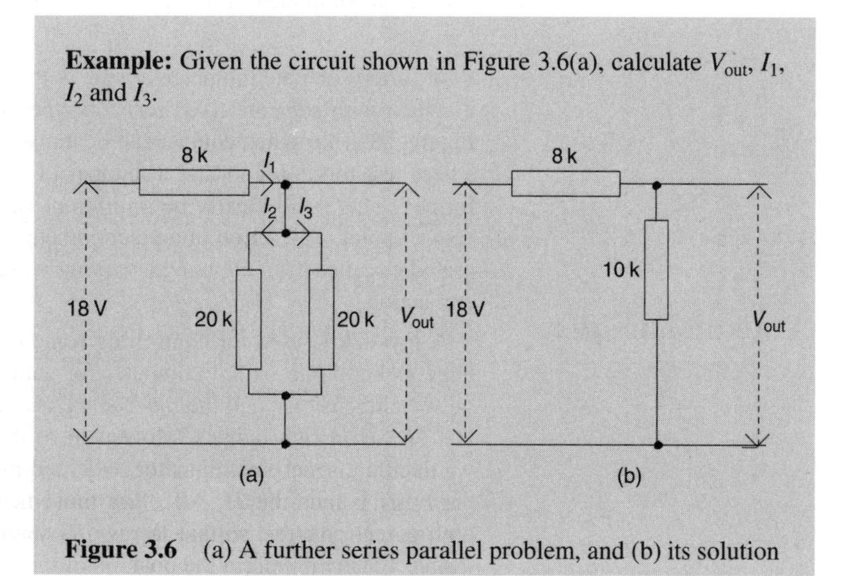


**Figure 3.6** (a) A further series parallel problem, and (b) its solution

**Solution:** Start with the parallel, equal 20k resistors. Use either the product/sum rule or recognize that two equal parallel resistors have a value equal to half that of one, so that the circuit becomes that shown in Figure 3.6(b). The total resistance is therefore 10k + 8k = 18k. Now 18V across 18k will cause a total current $I$ = 1 mA. This current through 10k will produce a $V_{out}$, = 10V. This is the voltage across both 20k resistors, so that $I_1$ and $I_2$ are both equal to 10/20k = 0.5 mA.

## The potentiometer

Figure 3.7 shows the working principle, symbol and a typical photograph for the component called the **potentiometer**, which is in effect a variable potential divider. A sector of a circle, or of a straight strip, of resistive material is connected at each end to fixed terminals, and a sliding contact is held against the resistive material by a spring leaf. As the position of the sliding contact is altered, so will different values of resistance be created between the sliding contact and each of the two fixed terminals. The symbol is a reminder of the principle.

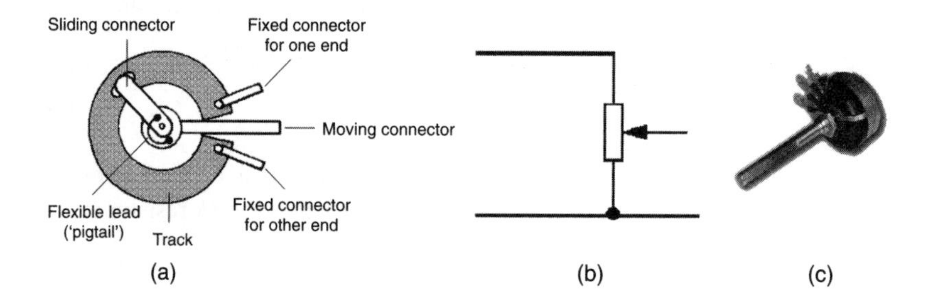


**Figure 3.7** (a) Circular potentiometer construction, (b) symbol, and (c) typical component [photograph: Alan Winstanley]

A variety of potentiometer circuit is the switched potentiometer, using a switch with separate fixed resistors, the principle of which is shown in Figure 3.8. The switch contact can be moved to connect to any of the points where resistors join, so that a number of potential divider circuits can be formed. There will clearly be a different value of voltage division at each contact point. The action of a potentiometer can also be carried out by integrated circuits (ICs), allowing a varying voltage to implement a potentiometer action.

A **joystick** is used for controlling some types of action, and is used to a large extent along with a computer for games. The joystick can be moved on two dimensions, and the mechanical action operates two potentiometers that are set at right angles. Movement of the joystick forwards and backwards affects one potentiometer only, and movement left and right affects the other potentiometer. All other movements of the joystick will affect both potentiometers, so that the two potentiometer output provide signals that accurately represent the position of the joystick.

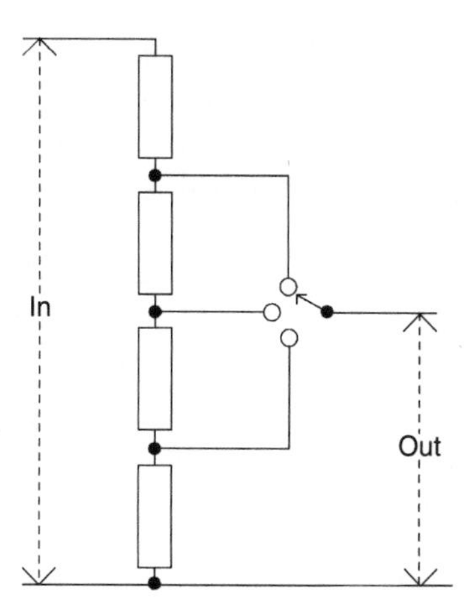


**Figure 3.8** A simple switched potentiometer

All resistors, whether fixed or variable, are specified by reference to their resistance value, their method of construction and their power rating. The abbreviation ww or WW is normally used to indicate wire-wound resistors.

## Measurements

Electrical quantities are measured by instruments such as **multimeters** and **oscilloscopes**. A multimeter (or multirange meter) is an instrument that is capable of indicating the values of several ranges of currents, voltages and resistances.

To measure a voltage, the multimeter is first switched to the correct range of voltage and the leads of the multimeter are clipped to the points across which the voltage is to be measured. One of these points will often be the chassis, earth or negative supply line. The circuit is now switched on. The voltage is then read from the correct scale (Figure 3.9). The meter is shown as a dial type, but modern digital meters produce a readout of digits.

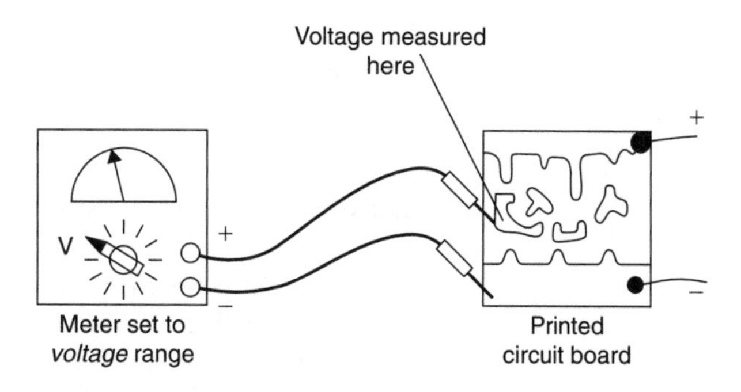


**Figure 3.9** Measuring voltage on a printed circuit board

To read current flow (Figure 3.10), the circuit is broken at the place where the current is to be measured; in this example, the current supplied to the whole circuit. The multimeter leads are then clipped on, one to each side of the circuit break. The meter is set to a suitable range of current flows, and the circuit is switched on. Current value is then read from the correct scale. It is seldom practicable to read current in this way, however, and circuits often contain small resistors so that current in a part of the circuit can be calculated by measuring the voltage across one of these resistors.

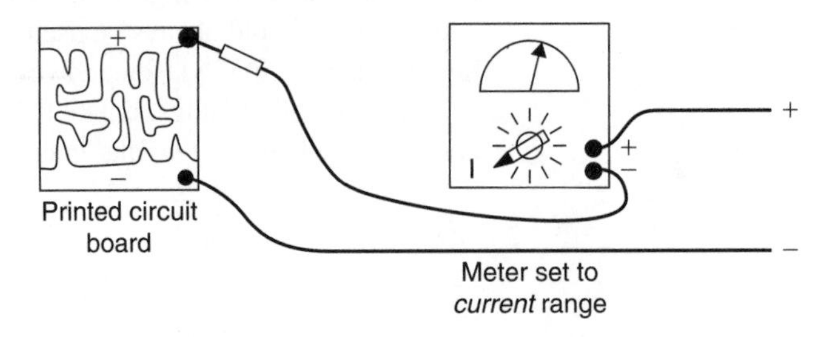


**Figure 3.10** Measuring current flow to a printed circuit board

Note that, for measurements of either voltage or current, the meter must be connected in the correct polarity, with the + (usually red) lead of the meter connected to a more positive voltage.

To read resistance values (Figure 3.11), the multimeter leads are first short-circuited. The meter is switched to the 'ohms' range, and the set zero adjuster is used to locate the needle pointer over the zero ohms mark. The leads are then reconnected across the resistor to be measured, and its resistance is read off the scale.

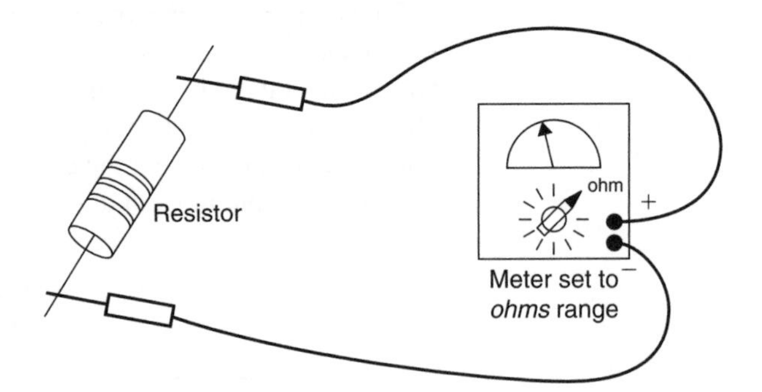


**Figure 3.11** Measuring resistance

If there is any doubt as to which sort of values of voltage or current will be encountered, the highest likely range of each must first be tried on the multimeter. If the circuit is powered from a battery, it can usually be assumed that no voltage will exceed the battery voltage, but this is not always true, because some circuits contain inverters that multiply the input voltage to a higher level. It is safer always to start with the meter set on its highest voltage range. The range switch is then used to select a lower range of values, until a range is found which gives a reading that is not too near either end of the scale. Note that the resistance ranges make use of a battery inside the meter, and if this battery is dead resistance readings cannot be made.

Multimeters always draw current from the circuit under test, so the values of voltage or current that are measured when the meter is connected are not necessarily the same as those that exist when the meter is not connected. Using modern digital meters overcomes this problem to a considerable extent. These instruments have very high resistance and so draw very little current from a circuit. They should always be used when measurements have to be taken on high-resistance circuits. Note, however, that they rely entirely on their battery to operate on all readings, unlike the older analogue type of meter. If you do not check the battery at regular intervals your readings will not be reliable.

When you have finished using a multimeter **always** leave it switched to its highest voltage range.

## Practical 3.2

Construct the circuit shown in Figure 3.12. With an analogue multimeter switched to the 10 V scale, connect the negative lead to the negative line. Measure the supply voltage, and adjust it to exactly 9 V. Measure the voltages at points X and Y, and record the two values.

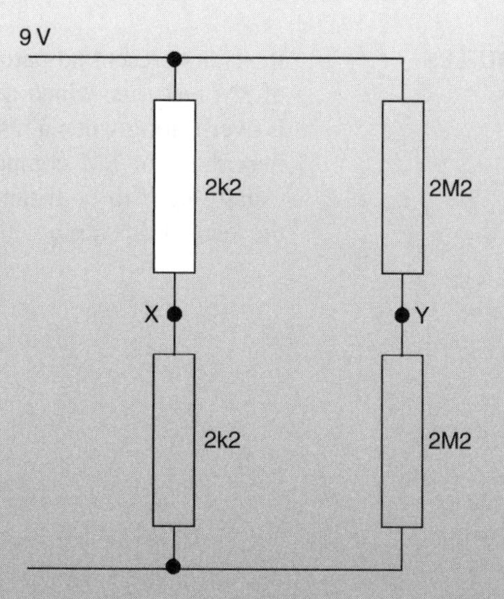


**Figure 3.12** Circuit for Practical 3.2

Now use the potential divider formula to calculate what voltages would be present at X and at Y if no meter were connected. Use a digital voltmeter to read the voltage at point Y. Which type of meter is preferable for measuring the voltage at Y, and why?

**Examples:** These examples should be solved by mental arithmetic, but you can use a calculator to check your answers if necessary. In each question, two quantities are given and you should calculate the third.

| | | | | | |
|---|---|---|---|---|---|
| $V = 9\,V$ | $V = 6.6\,V$ | $V = ?$ | $V = 27\,V$ | $V = 12\,V$ | $V = ?$ |
| $R = 18k$ | $R = ?$ | $R = 10k$ | $R = 1k8$ | $R = ?$ | $R = 2k2$ |
| $I = ?$ | $I = 2\,mA$ | $I = 3\,mA$ | $I = ?$ | $I = 4\,mA$ | $I = 3\,mA$ |

**Insulation testers** are used to measure the very high resistances of insulators. Testers such as the well-known Megger™ operate by generating a known high voltage and applying it through a current meter to the insulator under test. Any current flowing through the insulator causes the meter needle to deflect, and the scale indicates the resistance value of the insulator in megohms. Typical proof test voltages are 250 V or 500 V d.c. and 500 V a.c. Megger is a registered tradename, but it is widely used to mean any meter for high resistance measurements.

For measurements of very low resistance, such as connectors, printed circuit board tracks or soldered joints, the lowest range of a digital multimeter (typically 0.001 Ω minimum reading) can be used, or a digital milliohmmeter, many models of which can read down to 1 μ Ω.

## Resistor notes

Resistors present no particular handling problems apart from the small size of SM resistors, which need to be handled with tweezers. The main hazard is overheating when a resistor is soldered into place, and this is particularly important for SM components. For a conventional resistor the leads will usually need to be trimmed to the required size, and you may have to tin the leads if this has not already been done.

## Multiple-choice revision questions

3.1 In the three-band resistor colour code, the third band represents the:
(a) first significant figure
(b) second significant figure
(c) tolerance
(d) number of zeros.

3.2 Two resistors connected in parallel will have a resistance that is:
(a) higher than either single resistor
(b) lower than either resistor
(c) equal to the sum of the resistance values
(d) higher than the sum of the resistance values.

3.3 The resistance of a metal increases when it is heated. Its temperature coefficient is:
(a) positive
(b) negative
(c) zero
(d) infinite.

3.4 A resistor is rated at 1/4 W. With 6 V across it would be just about overloaded if the current flowing were:
(a) 5 mA
(b) 10 mA
(c) 20 mA
(d) 40 mA.

3.5 A reading on a potentiometer circuit is 5 V, but a calculation suggests that it should be 8 V. This is because:
(a) the potentiometer circuit is faulty
(b) the resistors in the circuit are of low wattage
(c) the meter used has a high resistance
(d) the meter used has a low resistance.

3.6 Generally, a circuit has to be broken to measure:
(a) insulation
(b) voltage
(c) current
(d) power.

# Unit 2
## A.c. technology and electronic components

### Outcomes

1. Demonstrate an understanding of electromagnetic devices and capacitors and apply this knowledge in a practical situation
2. Demonstrate an understanding of alternating current and voltage and apply this knowledge in a practical situation
3. Demonstrate an understanding of a.c. mains supply safety and distribution and apply this knowledge in a practical situation.

Note: Because it is more logical to treat alternating current ahead of electromagnetic devices and capacitors, Outcome 2 has been dealt with first.

# 4 Magnetism

## Flux patterns

When iron filings are sprinkled on to a sheet of paper that has been placed over a bar magnet, the filings will arrange themselves into a definite pattern, which is called a **field** or **flux pattern**. This flux pattern (Figure 4.1) is a map of the direction and strength of the forces which are exerted on the iron filings because they are close to the magnet. The lines that the filings set into indicate the direction of the force. The clumping together of filings is some indication of the strength of the force. The greatest amounts of clumping exist at the ends of the magnet, and these ends are called the **poles** of the magnets, the places where the magnetic effect is strongest.

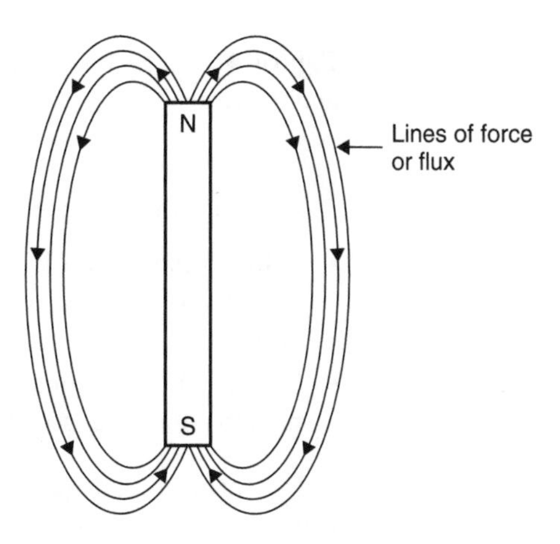


**Figure 4.1** A typical flux pattern around a bar magnet. The arrows indicate the direction that a compass needle would point to at that position in the field

These forces exist and can be detected without any physical contact between the magnet and the iron filings. The general name for a force of such a type is a **field force**. The flux pattern, therefore, is a map of the **magnetic field** of the magnet, and it indicates the direction and strength of this field.

The pattern of the field that exists around a magnet can be changed in several ways. Figure 4.2 shows how the flux pattern appears when two magnets are placed close to one another. When both magnets are aligned in the same north–south direction, the pattern is as shown in the upper diagram. When one of the magnets is reversed, the pattern alters to that shown in the lower diagram.

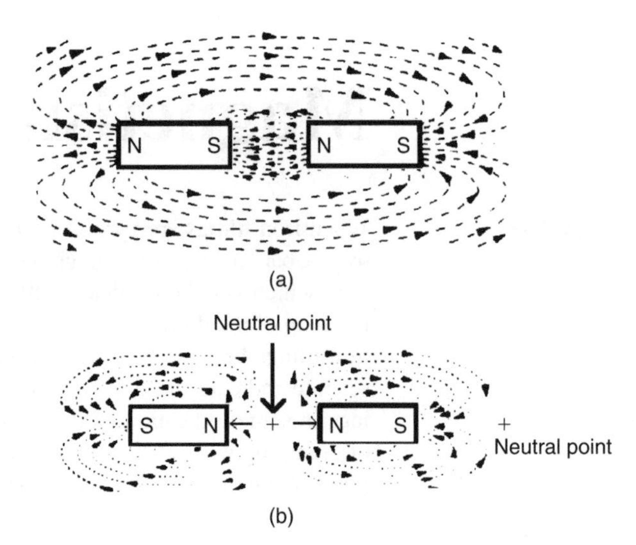


**Figure 4.2** The flux pattern around two adjoining magnets: (a) N to S, and (b) N to N

The shape of a flux pattern is governed by two rules:

- A line behaves as if it were carrying a magnetic 'current' (a **flux**) in the direction of the line. This direction can also be indicated by a compass needle.
- The flux lines become closely spaced together where there is a strong field, but they are more widely spaced where the field is weaker.

The shape of a flux pattern is not noticeably altered by being close to non-magnetic materials such as plastics, copper or aluminium. Placing magnetic materials in a magnetic field, however, causes the flux pattern to change as shown in Figure 4.3(a). The magnetic material behaves as an easy path for the lines of flux, just as a metal offers an easy path for current. This allows us to manipulate and control lines of flux in a **magnetic circuit**, a topic that is important for all magnetic devices such as motors, relays and solenoids.

Some metal materials, called soft magnetic materials, can concentrate flux lines, although they are not themselves permanent magnets. These metals (such as soft iron, permalloy and mumetal) can be used as magnetic screens because flux will flow along the material rather than through it. Figure 4.3(b) shows that a cylinder of soft magnetic material will have no flux pattern inside the cylinder even when there is a strong field outside it. Such screening materials are used to shield cathode ray tubes and other components from stray magnetic fields.

A **soft magnetic material** is one that is easily magnetized by another magnet, but it loses that magnetism equally easily. A **hard magnetic material** (such as steel and several steel alloys with cobalt and nickel) is difficult to magnetize, but it will retain its magnetism for long periods unless it is heated, struck with a hammer or demagnetized by alternating fields. Hard

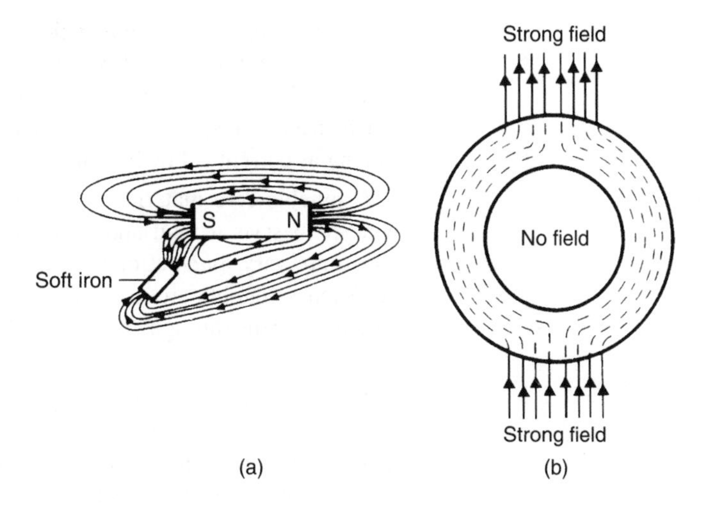


**Figure 4.3** (a) Flux lines concentrated by presence of a piece of soft iron, and (b) magnetic screening

magnetic materials are used for making permanent magnets; soft magnetic materials are used for electromagnets and magnetic shields.

## Electromagnets

Permanent magnets are not the only source of magnetic flux. When an electric current is passed through any conductor, a magnetic flux pattern is created around the conductor. The shape of this flux pattern around a straight conductor (Figure 4.4a) is circular, unless it is distorted by the presence of other magnetic material, and the pattern has no start or finish points. This leads us to the conclusion that all flux lines are, in fact, closed lines without either a start or a finish.

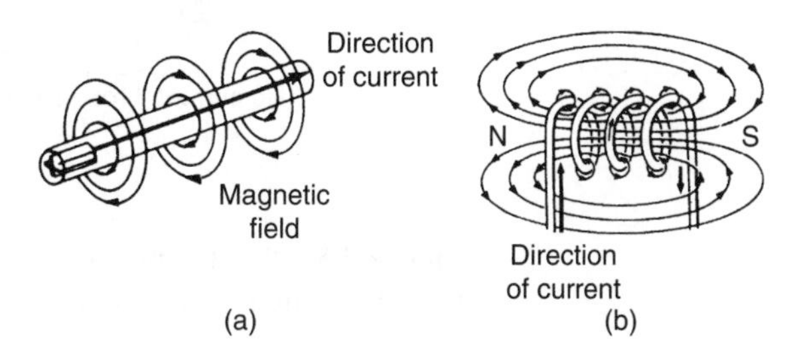


**Figure 4.4** The flux pattern around (a) a wire carrying current, and (b) a solenoid carrying current

A conductor that is wound into the shape of a coil is called a **solenoid**. When current flows through the wire of a solenoid, the flux patterns of each part of the wire in the solenoid add to one another, so producing the pattern shown in Figure 4.4(b). This pattern is similar to that of a bar magnet, except that the shape of the flux pattern inside the coil can be seen.

The addition of a soft magnetic material as a **core** inside a solenoid coil greatly concentrates the flux pattern inside the core and therefore also at its ends. The concentration of the flux pattern means that a coil becomes a much stronger magnet when it has a soft magnetic core, although only for as long as current is flowing through the wire of the coil.

A suitable core material can in practice increase the strength of a magnetic field several thousand-fold. This allows us to make electromagnets which are very much stronger than any permanent magnet could possibly be. Such solenoids are used to move soft iron pieces (armatures) so that passing current through the solenoid will cause mechanical movement of the armature.

All magnetic flux is caused by the movement of electrons. This can be the movement of electrons through a conductor, called **electric current**, or it can be the spinning movement of electrons within the atoms which occurs in permanent magnetic materials. These spinning movements balance each out in most materials (so that most materials are only very faintly magnetic), but the materials that are classed as strongly magnetic (called **ferromagnetic** materials) are made up of atoms that have more electrons spinning in one direction than in the opposite direction.

## Motor effect

Another important type of alteration in the flux pattern of a magnet occurs when a wire is placed in a magnetic field and has a current flowing through it. In Figure 4.5, the points marked N and S are the north and south poles, respectively, of two bar magnets. The little circle with a cross inside it which lies between them represents a conductor wire seen end on, that is to say, with one of its ends running into the paper away from you, and the other end running out of the paper towards your eye.

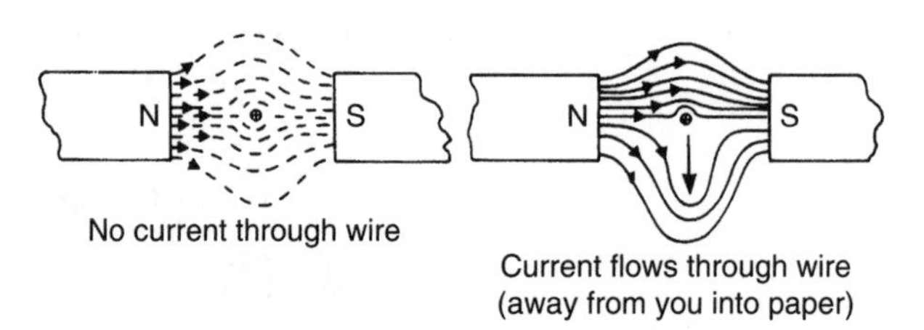


**Figure 4.5** What happens to the flux lines of magnets when current flows through a wire in a magnetic field

When an electric current is passed through the wires the flux pattern between the magnets becomes distorted and this pattern can be revealed using iron filings or small compass needles.

When current flows and causes this distortion of the field, a magnetic force is exerted on the wire itself, acting at right angles to the wire and also at right angles to the flux lines. The direction in which this force is exerted can be remembered by using the **left-hand rule**, as pictured in Figure 4.6. The thumb and first two fingers of the left hand are extended at right

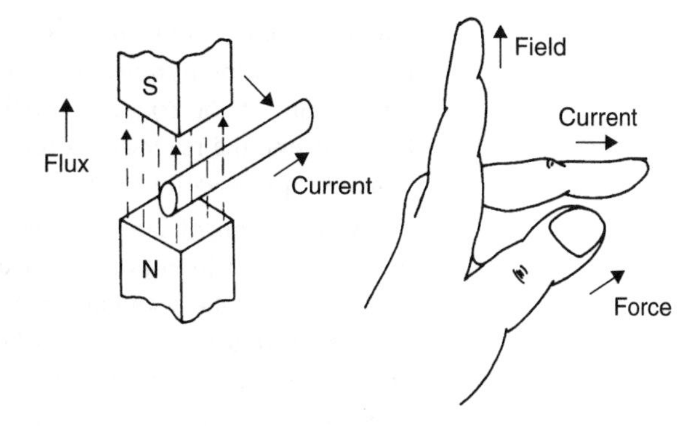


**Figure 4.6** The left-hand rule for finding the direction of force on a conductor

angles to one another. The **F**irst finger then points in the direction of the **F**lux (N to S direction); the se**C**ond finger shows the direction of **C**urrent flow (positive to negative); while the thu**M**b points along the direction of the **M**agnetic force exerted on the wire. Unless the wire is held firmly, it will move in the direction of this force. A familiar illustration of this effect is the way that the cables to an electric welder jump apart when the arc is struck: each cable generates a magnetic field and so exerts a force on the other cable.

The force that exists between a magnetic flux pattern and a wire carrying current is the principle of the electric motor. When current flows through the coil of wire, shown in cross-section in Figure 4.7(a), the ends of the coil will experience forces that are equal in size but opposite in direction. The coil will therefore rotate about its central axis. This rotation will only continue, however, until the plane of the coil lies in line with the flux (in what is called its neutral position). In this position no force is acting to cause rotation because the coil is now aligned with its flux pattern. Another way of saying this is that the coil axis no longer cuts across the flux lines.

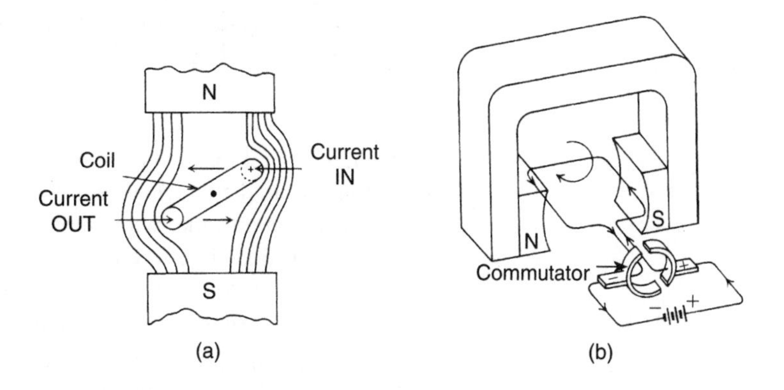


**Figure 4.7** (a) Forces acting on a coil that carries current in a magnetic field, and (b) electric motor principle

Continuous rotation is possible if the direction of current flow through the coil of wire is reversed just as the neutral position is reached. The momentum of the rotating coil will carry it a little way past the neutral position, and the reversed current will exert a force on both ends of the coil which keeps the coil rotating.

The direction of current flow through the wire can be reversed at the correct point in its rotation by using a rotary switch called a **commutator** (Figure 4.7b). The segments of the commutator are made of copper, and the contact (or **brush**) is of soft carbon, which conducts current but allows the movement of the commutator segments with fairly low friction.

An electric motor with a single coil and a two-segment commutator does not run at a perfectly uniform speed. The coil moves most rapidly when it is at 90 degrees to the plane of the flux and most slowly as it passes the neutral position. In practice, d.c. motors use several coils and corresponding pairs of commutator segments. This provides smoother running and avoids the problem of 'sticking' when the motor will not start if the coil happens to be at rest near its neutral position. Small motors usually have a twin coil arrangement with four commutator segments.

The set of revolving coils is known as the **armature** of the motor, and each end of a coil is called a **pole**, so that a small motor can be described as having, for example, a four-pole armature. The magnet (which may be either a permanent magnet or an electromagnet) is known as the **field** or **field winding**. Direct current motors are easy to control and their direction of rotation can be reversed by reversing the direction of current in either the armature or the field, but not both together.

## Practical 4.1

Connect up a small electric motor so that you can measure the current through the armature winding. Measure this current (a) when the armature is revolving freely, (b) when there is a load (clamp the shaft lightly between your fingers), and (c) when the shaft cannot move (this reading must be taken quickly to avoid damage to the armature and brushes).

The moving coil meter (Figure 4.8) uses the same principle as the electric motor to measure the direct current flowing through the coil. In this case, the coil is wound on a former with a pointer attached and mounted in bearings. The coil is positioned between the shaped poles of a powerful electromagnet. Current is fed to the coil through a pair of spiral springs which act to cause the coil to take up a zero position when no current is flowing.

When a current flows through the coil, the magnetic field of the coil works with the flux of the permanent magnet to turn the coil. The amount of turning is proportional to the amount of current flowing in the coil because the pointer stops at a position where the magnetic force is exactly

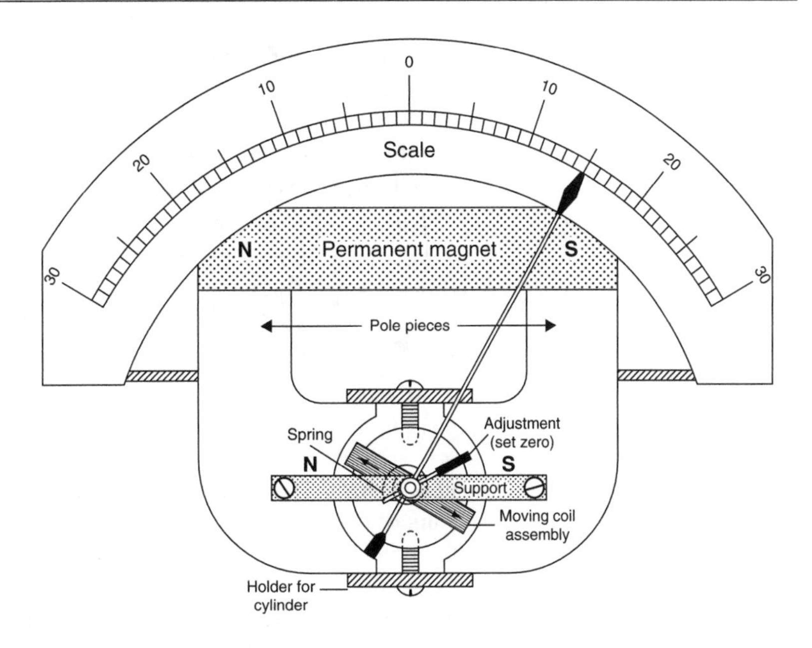


**Figure 4.8** Moving coil meter principle

balanced by the mechanical force of the springs. Moving coil meters have now been replaced for most measuring purposes by digital meters, which use an entirely different principle.

## Practical 4.2

Connect up a voltage source with a low-value wire-wound variable resistance and an ammeter in series with a solenoid (which can be the winding of a relay). Find the minimum value of current that will just pull in the armature of the solenoid.

## Electromagnetic induction

The reverse of motor action is called **electromagnetic induction**. When the flux pattern around a conductor (with no current passing) is changed by moving the wire through the flux, a voltage is generated between the ends of the wire. This induced voltage is greatest when the direction of the flux and the movement of the wire are at right angles to one another. This is the basic electrical generator principle.

The polarity of the induced voltage can be remembered by using the **right-hand rule** illustrated in Figure 4.9. The first two fingers and the thumb of the right hand are extended at right angles to one another. The first finger is pointed in the direction of the field of the magnet (N to S), and the thumb in the direction of the motion of the wire. The second finger then indicates

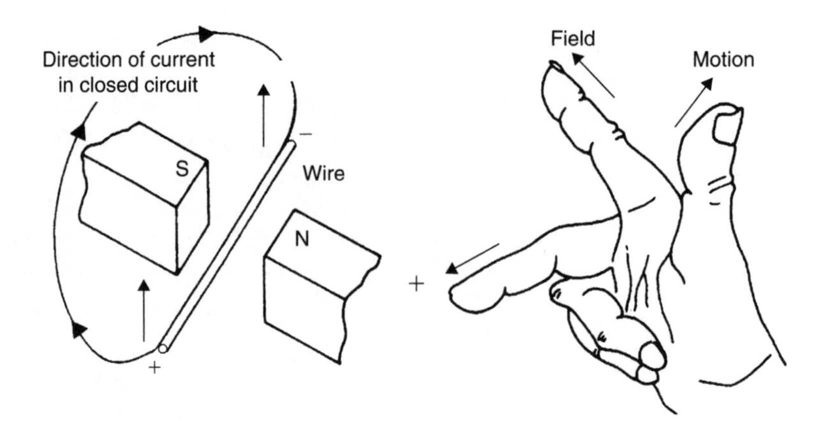


**Figure 4.9** The right-hand rule for finding the polarity of an induced voltage

the end of the wire which is positive. If the ends of a wire are connected to a load, current will flow. Remember that the **F**irst finger points in the **F**ield direction, the **M**otion is in the thu**M**b direction, then the **I**ndex finger points to the direction of current (**I**).

This principle of motional induction is used to generate a.c. in an alternator. Figure 4.10 shows a simple alternator using the same construction as an electric motor but with the commutator replaced by **slip-rings**. These slip-rings do not reverse the connections to the coil. The voltage produced when the alternator is set spinning is not a steady (d.c.) one but takes the shape of a wave, a sine wave (more details later), whose voltage reaches maximum when the wire of the coil is momentarily moving at right angles to the flux, and zero at the instant when the wire of the coil is moving in line with

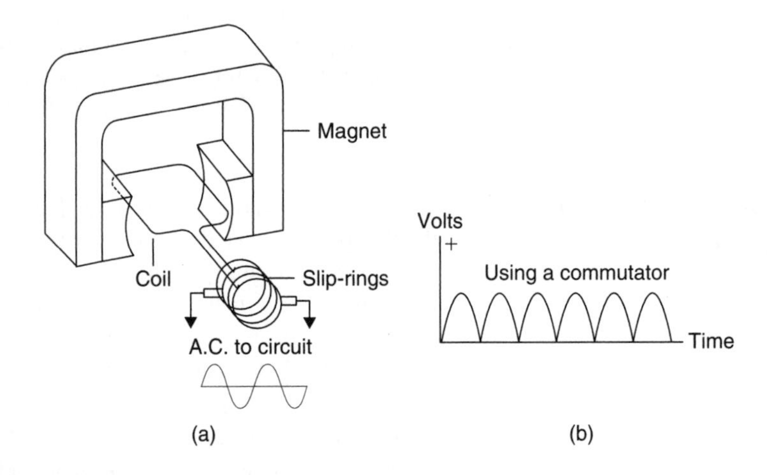


**Figure 4.10** (a) The principle of the a.c. generator, using slip-rings, and (b) output wave when a commutator is used in place of slip-rings

(parallel to) the flux direction. The reversal of polarity is caused by the fact that the wire during its rotation must cut across the flux lines in one direction for half of its rotation and in the opposite direction for the other half.

The voltage and the frequency of a.c. produced by an alternator both increase as the speed of rotation increases. Large voltage outputs can be obtained by using large magnetic fields and many turns of wire in the rotating coil. The frequency of the a.c. output is determined entirely by the rotating speed of the spinning coil and the number of poles of the magnet.

If the slip-rings are replaced by a commutator, the waveform generated will be of the unidirectional form (Figure 4.10b). This is close enough to d.c. to be useful for some purposes, such as battery charging. Alternators for cars use a three-pole arrangement with a set of six diodes rectifying the output to a reasonably smooth d.c.

A d.c. generator that uses several coils and commutator segments, however, produces a reasonably smooth d.c. output, although the switching at the commutator segments produces sparks, causing spikes of voltage to appear. This causes interference, called *hash*, on car radios, and the use of alternators and diodes on car electrical systems was a major step forward in improving car radio reception.

## Inductors

It is also possible to generate a voltage by induction without any mechanical movement. If the strength of the flux that cuts across a conductor is varied, a voltage will be induced in the conductor, just as if the variation of flux had been caused by movement. This effect, called **transformer induction**, is responsible for the very important electrical effects known as **mutual inductance** and **self-inductance**.

When a current starts to flow through a coil of wire, or is switched off, the flux lines around the coil will respectively expand or collapse. This variation in the flux lines induces a voltage in the coil itself and this induced voltage is always in a direction that will **oppose** the change of current flow. When, for example, a coil is suddenly connected to a battery, the induced voltage acts to oppose the battery voltage so that current changes more slowly. When the coil is disconnected from the battery, the induced voltage will aid the battery voltage, trying to keep the current flowing and generating a brief pulse of voltage.

Both of these effects, which are illustrated in Figure 4.11, are momentary; they last only for the brief period in which current flow through the coil is changing. If the current through the coil is an alternating current (which is continually changing) an induced a.c. voltage will always be present. This induced voltage is also alternating, and it acts to oppose the flow of current, making the coil behave as if it possessed greater resistance to a.c. than for d.c.

Because it is the coil itself that induces the opposing voltage, this opposing voltage is called an electromotive force (emf) of **self-induction** or **back emf**. The size of the induced voltage depends on the rate at which the flow of current changes, and on the shape and size of the coil. These geometrical factors of shape and size are measured by the quantity called **self-inductance**, which is measured in units of henries (H). A coil has an

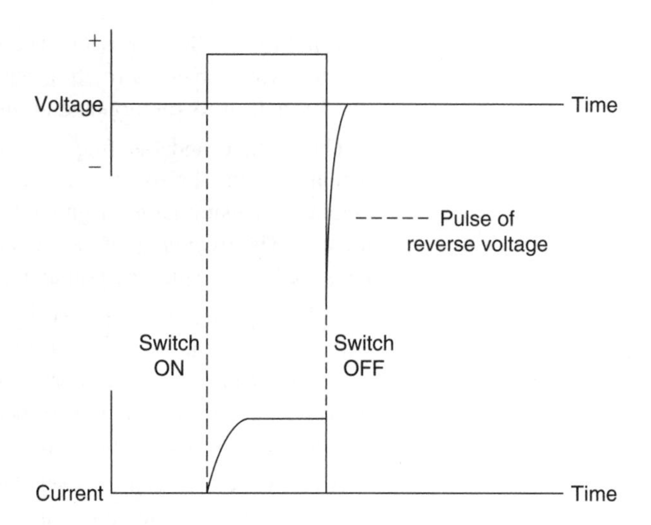


**Figure 4.11** The effect of inductance when current through an inductor is switched ON and OFF

inductance of 1 H when a rate of change of current of 1 A per second causes an induced voltage of 1 V.

In practical work, the smaller units of millihenries (mH) and microhenries (μH) are more useful. A few centimetres of straight wire will have an inductance of less than 1 μH. A coil used to tune a medium-wave radio will only have an inductance of some 650 μH. An iron-cored choke which is intended to act as a high impedance to audio signals might typically have an inductance of 50 mH. Only a very large coil wound on a massive laminated iron core would have an inductance that needed to be measured in units of henries.

## Practical 4.3

Connect up a solenoid or other large inductance winding to a suitable voltage source, and connect a neon lamp (which lights at 80 V applied) across the solenoid. Does the neon flash when you abruptly cut off the current? This indicates the back emf.

## Transformers

The changing flux around a coil that carries a changing current will also induce a voltage in another coil. This is the principle of the **transformer**, a device that is used extensively in electricity and electronics. A simple transformer, pictured in Figure 4.12(a), consists of two coils that are wound on a common magnetic core. When the current through one coil (the **primary** winding) changes, a voltage is induced in the other winding (the **secondary** winding). The symbol (b) indicates the construction.

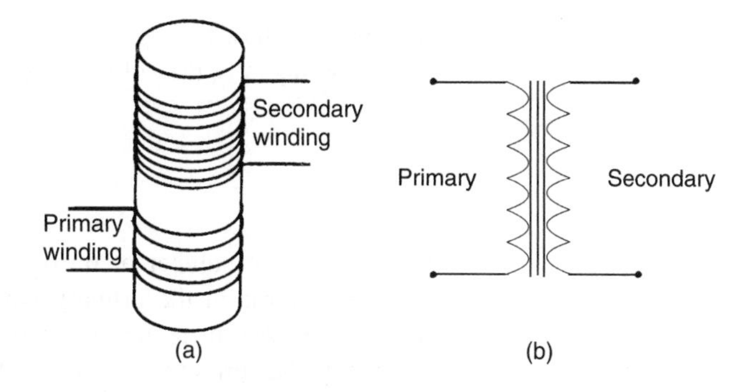


**Figure 4.12** (a) Transformer principle, and (b) circuit symbol

Remember that a single inductor has self-inductance. When inductors are wound on the same core, or close together, each winding still has **self-inductance**, and there will also be **mutual inductance** between windings. The mutual inductance that exists between the windings of a transformer means that a changing *current* passed through the secondary winding will cause a *voltage* to be induced in the primary winding. The construction of a transformer will be more easily understood if you look at the cross-section of a strip of transformer core and its windings illustrated in Figure 4.13(a). The primary and secondary windings of a transformer are always arranged to be as close to each other as possible. Figure 4.13(b) shows a typical small transformer.

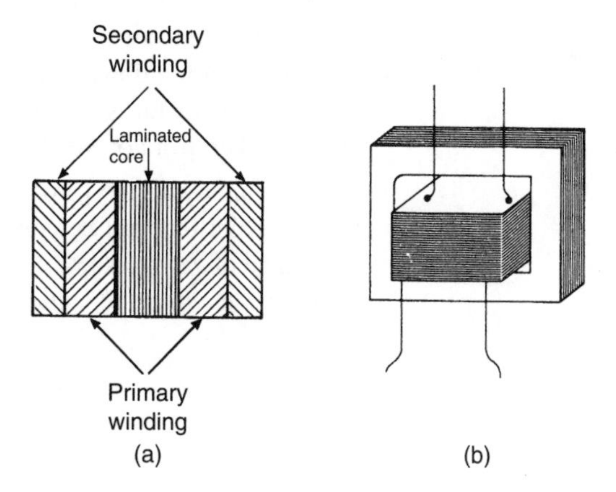


**Figure 4.13** (a) Cross-section of windings on a transformer core, and (b) a typical small transformer

One of the principal actions of a transformer is to pass signals from one circuit to another without any connection other than the magnetic flux existing between the two circuits. The transformer therefore passes a.c. signals but blocks d.c. A transformer can also achieve a **step-up** or **step-down** of alternating voltage or current depending on the ratio of the number of turns of wire in each of its two windings. For a given a.c. voltage across the primary winding ($V_p$), the voltage across the secondary winding will be

$$V_s = V_p \frac{N_s}{N_p}$$

where $N_s$ is the number of turns in the secondary winding and $N_p$ is the number of turns in the primary winding of the transformer. If the transformer is perfect, meaning that there is no power loss between primary and secondary, the current ratio $I_p/I_s$ will be equal to $N_s/N_p$, the turns ratio.

The construction of a transformer core depends on the frequency range of the signals that the transformer is designed to handle. For the lowest audio frequencies, a large core of soft magnetic material in the form of thin laminations is enough. At higher frequencies, ferrite dust cores must be used to reduce as far as possible the magnetic losses (**eddy-current** losses) that arise in metal cores when high-frequency signals pass through them. At the highest frequencies for which coils are used, no metal core material is acceptable. Coils either have to be wound on plastic formers, or they are made self-supporting. For the ultra-high frequencies (UHF) used for television transmission, short, parallel metal strips carry out transformer action.

The **autotransformer** uses a single winding to act as both primary and secondary of the device; it amounts to a tapped inductor. The autotransformer behaves exactly as does the transformer just described, except that it provides no d.c. isolation between its primary and secondary windings. The ratios of voltage, current and resistance (impedance) in the autotransformer are identical to those in the ordinary transformer, and the ratio of turns is worked out in exactly the same way. Many autotransformers are used for stepping down voltage, so that the primary number of turns is the total number of turns on the core, and the secondary number is the small amount taken from one end of the winding to the tap point.

A commonly used type of autotransformer, called a Variac™, has a sliding tapping point that can be continuously varied to provide an output (secondary) range of zero to some 130% of the nominal primary voltage. It is useful in circuits in which the voltage needs to be set precisely to a known value.

## Multiple-choice revision questions

4.1 The cores of electromagnets are usually made from:
(a) steel
(b) nickel
(c) soft iron
(d) hard iron.

4.2 When two permanent magnets with like poles are placed close to each other, the effect is to:
(a) attract each other
(b) repel each other
(c) induce a current into each other
(d) generate heat.

4.3 When a current flows in a wire that is placed in a magnetic field there is a mechanical force exerted on the wire. The direction of this force is:
(a) the same as that of the magnetic field
(b) the same as the current
(c) at right angles to both field and current
(d) opposite to the direction of the current.

4.4 A sine wave voltage is generated when a coil :
(a) revolves in a magnetic field
(b) moves in line with the magnetic field
(c) moves in a straight line across the magnetic flux
(d) is held stationary.

4.5 When a current that has been flowing through an inductor is suddenly cut off, there is a:
(a) large pulse of voltage
(b) small pulse of current
(c) large pulse of current
(d) continuous voltage.

4.6 A transformer has 1000 turns in its primary winding and 2000 turns in its secondary. It will give:
(a) 20 V d.c. output for 10 V d.c. input
(b) 10 V d.c. output for 20 V d.c. input
(c) 20 V a.c. output for 10 V a.c. input
(d) 10 V a.c. output for 20 V a.c. input.

# 5 Capacitance and capacitors

## Capacitance

A **capacitor** is a component that will store electric charge. A simple capacitor can be made by using two conducting plates separated from each other by a (non-conducting) gap. When the plates are connected to a battery, one plate to positive, the other to negative, a (small) current will flow, dying down to zero in a very short time. The action of the battery has been to pump electric charge (electrons) from one of the plates on to the other until the latter is so negatively charged that it will take no more. It is this pumping action, a movement of electrons caused by the potential of the battery, that is detectable as momentary current flow. Because one plate is now negatively charged the other is positively charged, because electrical charges always exist in pairs.

If the capacitor is now disconnected from the battery, the electrons that have been moved from one of its plates to the other will still remain out of place, but will return to the more positive plate through any conducting path that connects the plates. This return movement will cause another momentary current to flow, this time in the opposite direction.

While it is connected to the battery and from a short time after it has been disconnected from the battery, the capacitor is said to be **charged**. The amount of charge that is stored depends on the voltage of the battery (or other supply) that is used to do the charging, and on another quantity called **capacitance**, which is a measure of the ability of the capacitor to store electric charge.

While a capacitor is being charged or discharged, the ratio of the charge transferred between the plates to the voltage across the plates remains constant, so that a graph of charge plotted against voltage is a straight line (Figure 5.1). In symbols, $C = Q/V$, where $Q$ is the charge measured in coulombs and $V$ is the voltage in volts. The constant $C$ is the capacitance which is measured in farads (F) when the other units are quoted in terms of coulombs and volts. This capacitance is constant for a given capacitor and its size depends on dimensions and materials.

The **coulomb** is a very large unit. It is the amount of charge carried by $6.28 \times 10^{18}$ electrons and so the **farad** is also a very large unit. For this reason, the subdivisions of the farad called the microfarad (μF), the nanofarad (nF) and the picofarad (pF) are used in electronic work: $1\,\mu F = 10^{-6}\,F$ (one-millionth of a farad), and $1\,nF = 10^{-3}\,\mu F$ (one-thousandth of a microfarad). The picofarad is one-thousandth of the nanofarad, and thus one-millionth of the microfarad, equal to $10^{-12}\,F$.

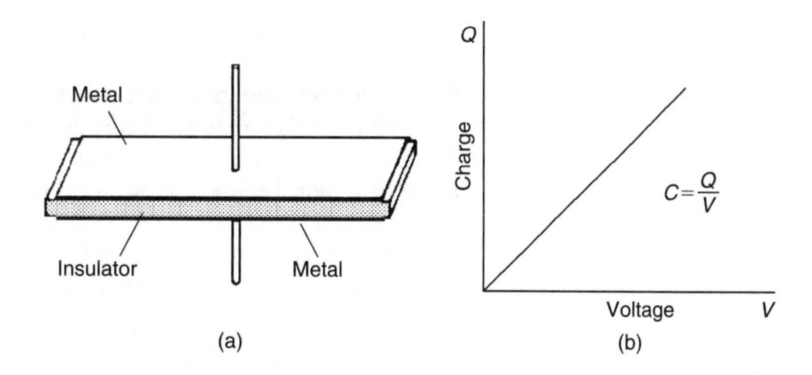


**Figure 5.1** (a) Simple capacitor construction, and (b) charge, voltage and capacitance

The value of the capacitance that can be achieved by the arrangement of two parallel conductors depends on three factors:

- the overlapping area of the conductors
- the distance they are apart
- the type of insulating material (or dielectric) that is used to separate them.

These factors affect the capacitance value in the following ways.

Capacitance is proportional to the area of the conductors that overlaps; this is their effective area for purposes of capacitance (see Figure 5.2). The greater the area of overlap, the greater the capacitance. Some types of capacitor form comparatively large values of capacitance by using conductors in the form of strips which can be wound into rolls, so that a large amount of capacitive area is obtained in a small space.

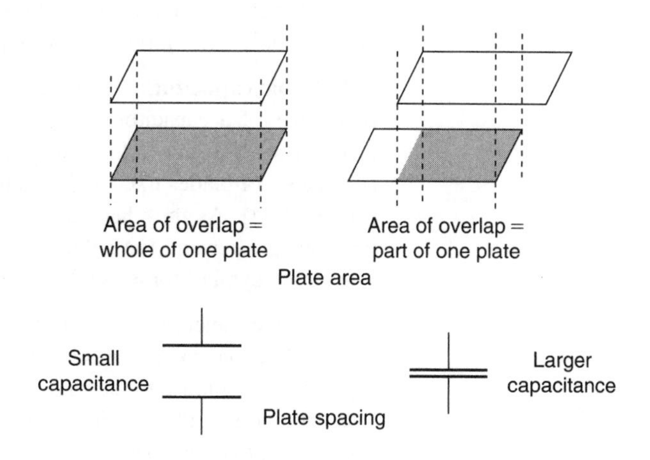


**Figure 5.2** Capacitor plate overlap and spacing

Capacitance is inversely proportional to the spacing between the conducting plates. This means that, for a given pair of plates, a reduction in the distance between them increases their value of capacitance. High-value capacitors therefore need insulators that are as thin as possible. Thin insulators, however, are easily broken down if the voltage across them is too high. Every capacitor therefore carries a **maximum voltage rating**, which must not be exceeded. Failure of the insulating material in a capacitor is called **dielectric breakdown**. It results in the failure of the capacitor, often producing a short-circuit between the terminals.

The type of material that is used as an insulator in a capacitor also contributes to its value of capacitance. When air or a vacuum is used as an insulator, capacitance between a given pair of plates is at its lowest possible value. Using materials such as waxed paper, plastics, mica and ceramics can increase the capacitance up to 25 times the value given by the same thickness of air or vacuum. This greater efficiency is caused by an effect that is measured as the **relative permittivity** of the material.

The formula for relative permittivity is:

$$C = \frac{\varepsilon \times \text{Area}}{\text{Spacing}} \text{ or } C = \frac{\varepsilon A}{d}$$

The formula above summarizes the three factors that govern a capacitor's maximum value of capacitance. The units are capacitance in farads, area in square metres and spacing in square metres. The factor $\epsilon$ (Greek epsilon) is the **permittivity**. This can be put into more practical units, so that for plates separated by air the formula becomes as shown above, with $A$ in square metres, spacing $d$ in millimetres, and $\epsilon$ equal to 0.0088.

Many materials and methods of construction have been used to make capacitors, but only a few representative types can be usefully considered here.

**Variable capacitors** operate by changing the area of overlap between the two plates of a capacitor, or by altering the spacing between the capacitor plates. The old-fashioned multiple plate variable capacitor (seldom seen now) has one set of blades fixed and an interleaving set mounted on a rotating shaft (Figure 5.3a). As the plates are more fully meshed, so the area of overlap becomes greater and the value of capacitance is increased (and vice versa). The circuit symbol for a variable capacitor is illustrated in Figure 5.3(b).

When the dielectric is air, a physically large capacitor is needed even for a variable capacitance of only 500 pF maximum, so for making miniature capacitors the plates are separated by sheets of solid dielectric. In compression trimmers (Figure 5.3c), most of which have a capacitance of some 10–50 pF maximum, sprung plates are separated by solid dielectric sheets. An increase in capacitance is obtained by compressing the plates towards each other by means of a screw. Variable capacitors such as these are designed with the moving (adjustable) plates always connected to earth, so

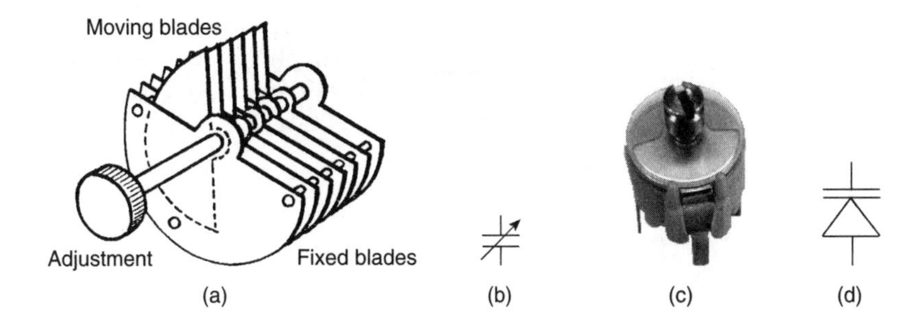


**Figure 5.3** (a) An old-style variable capacitor using meshing plates, (b) symbol, (c) modern trimmer [photograph: Alan Winstanley], and (d) varicap diode symbol

as to avoid the changes of capacitance that would otherwise occur when the control was touched; an effect called hand or body capacitance.

Another type of variable capacitor is the varicap diode, the symbol for which is illustrated in Figure 5.3(d). This device is a semiconductor diode whose capacitance is varied by changing the voltage across it. This is used now to allow capacitance changes by remote control. See also Chapter 7.

## Practical 5.1

Use a capacitance meter set so as to measure values of up to 1000 pF, and measure the capacitance between two metal sheets separated by insulators of different thickness. Use sheets of different area to show that the capacitance value depends on area. Try to keep the arrangement of wiring unchanged to avoid effects of stray capacitance.

Small (picofarad and lower nanofarad ranges of value) capacitors of fixed value (Figure 5.4a) are simply constructed, with the plate and the insulator arranged parallel to one another. Their insulators are often thin sheets of mica, which can be metal coated on both sides to form a mica capacitor. This is a type that is chosen when stability of capacitance value is important. Thin sheets of porcelain ceramic are also used now, and these capacitors are replacing the mica type. The complete capacitor, with its leadout wires attached, is then dipped in wax or plastic to insulate the plates and protect the assembly from moisture.

There is some confusion over names of mica capacitors. Mica has a silvery appearance and a mica sheet is often referred to as called **silver mica**. A capacitor may also be formed using mica plates that have been silvered on both sides, a **silvered mica** capacitor. In practice, both terms mean the same.

An alternative construction is the **tubular** type, in which the inside and outside surfaces of a ceramic tube are coated with metal to form a capacitor (Figure 5.4b). This type is also coated in wax or plastic and is used for ranges covering picofarad and lower nanofarad values.

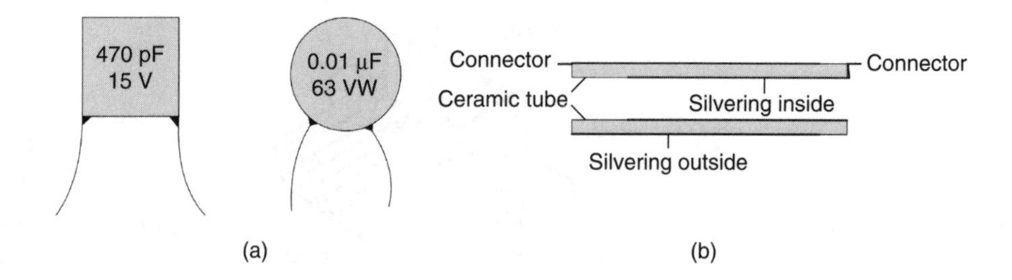


**Figure 5.4** (a) Small fixed-value mica capacitors, and (b) tubular ceramic capacitor construction

For wound (or rolled) capacitors, used for larger values of capacitance, the conductors take the form of foil sheets. Typically, these are formed by evaporating metal on to both sides of a long strip of insulator, placing an insulating sheet along one of the sides to prevent short-circuits, and then rolling the whole assembly up like a Swiss roll. Separate metal foils can also be rolled together in this way, as shown in Figure 5.5.

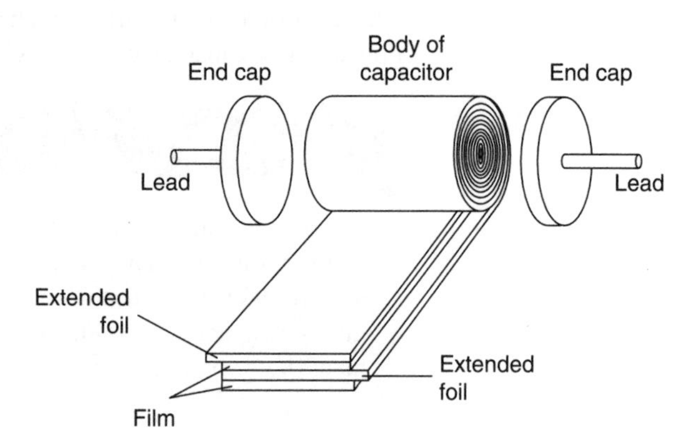


**Figure 5.5** Construction method used for typical modern rolled plastic dielectric capacitors

Rolled capacitors using insulators that can withstand high voltages (high dielectric breakdown values) are used for manufacturing capacitors that must be of fairly high capacitance values and also withstand high voltages. The insulators are nowadays plastics of various kinds, such as polycarbonate, polythene or polyester. This type of construction can be used for values of 1 nF to 1 μF or more, but for larger values of capacitance rolled capacitors are bulky.

**Electrolytic capacitors** are constructed differently using thin films of aluminium oxide as the insulator. A typical electrolytic capacitor is made using sheets of aluminium foil that are separated by a porous material soaked in a slightly acid solution (the **electrolyte**). When a voltage is first applied across the plates a current flows briefly until one plate is covered with an invisibly thin film of aluminium oxide. This process is called **forming**. The oxide film is a good insulator, and since it is thinner than any other usable solid material, electrolytic capacitors can have very large values of capacitance. In addition, the area of the foil can be greatly increased by corrugating it or, better still, by etching it to a rough finish. The voltage that can be applied across the film is, however, limited; and it must be applied in the correct polarity. Typical values are 1 μF to more than 1 F.

One of the terminals of an electrolytic capacitor is marked (+), and the other (−); and this polarity **must** be observed. Connecting an electrolytic capacitor the wrong way round would cause the oxide film to be broken down, and large currents would flow. This could cause the casing to burst, spraying both the operator and the rest of the circuit with corrosive material. Electrolytic capacitors used to smooth power supplies are particularly at risk in this respect, for very large currents would flow in the event of an internal short-circuit.

Figure 5.6 shows two typical electrolytic capacitor shapes, one with radial leads and one with axial connections. In general, the radial lead types are used with the larger components, while the axial leads need to be bent and cropped before insertion into a printed circuit board (PCB).

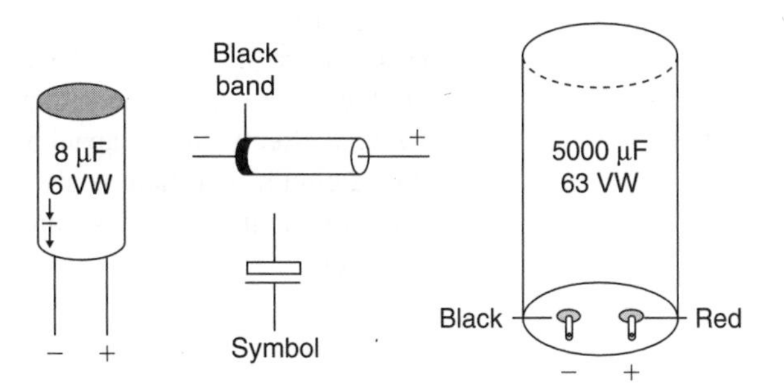


**Figure 5.6** Typical electrolytic capacitor shapes, and their symbol

Electrolytic capacitors are used for power supplies (see Chapter 7). They are also used as coupling and decoupling capacitors. In both of these applications, the capacitor allows a.c. (signal) currents to pass freely, even if the signal frequency is low, while preventing the passage of all but a very small d.c. leakage current. If any d.c. leakage current is undesirable, then capacitors of a different construction must be used. Electrolytic capacitors should not be connected in series, because the leakage currents are likely to be different, and this may cause the voltage across one capacitor to be much larger than the voltage across the other. In the few examples where this has to be done, the capacitors will have resistors connected in parallel.

For some purposes, a useful alternative to the aluminium type is the tantalum electrolytic (using tradenames such as **tantalytic** or **tantalix**), which permits very much less leakage current but is a more expensive component.

Capacitors nowadays are regarded as the least reliable components in electronic circuits, and they are the first items to be checked when a problem arises. Electrolytic capacitors are particularly likely to cause trouble, and some servicing procedures specify that all electrolytic capacitors are replaced on servicing just as a matter of course. Failure of the power supply smoothing capacitor is a common source of breakdowns of devices ranging from microwave ovens to computer power supplies.

Wound capacitors with values in the range 1 nF to 1 F are used in audio coupling, decoupling and filtering, in radio frequency (RF) decoupling and in low-frequency oscillators. Ceramic capacitors are used in RF decoupling, and silver mica capacitors in RF tuned circuits. Variable capacitors are used in oscillator, tuned amplifier and tuned filter circuits.

## Stray capacitance

In addition to any capacitors that may be deliberately placed in a circuit, a stray capacitance exists between any two parts of a circuit that are physically close to one another but not connected. Stray in this sense means unintentional and unplanned. When the circuit is used with low-frequency signals, this stray capacitance is not very important, but things can be very different in circuits working at high frequencies. If, for instance, the circuit shown in Figure 5.7 were required to handle high-frequency signals, the stray capacitances that are indicated by dotted lines could cause unwanted effects. Stray capacitances can often be minimized by careful layout of a circuit, but there is always an unavoidable stray capacitance across each individual component and this amount is not reduced by altering the layout.

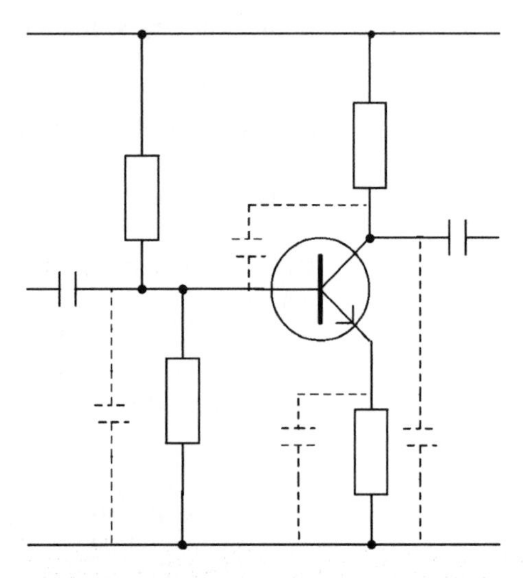

**Figure 5.7** Stray capacitances, shown dotted here in a simple amplifier circuit, can be significant at high frequencies

## Practical 5.2

Use a capacity meter to show the effects of stray capacitance. Observe the changes in capacitance that appear when the meter is connected between a metal plate and earth, and you place your hand near the plate.

**Screening** makes indirect use of the effects of stray capacitance to shield one part of a circuit from signals radiated by another part. An electrostatic screen removes unwanted signals that could be passed by stray capacitances by enclosing the circuit that could be affected inside an earthed metal box. This box need not be solid, but can be made using metal mesh. Any stray signal current reaching the box returns to earth through the metal casing, and is so prevented from affecting the circuits being screened. A different form of screening is needed for magnetic signals, as we have seen in Chapter 4.

# Capacitor combinations

Capacitors, like resistors, can be connected in series, in parallel, or in series–parallel combinations. Whatever the method of connection, any combination of capacitors can be replaced, as far as calculations are concerned, by a single capacitor of equivalent value. The size of this equivalent capacitance can be calculated from first principles. It is equal to the total charge stored by the capacitor network, divided by the total voltage across the network.

This method of calculation must be used in complex capacitor circuits, but simple formulae can be used to calculate the resultant capacitance when capacitors are connected either wholly in series or wholly in parallel (see Figure 5.8).

When capacitors are connected in **parallel**, their equivalent value is found by adding their individual values of capacitance. Thus,

$$C_{\text{total}} = C_1 + C_2 + C_3 + \cdots$$

When capacitors are connected in **series**, their equivalent value is found by adding the inverses of their individual values of capacitance. Thus,

$$\frac{1}{C_{\text{total}}} = \frac{1}{C_1} + \frac{1}{C_2} + \frac{1}{C_3} + \cdots$$

or for two capacitors,

$$C_{\text{total}} = \frac{C_1 \times C_2}{C_1 + C_2} \text{ (product divided by sum)}$$

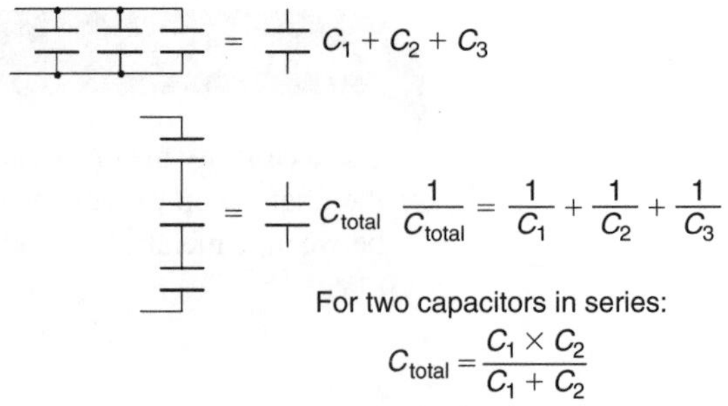

$$= C_1 + C_2 + C_3$$

$$= C_{total} \quad \frac{1}{C_{total}} = \frac{1}{C_1} + \frac{1}{C_2} + \frac{1}{C_3}$$

For two capacitors in series:

$$C_{total} = \frac{C_1 \times C_2}{C_1 + C_2}$$

**Figure 5.8** Capacitors in parallel and in series

These formulae for capacitors in combination are at first sight similar to those governing resistors in combination. The important difference is that the formula for adding resistors in series is the same as that for adding capacitors in parallel, while the formula for adding resistors in parallel is the same as that for adding capacitors in series.

## Capacitance meters

Capacitance meters, once expensive, are now available at reasonable prices. One type makes use of a set of standard capacitors to compare values, using the type of circuit called a bridge (Figure 5.9). The capacitor is connected to two terminals, and the instrument set for capacitance readings using one of the set values. A dial is turned until a meter or another display reads zero, and the value of capacitance is found by multiplying the dial reading by the value of standard capacitance selected. If no zero point can be found, another set value needs to be used.

More commonly now, direct-reading capacitance meters are used. These alternately charge and discharge the capacitor from a set voltage and measure the current that flows, calculating the capacitance from the current value and the rate at which the capacitor is being charged and discharged. In practice, the capacitor is connected between terminals, and the instrument switched on. The capacitance is then read from a digital display.

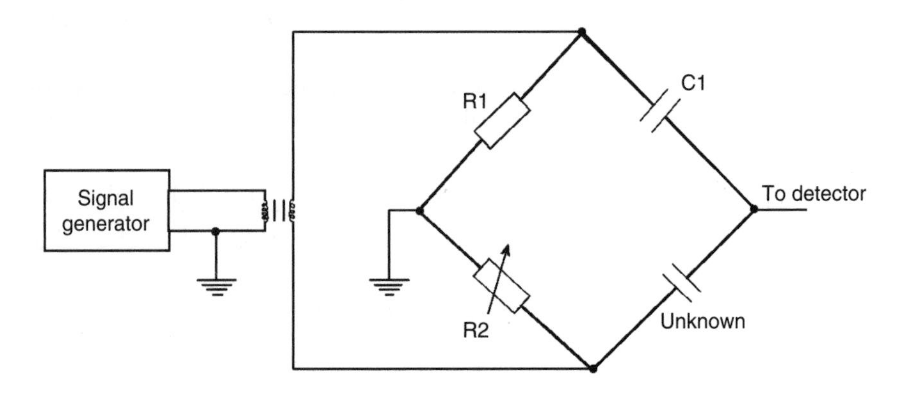


**Figure 5.9** Principles of a capacitance bridge

## Practical 5.3

Assemble a network (filter network) containing capacitors and resistors. Test the network for continuity and resistance.

## Multiple-choice revision questions

5.1 Two metal plates are insulated from each other. The capacitance between them will increase if the:
(a) plates are brought closer
(b) plates are moved farther apart
(c) area of the plates is reduced
(d) insulation is removed.

5.2 A 1 F capacitor is found to have a potential of 10 V across its terminals. The charge on the capacitor is:
(a) 0.1 coulomb
(b) 1.0 coulomb
(c) 10 coulombs
(d) 100 coulombs.

5.3 A d.c. voltage applied to an electrolytic capacitor must be:
(a) accompanied by a.c.
(b) of correct polarity and voltage
(c) of correct polarity and any voltage
(d) of correct voltage and any polarity.

5.4 Total capacitance is increased when capacitors are connected in:
(a) bridge form
(b) series
(c) series parallel
(d) parallel.

5.5 The reactance of a capacitor to a signal frequency will be large if the capacitance is:
(a) small and the frequency is high
(b) large and the frequency is high
(c) large and the frequency is low
(d) small and the frequency is low.

5.6 A capacitor is used in a radio circuit for tuning, so that it must have stable characteristics. The most suitable of the types listed below is:
(a) paper
(b) electrolytic
(c) porcelain
(d) ceramic tubular.

# 6 Waveforms

## Measuring waves

A **waveform** is the shape of a wave of alternating voltage (or current). Since electrical quantities are invisible a waveform is most easily represented by a graph of either voltage or current plotted against time. The waveform can be made visible by using the instrument called the **cathode-ray oscilloscope (CRO)**. This instrument (see later) can also be used to make measurements on waveforms, mainly of voltage **amplitude** and time **period** (or wave duration). For test purposes, waveforms can be generated by another indispensable instrument, the **signal generator**, which has controls for setting the waveform, frequency and amplitude of its output signals.

The amplitude of a waveform is the value of the voltage or current, and it varies in value during one cycle of a wave. The **peak-to-peak** (p-p) amplitude of a wave is easily measured on the face of a CRO tube (Figure 6.1), where it is found by measuring the vertical distance in centimetres between peaks of the wave and then multiplying this distance by the settings of the Volts/cm switch of the oscilloscope.

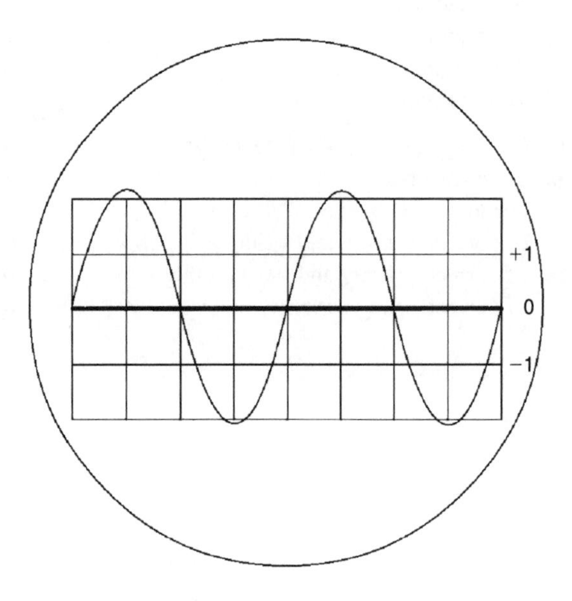


**Figure 6.1** Peak-to-peak amplitude being measured on an oscilloscope

For example, a p-p distance of 3 cm at a setting of 5 V/cm corresponds to $3 \times 5\,\text{V} = 15\,\text{V}$ p-p. On a graph, the p-p amplitude can be read from the calibrated vertical scale of the graph.

**Peak amplitude**, as opposed to peak-to-peak amplitude, is sometimes used as a measurement when the waveform is symmetrical. A symmetrical waveform has exactly the same shape above and below its centre line. Figure 6.2 shows a sine wave (which is always symmetrical) and a square wave (which is not always symmetrical, but is in this example).

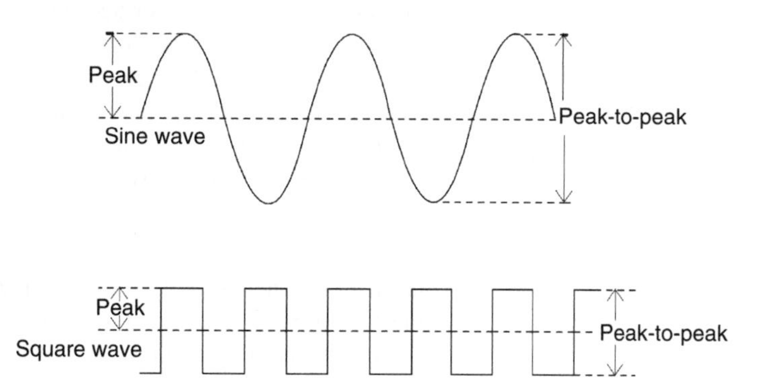


**Figure 6.2** Peak amplitude and peak-to-peak amplitude on symmetrical waveforms

Meters that are used to read values of wave amplitude do not normally measure the p-p amplitude. A meter that is set on its direct current (d.c.) voltage (or current) range will usually read zero when it is used for alternating current (a.c.). An a.c. meter is usually calibrated to read a quantity called **root mean square (r.m.s.)** value (see later). A waveform will have an **average value** of voltage or current flow in one direction if the area of the graph of the wave above (or below) the centreline of the graph is greater than the area of the graph of the wave in the other direction (Figure 6.3). This would produce a reading on a d.c. meter as well as on an a.c. meter, but these reading amounts would not be the same. When a mixture of d.c. and a.c. is present in a circuit, the average value as measured by a d.c. meter shows the amount of d.c. present.

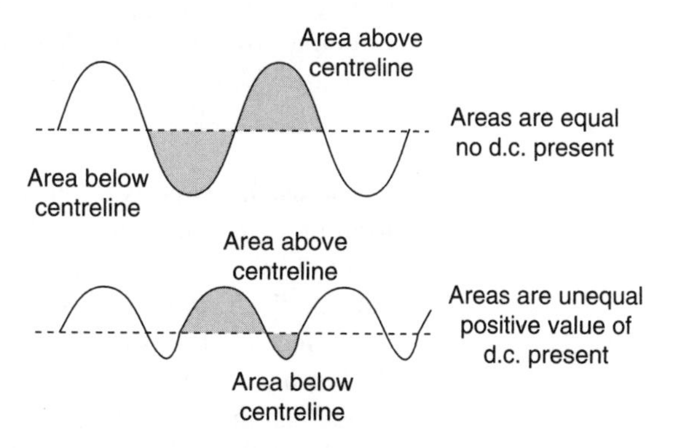


**Figure 6.3** The d.c. component of differing waveforms

Another method of measuring amplitude is used mainly for sine waves. It is called the **r.m.s.** When a sine wave has a value of current flow of 1 A r.m.s., it means that it will produce the same amount of power in a resistor as will a steady (d.c.) current of 1 A.

Equally, 1 V r.m.s. of a sine wave applied across a resistor will produce the same amount of power dissipation as will 1 V of d.c. When a waveform of a.c. is a sine wave, the r.m.s. value of its amplitude will be 0.707 times its peak value (which is the same as the peak value divided by the square root of 2). Root mean square values are particularly useful if we need to calculate the power dissipated by a.c., because we can use the same $V^2/R$ or $I^2R$ or $V \times I$ formulae as we use for d.c.

Waveforms of other shapes have different relationships between their r.m.s. and peak values; so meter readings of r.m.s. voltages and currents should be taken only when sine waves are being measured. Some specialized instruments (such as 'intelligent' digital multimeters) exist which measure the r.m.s. value of any waveform, but you will not find them much used in servicing applications.

Root mean square quantities are important for power calculations in circuits whose signals are sine waves, because they can be used for a.c. in exactly the same way as the d.c. voltage and current reading can be used for d.c. For example, the equations for power dissipated in a resistor $R$ with current $I$ flowing and voltage $V$ across the resistor will be: $P = V.I$ or $P = V^2/R$ or $P = I^2R$ whether $V$ and $I$ are d.c. values or a.c. r.m.s. values.

Measurements of voltage and current for mains-frequency a.c. are made using the a.c. scales of the multimeter just in the same way as measurements on d.c. circuits. The readings are always, unless you are using a specialized instrument, in r.m.s. units, which is normally what is required for a.c. mains readings. Measurements of voltage and time period for signal frequencies are carried out using the oscilloscope.

## Practical 6.1

Use a multimeter set to a suitable a.c. voltage range to measure the output from a nominal 6 V transformer. Ensure that no contact can be made with the primary connections.

The **oscilloscope** is an instrument that is capable of displaying a waveform on a specially produced glass screen. This screen is engraved with a set of divisions (such as centimetres and millimetres) referred to as a **graticule**. At the heart of the display system is a sawtooth waveform that is used to produce the horizontal beam deflection across the tube face. This causes the beam to traverse the screen relatively slowly during the forward writing period and then rapidly fly back to repeat the process in a continuous manner. While this is in progress, the signal to be examined, the work signal, is applied to the vertical deflection system so that a two-dimensional pattern that shows its amplitude, periodic time and general shape can be displayed.

The signal to be examined is connected as an input to the Y amplifier section by way of a switched attenuator which typically provides an input impedance ranging from 1 to 10 M in parallel with a capacitance of 10–20 pF. This attenuator is calibrated in terms of volts per centimetre of vertical graticule distance. The horizontal (X) deflection system is driven from a sawtooth generator stage that provides a waveform with very low distortion. This signal is further amplified before being used to drive the X deflection system. The timebase sawtooth can be synchronized to the input (Y) signals.

Like all test instruments, it is important to be able to rely on the accuracy of the readings obtained. The CRO usually has a built-in calibrator that needs to be checked against some standard. The substandard typically consists of an internally generated square wave with a frequency of 1 kHz and an amplitude of 1 V. When used as the work input, the gain of the Y amplifier can be preset to indicate an amplitude of 1 V. The periodic time for this waveform is 1 ms and thus the duration of one timebase scan can be similarly adjusted.

To measure the peak-to-peak amplitude of a signal, the vertical distance, in cm, between the positive and negative peaks must be taken, using the graticule divisions. This distance is then multiplied by the figure of sensitivity set on the **Volts/cm** input sensitivity control.

To measure the duration of a cycle of an a.c. signal, its periodic time, between successive positive or negative peaks, the horizontal distance between them is taken, using the graticule scale. This distance in centimetres is then multiplied by the time value read off the **Time/cm** switch scale. The frequency of the wave can then be calculated from the formula:

$$\text{frequency} = 1/(\text{periodic time}).$$

## Practical 6.2

Use a signal generator connected to an oscilloscope to view and measure sine and square waveforms.

The **periodic time** (or duration) of a waveform is another important quantity which can be measured by using an oscilloscope, or which can be read from a graph. The period of a wave is the time taken for a complete cycle of the wave, as illustrated in Figure 6.4.

To measure this quantity on the CRO display, the waveform must be **synchronized** or **locked**, which means that the display must be steady on the screen, not drifting across it. The distance in centimetres between corresponding points is then measured. The corresponding points may be the positive or the negative peaks, or the zero voltage levels at the points where the voltage is changing in the direction, or any other features which repeat once and once only, per cycle. The measured distance in centimetres is then multiplied by the setting of the Time/cm switch to give the time period in seconds, milliseconds or microseconds. Special care is needed when this

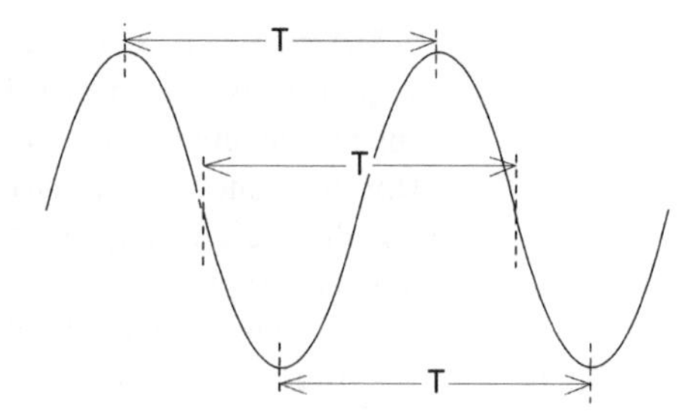


**Figure 6.4** Measuring periodic time for a waveform. The time is always measured between corresponding points (three possible sets of points shown here)

reading is taken because some models of oscilloscope incorporate potentiometer control of time/cm, in addition to the switch, and this potentiometer must be set to one end of its travel before the time reading is taken.

The **frequency** of a wave is the number of complete cycles of the wave accomplished per second and is equal to 1/period. If the period is measured in seconds, 1/period gives the frequency in units of hertz (Hz), or cycles per second. If the period is measured in milliseconds (ms), the frequency value will be in kilohertz (kHz). If the period is measured in microseconds (μs), the frequency value will be in megahertz (MHz).

The frequencies used in electronics are often identified by their *range* of value rather than by quoting exact frequencies. Thus, the expression low frequency (LF) covers all frequencies below 50 Hz extending right down to d.c. Audio frequency (AF) covers the range of frequencies from about 40 Hz to 20 kHz, which is about the range of sound wave frequencies which can be detected by the (younger) human ear.

Frequencies higher than the audio range are collectively known as radio frequencies (RF), but certain ranges of RF have distinctive names. For example, the range 460–470 kHz is called intermediate frequency (IF) because of its special use in AM superhet receivers (see Chapter 17). Frequencies in the range 30 MHz to about 300 MHz are called very high frequency (**VHF**), and frequencies above about 300 MHz ultra high frequency (UHF). The UHF range is often taken as 300 MHz to 3 GHz (3000 MHz), with the range 3–30 GHz labelled as **SHF** (super high frequency), and frequencies above 30 GHz as extremely high frequencies (**EHF**). These names are misleading and difficult to remember, and it is always better to refer to the frequency range in MHz or GHz. More often now the term **microwave frequencies** is used to mean the frequencies from around 1 GHz upwards. In this range, the range of around 10–13 GHz is used for satellite television transmission, and Global Positioning Satellite (GPS) systems use a frequency around 1.5 GHz. Mobile phones use two bands of frequencies in the range 890–960 MHz. Radar uses frequencies ranging from 15 GHz upwards.

## Phase

Many circuits can cause the phase of a sine wave to shift. This means that the peak of the waveform at the output does not occur at the same time as the peak of the wave at the input (Figure 6.5). Phase shift is measured in degrees, with 360° representing one complete cycle of difference between input and output waves. On this scale, 180° represents half a cycle of difference, and 90° a quarter of a cycle.

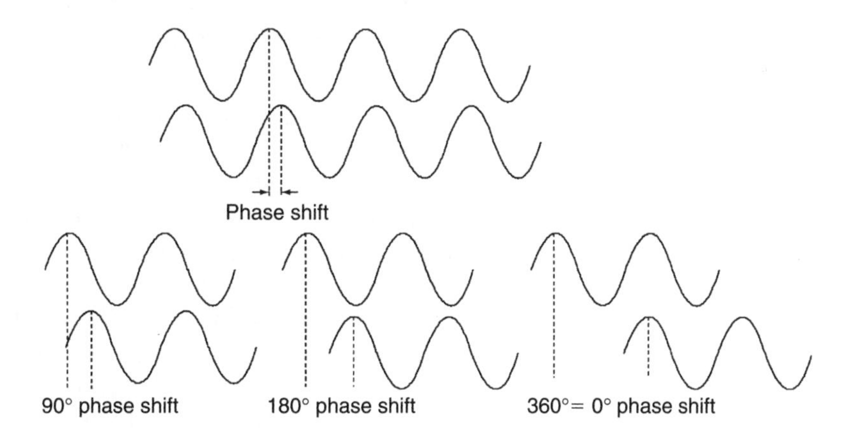


**Figure 6.5** Phase shift illustrated

A double-beam oscilloscope, which uses two separate Y (vertical deflection) amplifiers and a common timebase to control two CRT beams, can be made to show phase shift by displaying both input and output waveforms on the same timebase. When the time difference $t$ between the waves is measured, the phase shift is given in degrees by the expression: $t/T \times 360°$, where $T$ is the time period or duration of a complete cycle of the waveform.

**Example:** The time duration between the peaks of two sets of waves of the same 50 μs period is 10 μs. What is the phase shift in degrees?

**Solution:** Substitute the data in the equation $t/T \times 360°$, then $(10/50) \times 360 = 72$, a phase shift of 72°.

**Example:** The phase shift between two waves is 45°. If the frequency of each wave is 5 kHz, what will be the time difference between peaks as measured by the oscilloscope?

**Solution:** Using $t/T \times 360 =$ shift and remembering that 5 kHz is 5000 Hz with a period of 1/5000 seconds = 0.0002 s, we get $t/0.0002 \times 360 = 45$, so that $t = (45 \times 0.0002)/360$, a time of $2.5 \times 10^{-5}$ or $25 \times 10^{-6}$ seconds, 25 μs.

## Capacitors and inductors

Resistors behave to a.c. as they behave to d.c., dissipating heat when current flows, and obeying Ohm's law. The expressions for power dissipated, as we have noted earlier, are the same as for d.c. provided that we work with r.m.s. a.c. values. Capacitors and inductors behave quite differently.

As far as a d.c. circuit is concerned, a capacitor is an insulator. By contrast, an inductor is simply a resistor. In an a.c. circuit, both capacitors and resistors act to pass a.c., but will limit the amount of current that an applied a.c. voltage can cause, just as a resistance limits the amount of current that a d.c. voltage will cause.

There is, however, an important difference. When an a.c. voltage is applied to a capacitor or an inductor, the **phase** of the voltage wave across the component is not the same as the phase of current through the component.

To be precise, a perfect capacitor and a perfect inductor will cause a 90° phase shift of voltage compared to current. Practical capacitors are almost perfect in this respect, but inductors (because they have some resistance) are not, and the phase shift in an inductive circuit can never be exactly 90°.

The other difference between a capacitor and an inductor is that the phase shift is in opposite directions. A capacitor will cause the voltage wave across it to be 90° later than the current wave. An inductor will force the voltage wave to be 90° ahead of the current wave. There is a convenient way of remembering these phase shifts: **C-I-V-I-L**, which you can think of as meaning **C**: **I** before **V**; but **V** before **I** in **L**. These phase shifts can be represented in a phasor diagram (Figure 6.6a). In this type of diagram, the phase of current is represented by a horizontal line. The phase of voltage across a capacitor is indicated as a line at 90° downwards, and the phase of voltage across an inductor as a line vertically upwards.

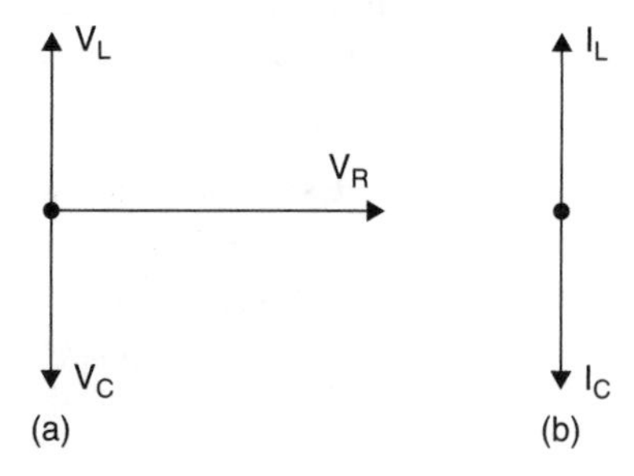


**Figure 6.6** (a) Phasor diagram for a circuit with L, C and R, and (b) phasor diagram for a circuit with C and L only

In case you think this is just academic, think what can happen in a circuit that contains a capacitor in parallel with an inductor (Figure 6.6b). In a circuit like this, the voltage across both components is the same, and can be represented as a horizontal line. The phase of current through the capacitor will then be represented by a line vertically upwards (think of the V-I example for a capacitor and turn it around to make the voltage line horizontal). The phase of current through the inductor will be represented by a line vertically downwards. The current that flows through the combination of C and L will be the difference between these two current values (because they are in different directions).

Now we know that the reactance of a capacitor or an inductor changes as the frequency of the applied voltage or current is changed. The changes are in opposite directions, so the reactance of the inductor increases as we increase frequency, and the reactance of the capacitor decreases as we increase frequency. There must be some frequency at which these reactance values are exactly equal in size, but in opposite phase.

This condition is called **resonance**, and at resonance the currents in the inductor and the capacitor are exactly equal and opposite, so that no current flows in the rest of the circuit. As far as the rest of the circuit is concerned, the parallel combination of capacitor and inductor has a very high resistance so that very little current flows through the combination. In practice, because the inductor has resistance and its phase angle is not exactly 90°, the resistance of the circuit at resonance is not infinite, but it is very much higher than it is at any other frequency. The importance? This is how every radio, television, or any other communication device selects one frequency out of all the others.

### Practical 6.3

Assemble CR and LR circuits and use a double-beam oscilloscope to show the phase shift in a circuit containing a resistor and a capacitor and in a circuit containing a resistor and an inductor.

## Block diagrams

A block diagram of an electronic circuit is used to show what the circuit does rather than details of how it does it. A block diagram simplifies fault diagnosis, because it is often easier to diagnose, from a block diagram, the general area of the circuit that might be faulty. This type of diagnosis can be much more difficult if you use a full circuit diagram. In addition, many systems that use integrated circuits can be illustrated only by using block diagrams.

Another reason for the use of a block diagram is that the full circuit diagrams of some electronic devices, such as colour television receivers and particularly computers, are far too large and complex to draw on a single sheet of paper. The circuit is divided into blocks, with one sheet of paper showing the full circuit diagram of one block, with all (or most) of the components and connections drawn in. The relation of each of these blocks to one another is then indicated on a full block diagram of the whole device.

Block diagrams can also be used to show the waveforms that ought to be present at various points in a circuit. These waveforms can be made visible with the aid of a CRO. Any significant change in the shape or the amplitude of a waveform can indicate a fault in that particular part of the circuit.

Both block diagrams and waveforms showing typical signals will be used throughout this book. Their purpose and use should be thoroughly understood at this stage.

The shapes that you see in a block diagram are, not surprisingly (but not always), rectangular blocks. One notable exception is the amplifier block, a triangle with one side indicating input(s) and the opposite corner indicating output. Each block shows signals in and signals out, along with a name for the block and, very often, small sketches of the signal waveforms. All power supplies are ignored, as are individual components.

If you are to make sense of a block diagram, however, you must know what the circuit is intended for and how it deals with the actions. The block diagram for a computer monitor, for example, does not mean a lot until you know what input signals exist and how they are used to affect a cathode-ray tube or LCD screen. The point is important: words and pictures go together and both are needed; you cannot learn from block diagrams alone. A picture, in the old proverb, might be worth a thousand words, but you often need to use a thousand words to explain a picture, even if it is not a piece of modern art.

## Multiple-choice revision questions

6.1 The peak-to-peak amplitude of a wave could be measured using:
(a) a digital voltmeter
(b) an oscilloscope
(c) an analogue meter
(d) a multimeter.

6.2 If a wave has a d.c. component, then it must be:
(a) a sine wave
(b) a square wave
(c) a wave symmetrical about earth potential
(d) a wave asymmetrical about earth potential.

6.3 A 1MHz wave travelling at $3 \times 8$ m/s has a wavelength of:
(a) 3 cm
(b) 30 cm
(c) 300 cm
(d) 300 m.

6.4 In an a.c. series circuit, the sum of the voltages across the components is not the same as the applied voltage. This is because:
(a) only currents can be added
(b) there are phase differences between the voltages
(c) the instruments are not intended to measure a.c.
(d) the components are not intended to be used with a.c.

6.5 Radio signals are tuned using the principle of:
(a) resistance
(b) phase
(c) resonance
(d) amplitude.

6.6 In a block diagram, an amplifier is indicated by a:
(a) square
(b) triangle
(c) rectangle
(d) circle.

# Unit 3
## Electronic devices and testing

### Outcomes

1. Demonstrate an understanding of semiconductor diodes and d.c. power supplies and apply this knowledge safely in a practical situation
2. Demonstrate an understanding of semiconductor active devices and apply this knowledge in a practical situation while observing safe practices.

Note: This unit calls for practical work in assembling and testing transistor and IC amplifier circuits. This work is dealt with in Chapters 6 and 11 because it calls for knowledge of waveforms that has not yet been established.

# 7 Semiconductor diodes

Semiconductors are not simply materials whose resistivity is somewhere between that of a conductor and that of an insulator. Although a pure semiconductor will have a resistivity value that is not as high as that of an insulator, it certainly does not approach the low value that we would expect of a conductor. The two items that class a material as a semiconductor are the effect of temperature and the effect of impurity, and both of these effects are closely related.

Suppose, for example, that we have a specimen of pure silicon. Its resistivity is very high, so this material would normally be thought of as an insulator. When the pure silicon is heated, however, its resistivity drops enormously. Although the drop is not enough to place hot silicon among the ranks of good conductors, the contrast with any other insulators is quite astonishing.

The effect of impurities is even more amazing. Even tiny traces of some impurities, one part per hundred million or so, will drastically change the resistivity of the material. It is because of this remarkable effect of impurity that it took so long to discover semiconductors; it was only in the twentieth century that methods were discovered of purifying elements such as silicon to the extent that the resistivity of the pure material could be measured.

The main difference between a semiconductor element and any other element is that a semiconductor can have almost any value of resistivity that you like to give it. It is, in other words, a material that can be engineered to have the electrical characteristics that you want of it. The manipulation is done by adding very small quantities of other elements.

The first semiconducting materials to be used were the metallic elements germanium and silicon, and silicon is still the most used semiconductor material. The compound called gallium arsenide is also a semiconductor, however, and it can be used to form diodes and transistors that will operate at very high frequencies, well into the microwave region. Gallium arsenide transistors are used in the low noise block (LNB) circuit that is part of a satellite television aerial system; this amplifies the feeble microwave signals and converts them to lower frequencies. Gallium arsenide is also used to form light-emitting diodes (LEDs) (see later in this chapter).

There are two ways that electric current can be carried in solid materials, and this applies particularly to crystals of solids. Crystals are never perfect, and when a crystal of an almost pure material has been deliberately made impure (or **doped**), the crystal contains atoms of a different type. If these atoms possess more or fewer electrons in their outer layer than the normal atoms of the crystal, then the electrical characteristics will also change.

One way of changing the characteristics is to release more electrons. The other way is to release more holes. A **hole** is a part of a crystal that lacks an electron. Because of the structure of the crystal, a hole will move from atom to atom and when it does, it behaves just as if it were a particle with a positive charge. Within the crystal, the hole has a real existence; its charge and even a figure for its mass can be measured. The important difference is that the hole is a gap in a crystal, it does not exist outside the crystal. The electron, by contrast, can be separated from the crystal, and can even move in a vacuum, as in a cathode-ray tube.

By adding impurities to pure semiconductor materials, then, we can create the value of resistivity we want, and it can be arranged so that most of the current is carried either by electrons or by holes. If most of the current is carried by electrons, the material is called **N-type**, and if most of the current is carried by holes it is called **P-type**. Note that it is only the current carriers inside the material whose positive or negative charges are important; the numbers of these positive and negative charges in the crystal are equal, and the material as a whole has no charge.

A **junction** is an area inside a crystal where P-type material meets N-type material.

You cannot create a junction by taking a piece of N-type semiconductor and placing it in contact with a piece of P-type material, because you could never get the pieces close enough. The junction has to be created so that these opposite types meet within the crystal. By connecting a wire to the N-part and one to the P-part of the crystal, the component called a **diode** is created. The symbol for a diode is illustrated in Figure 7.1(a). For such a diode, the portion using P-type material is the **anode** and the portion using N-type material is the **cathode**.

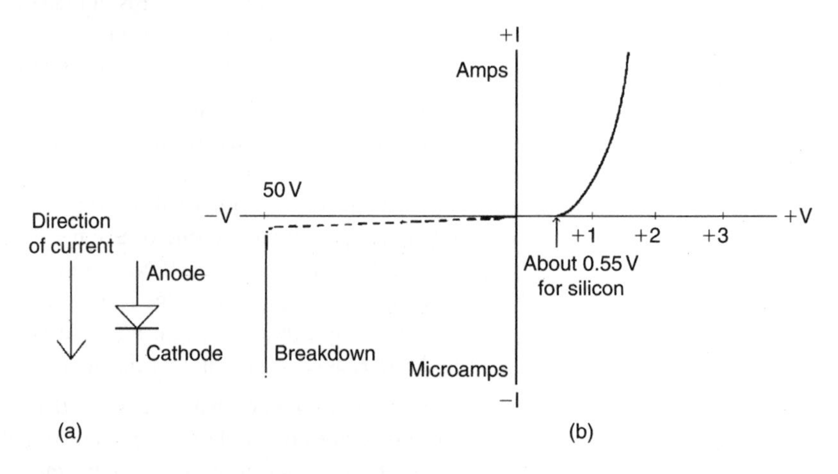


**Figure 7.1** (a) Diode symbol, and (b) a typical characteristic for a diode using P and N silicon (a silicon diode). Note the different scales used for voltage and current

A diode does **not** obey Ohm's law, so we cannot use $V = RI$ or any other simple formula to find a *constant* relationship between current and voltage. The relationship $V/I$, which is the resistance at each different value of current, must be read from a graph, and a graph of this type is called a **characteristic**. A typical characteristic for a semiconductor diode is illustrated in Figure 7.1(b), assuming a diode made from P-type and N-type silicon. This shows that a diode is a one-way path for current. The resistance of the diode is different for each part of the graph.

The characteristic is really made up from two separate graphs. The right-hand side is the graph for current flow, meaning that the diode has its anode (P-type material) connected to the positive side of a supply. This is referred to as the **forward** direction, and the external voltage is called a **forward bias**.

The graph axes are scaled in terms of small voltages and currents, and the graph shows that no current flows until the forward voltage is about 0.6 V. When the diode starts to conduct, the graph shape is not the straight line that would be expected of a resistor, but a curve (an exponential curve). The curvature is upwards, so that the current value is multiplied for each small increase in voltage. There is no constant value of resistance; the behaviour is as if the resistance decreased as the voltage was increased.

When the voltage is applied in the reverse direction, so as to make the junction non-conducting, the voltage scale on the graph has to be changed to show anything happening. For this particular diode, nothing is measurable until the voltage reaches a high value, around 100 V in this example, at which point the diode becomes conducting with a very low resistance. Unless there is some means of limiting the current flow, such as a resistor connected in series, the current that passes in this condition will destroy the junction. The reverse voltage that is needed to cause this effect is called the **breakdown voltage**. Applying breakdown voltage to a diode is harmful only if excessive current is allowed to flow, and the principle of reverse breakdown is used deliberately in Zener diodes (see later in this chapter). The Zener is just one of a large set of diodes that are intended for specialized purposes.

A diode can be tested simply by using a multimeter to find which way round the diode has to be connected to register low resistance, and it should have a high (unreadably high) resistance to the flow of current in the opposite direction. Some meters have special diode-testing ranges, but the resistance range of an ordinary meter can also be used for this purpose.

When some meters are switched to the resistance range, the red (positive) lead becomes negative, and the black (negative) lead becomes positive. The diode conducts when the black lead is connected to the anode and the red lead to the cathode, and there is no reading when the connections are reversed. Note that some diodes, notably LEDs, can be damaged by the battery voltage of a resistance meter.

Diodes can fail open-circuit (o/c) or short-circuit (s/c), and either type of failure is easily detected by testing with a multimeter. An o/c diode will not pass current in either direction; it has a high resistance in both directions. An s/c diode will pass current equally easily in both directions.

A simple circuit that can be used for diode checking is illustrated in Figure 7.2(a). When current flows, the anode of the diode is connected as illustrated.

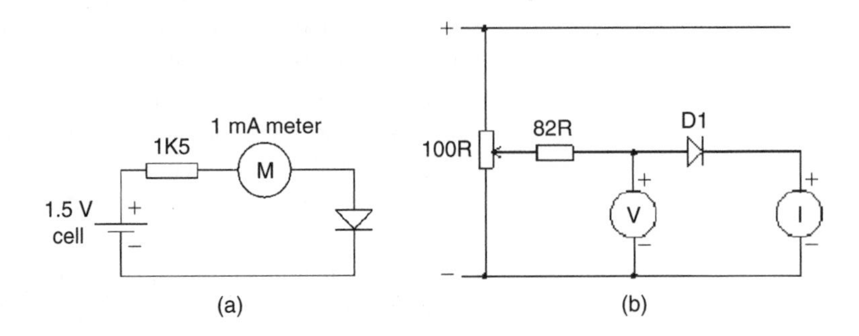


**Figure 7.2** (a) Testing a diode, and (b) circuit for Practical 7.2

## Practical 7.1

Use an ohmmeter to identify the anode and the cathode connections of an unmarked diode. A good diode should indicate a very high resistance in the reverse direction, and a low resistance in the forward direction. Remember that many multimeters operate with reverse polarity when the resistance range is selected, and be careful to avoid touching the leads when using the highest resistance ranges.

## Practical 7.2

Connect the circuit illustrated in Figure 7.2(b), using a silicon diode. Turn the potentiometer control so that the voltage across the diode will be zero when the circuit is switched on. Make sure that you know the scales of current and voltage you are using: a 1 V or 1.5 V scale of voltage and a 10 mA scale of current. The voltmeter should be of the high-resistance type.

Watching the meter scales, slowly increase the voltage. Note the reading on the voltmeter when the first trace of current flow is detected. This value of voltage is called the contact potential. Below this value, the diode does not conduct.

Note the voltage readings for currents of 1 mA, 5 mA and 10 mA. Draw a graph and ask yourself whether the diode obeys Ohm's law with a constant resistance value. Note that this circuit measures the voltage across the current meter, but this should be very small compared to the voltage across the diode.

## Diode types

Diodes of many types exist, and they can be classified roughly as signal, power, Zener, LED and photo types. **Signal diodes** are used for tasks such as demodulation (see Chapter 16), using a high-frequency signal into the diode. Signal diodes need not have a very low resistance when conducting, and it is more important to have very low values of stray capacitance between anode and cathode.

**Rectifier diodes** are used in power supply units (PSUs) (see later in this chapter). The principle is like commutation, changing over the connections to an a.c. supply twice on each cycle so that only the positive peaks are passed. Because the cathode of the rectifier diode is the positive d.c. output terminal, it is often marked with a + sign, or coloured red.

**Schottky diodes** use a different type of junction formed between aluminium and just one type of silicon. This has the advantage of conducting at a low forward voltage, and is used in modern power supplies, particularly for computers, to reduce heat dissipation.

**Zener diodes** are used in quite a different way. They are connected in the reverse direction, with the cathode connected to a positive voltage. A Zener diode has a low breakdown voltage, so that it must be used with a resistor in series to prevent damage. In this type of circuit (Figure 7.3a), the voltage across the diode is almost perfectly constant, even if the current varies considerably. This type of diode is used to obtain a steady voltage for regulators (see later). In the illustration, the input is a d.c. voltage higher than the Zener voltage, and the output is the Zener voltage, which does not change if the input voltage changes slightly. Zener diodes are available in a range of voltages, and the figures are written using a letter V in place of a decimal point, such as 5V6.

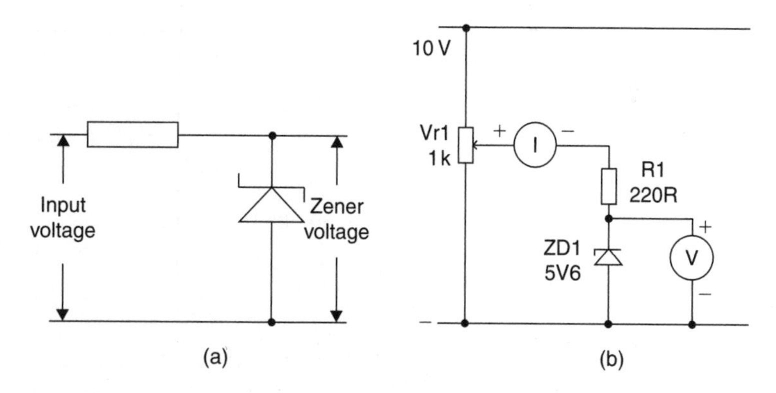


**Figure 7.3** (a) Using a Zener diode, and (b) testing a Zener diode

**LEDs** are diodes that emit light when the diode conducts in the normal forward direction. The usual colours are red or green, and some diodes can give a yellow light. There are also LEDs whose light changes colour when the voltage is changed. All LEDs have a very low breakdown voltage, so that even connecting then to a 1.5 V cell in the opposite direction can cause irreparable damage. LEDs are also easily damaged by excess current, so a limiting resistor must be placed in series between the LED and its supply voltage.

Information on diode connections and ratings can be obtained from manufacturers' data-books, from independently published data-books, or from the Internet, using a search engine such as Google.

## Practical 7.3

Connect a Zener diode in the circuit illustrated in Figure 7.3(b), making sure that the cathode of the diode is connected towards the positive pole of the supply. The voltmeter should be set to the 10 V range, and the milliammeter to the 10 mA range. Switch on, and adjust the potentiometer so that the voltage across the diode can be read for currents of 1 mA, 5 mA and 10 mA. Note the voltage readings for each of these current values.

If the first reading had been taken across a resistor, what would the reading for 10 mA have been? (Hint: find the resistance value and then use the potential-divider formula.)

## Practical 7.4

Connect an LED in the circuit illustrated in Figure 7.4. The voltmeter should be set to the 5 V range and the milliammeter to the 10 mA range. Be **very** careful to connect the LED with the correct polarity.

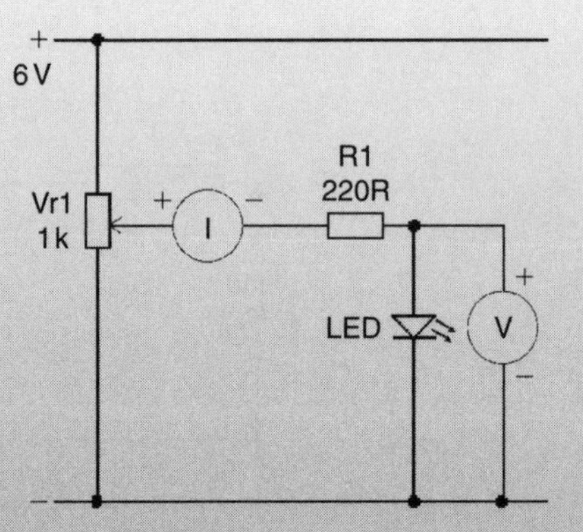


**Figure 7.4** Testing an LED

Starting with the potentiometer at its lowest setting, switch on and slowly turn up the voltage until the diode conducts. Note the voltage across the diode when the current first starts flowing. Examine the light output at current values of 1 mA, 5 mA and 10 mA and note the value of forward voltage at these currents.

**Photo diodes** are used with reverse bias, but they will conduct when they are struck by light. The currents that can pass are very small, a few microamps, so that amplification is needed. Photo diodes are used in light detectors when a fast response is needed; more sensitive devices are used if a slow response is acceptable.

**Varicap** or **varactor** diodes are used for tuning high-frequency circuits. These diodes are used with reverse bias, and are designed so that changing the amount of reverse voltage will alter the capacitance between anode and cathode. The varactor is used along with a fixed capacitor as part of a tuned circuit, so that altering the steady bias voltage will alter the tuned frequency.

Clipping, d.c. restoring and limiting are processes that can be carried out on waveforms of any shape, but which need diodes as well as the resistors, capacitors and inductors that are used in wave filters.

**Clipping** means the removal of part of either peak of a wave. A simple clipping circuit is shown in Figure 7.5(a). The silicon diode does not conduct when its anode is negative, and will conduct only when its anode voltage reaches about 0.5 V positive. The clipping circuit as a whole acts like a potential divider, with the diode resistance being very high for voltages of less than 0.5 V. The waveform is therefore clipped at about +0.5 V. Such a clipped waveform will have a d.c. component unless both positive and negative peaks are clipped equally. A circuit for doing this is shown in Figure 7.5(b).

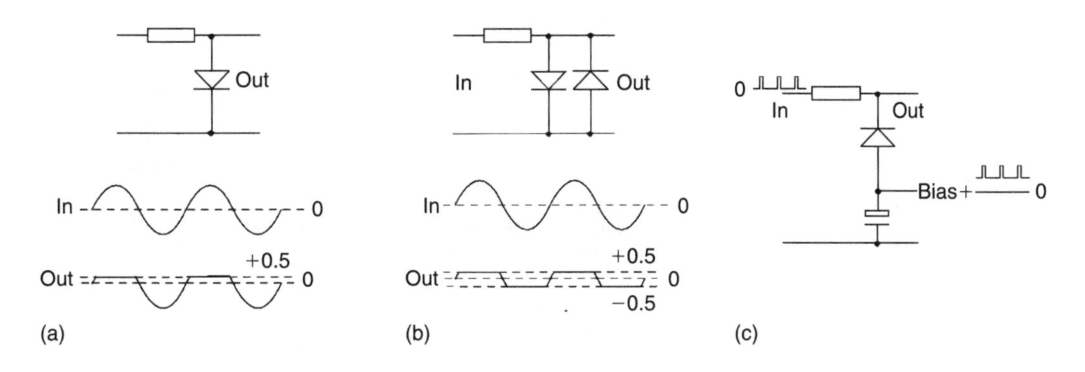


**Figure 7.5** (a) A simple clipping circuit, and (b) two-way clipping, (c) d.c. restoration

Restoration (or **d.c. restoration**) is the process of restoring to a signal the d.c. level which it will have lost if it has been passed through a capacitor or a transformer. Figure 7.5(c) shows a typical application. The unidirectional waveform has been passed through a capacitor, so losing its d.c. component. A simple d.c. restoring circuit then replaces the missing d.c. and the d.c. level is again present. The diode in the circuit does not conduct when the waveform is positive, but passes current when the wave swings negative relative to the zero (earth) line. When the diode conducts, capacitor C is charged positively and has a positive voltage between its plates when the diode ceases to conduct. This positive voltage is equal to the missing d.c. voltage, and the diode will thereafter conduct only as required on the negative peaks to keep the voltage across the capacitor at the correct level.

## Power supplies

The cheapest method of operating any electronic circuit whose power consumption is more than a few milliwatts is by means of a power supply unit (**PSU**) that draws its energy from the a.c. mains supply. The purpose of a PSU is to raise or, more usually, lower the mains supply voltage to an amplitude suitable for the application, and then to convert this a.c. voltage into a steady d.c. output. Sometimes additional circuits called **regulators** or stabilizers are needed to ensure that the d.c. output voltage remains steady even when mains voltage varies, or if the amount of current taken from the PSU is altered.

The a.c. voltage conversion is carried out by a transformer. The a.c. voltage from the mains is applied to the primary of the transformer, and induces a voltage between the terminals of the transformer secondary. The value of this secondary voltage depends on the relative number of turns in the two windings. Most modern transistor equipment requires a lower secondary voltage, usually within the range 6–50 V.

The first stage in the conversion of the low a.c. voltage into d.c. is carried out by a diode rectifier, or series of rectifiers, which alter the waveshape of the a.c. into one of the waveshapes shown in Figure 7.6, of which the full-wave (bridge) rectifier arrangement is preferred for most applications.

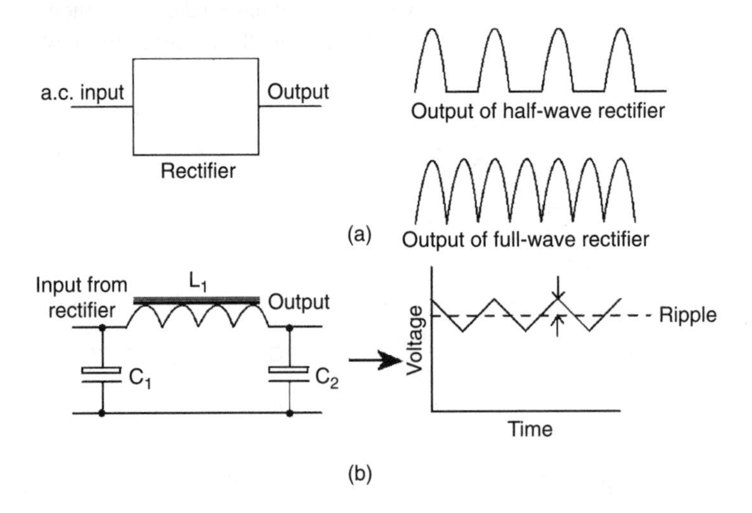


**Figure 7.6** (a) Rectifier action, and (b) smoothing. The dashed line shows average d.c. value

The result of rectification is to produce the waveforms shown on the right-hand side of Figure 7.6, which are of a voltage of one polarity only. This voltage will have an average value that is equivalent to d.c. of that same voltage, but is **not** a smooth, steady d.c. voltage.

To convert this output into smooth d.c., a reservoir capacitor and filter must be used. The reservoir capacitor ($C_1$) is charged by the current from the rectifier(s), and it then supplies current to the circuit every time the voltage output from the rectifiers drops. In this the waveform is smoothed into d.c. plus a small amount of remaining a.c. called **ripple**, indicated in Figure 7.6. Another capacitor (shown as $C_2$) can be used for further smoothing.

The ripple is almost zero when only a little current is taken from the PSU, but it increases considerably when more current is taken. If it becomes unacceptably high, a larger value of reservoir capacitor must be used, or additional filter stages must be added. Another way to reduce ripple is to add a **regulator** circuit (see later).

When no load current is taken from a rectifier circuit, the output from a half-wave circuit, as measured by a d.c. voltmeter, will be identical to that from a full-wave circuit supplied with the same d.c. voltage. When a load current flows, however, the output voltage of a half-wave rectifier circuit will drop by a greater amount than will that of the full-wave rectifier circuit, unless a much larger value of reservoir capacitor is used. Half-wave rectifier circuits are used in microwave cookers, but are seldom used in any other equipment.

The a.c. ripple that appears when the load current is taken from a rectifier circuit will be at mains frequency when the rectifier is a half-wave one, but at twice mains frequency (in the UK, 100 Hz) when the rectifier is a full-wave circuit. When the load current taken is very large, the voltage output from a half-wave supply will fall to about half the value of that from a full-wave supply connected to the same a.c. voltage. The most common type of rectifier circuit used now is the full-wave bridge type, which is usually obtained in a package form with two (a.c.) inputs and two (d.c. + and −) outputs. The bridge circuit and some typical bridge rectifier packages are illustrated in Figure 7.7.

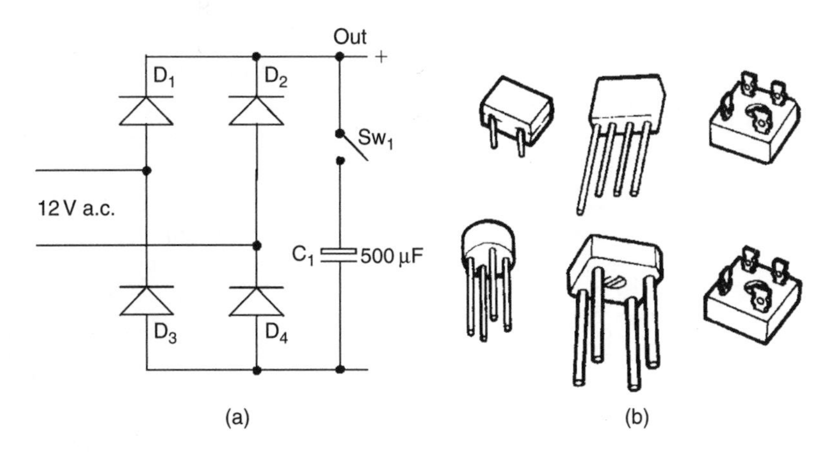


**Figure 7.7** (a) Bridge rectifier circuit, and (b) typical bridge rectifier packages

Neither type of supply has good regulation, although they are adequate for many purposes. Regulation is the quantity that measures the drop of output voltage that occurs when load current is taken from the supply circuit. A perfectly regulated supply would have constant output voltage whatever the noted load current; in other words, it would have a zero value of internal resistance.

Failure in any part of a conventional PSU is easy to diagnose. Symptoms are complete loss of output voltage, a drop in output voltage, excessive ripple or unusually poor regulation. Complete failure can be caused by a blown fuse or by an o/c transformer winding. A short-circuit in a half-wave rectifier will also bring output to zero. A short-circuit winding in the transformer will, in addition, generate excessive heat.

A drop in voltage, usually accompanied by excessive ripple, can be caused by the failure of one of the rectifiers in a full-wave set, or by an o/c reservoir capacitor. Poor regulation can also be caused by an o/c reservoir capacitor, or by a diode developing unduly high resistance.

## Practical 7.5

Connect the a.c. rectifier/reservoir circuits shown in Figure 7.8(a). (Note that these are intended for use with 12 V a.c. supplies only; they must never be connected to the mains.)

Use the a.c. voltage range of a multimeter to set the a.c. input to 12 V, and measure the output voltage of each circuit. Note down the readings under appropriate headings. Note also the output readings when the reservoir capacitors are temporarily disconnected, using the switch. Compare the ripple amplitude and frequency using an oscilloscope.

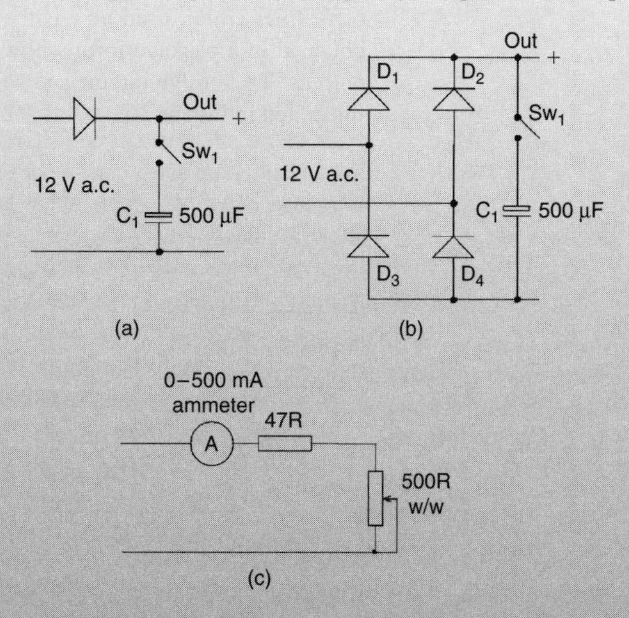


**Figure 7.8** Rectifier circuits: (a) half-wave, (b) full-wave bridge, and (c) load circuit for Practical 7.5

Now connect to each output in turn the load circuit (c). Measure the output voltages at various load currents from 50 mA to 200 mA for both rectifier circuits, and draw up a table of output voltage and load current. Note the effect on the output voltage and waveforms when any one diode is open circuited (so reverting to a half-wave circuit). Draw a graph of output voltage plotted against load current for each type of rectifier circuit.

## Regulation

A regulated power supply includes a transformer-rectifier circuit, with smoothing, and a regulator section. This is shown in block diagram form in Figure 7.9. The rectifier could be half-wave, bi-phase half-wave or bridge, but is almost always the bridge type. The reservoir capacitor is always a large-capacity electrolytic.

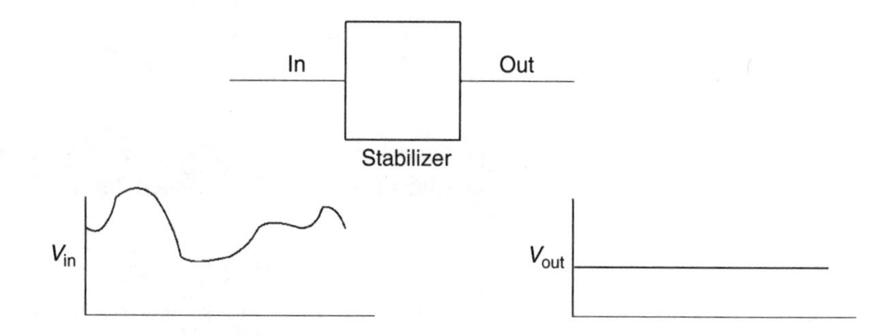


**Figure 7.9** Block diagram for a regulated supply. Note that the regulated output will be lower than the minimum level of the input, otherwise regulation would fail at this input voltage

The circuitry for a regulating portion of a PSU is so standardized that it is available in integrated circuit (IC) form, and this form is nearly always used in conventional (not switch-mode) PSUs. The regulator ICs can be obtained in a huge range of voltage output, current output and maximum dissipation values. If the output from a regulated supply appears unregulated or is very low this indicates failure of the regulator. This can be checked by disconnecting the connection between the reservoir capacitor and the regulator and testing the output from the rectifier unit. If this shows full voltage, the rectifier is all right but the regulator is not.

One important use for rectifier circuits is battery charging, and two main types of circuit are used. Lead-acid cells need to be recharged from a constant-voltage supply, so that when the cell is fully charged, its voltage is the same as that of the charger, and no more current passes. By contrast, nickel–cadmium cells must be charged at constant current, with the current switched off when the cell voltage reaches its maximum. Constant-current charging is needed so that excessive current cannot pass when the cell voltage is low. Simple transformer-rectifier circuits can be used for constant-voltage charging, but some form of regulation is preferable, particularly for sealed lead-acid cells.

Nickel–metal-hydride cells need a more complicated charger circuit, and one typical method charges at around 10% of the maximum rate, with the charging ended after a set time. Some types of cell include a temperature sensor that will open-circuit the cell when either charge or discharge currents cause excessive heating. For some applications trickle-charging at 0.03% of maximum can be used for an indefinite period.

Lithium-ion cells can be charged at a slow rate using trickle-chargers intended for other cell types, but for rapid charging they require a specialized charger that carries out a cycle of charging according to the manufacturer's instructions.

Note that there is no such thing as a completely universal battery charger and although several microprocessor-controlled chargers are available that can be used on all types of cells they cannot be used on sets of mixed cell types.

## Practical 7.6

Measure the internal resistance of a bridge rectifier circuit with an output of about 8 V. Now connect in a regulator IC whose regulated output is 5 V. Measure the internal resistance of the resulting circuit.

## Multiple-choice revision questions

7.1 The feature that is most typical of a semiconductor is:
(a) its resistivity value is between that of a conductor and an insulator
(b) its resistivity can be changed by adding tiny quantities of impurity
(c) it is composed of crystals
(d) it melts easily.

7.2 A diode:
(a) passes current in one direction only
(b) has a linear *V–I* graph
(c) becomes less conductive when it gets hot
(d) is unaffected by light or temperature.

7.3 Clipping is an action that is used:
(a) to make a waveform more linear
(b) to improve frequency range
(c) to restore the d.c. level of a signal
(d) to limit the amplitude of a signal.

7.4 An LED must be connected so that:
(a) only a very small current can flow
(b) only a very large current can flow
(c) the voltage across it is constant
(d) no reverse current can flow.

7.5 A full-wave bridge power supply:
(a) delivers perfectly smooth d.c.
(b) uses four diodes
(c) needs a special transformer
(d) has perfect regulation.

7.6 An IC regulator:
(a) needs an input voltage higher than the regulated output
(b) needs a high input current
(c) allows the output voltage to be varied
(d) does not reduce ripple voltage.

# 8 Transistors

## Bipolar junction transistor

Although transistors are seldom used as separate components in modern circuits, it is important to know how they work, because they form the basis of the integrated circuits (ICs) that are used in virtually all electronic circuits today.

The **bipolar junction transistor (BJT)** is a device that makes use of two junctions in a crystal with a very thin layer between the junctions. This thin layer is called the **base**, and the type of transistor depends on whether this base layer is made from P-type or from N-type material. If the base layer is of N-type material, the transistor is said to be a **P-N-P** (PNP) type, and if the base layer is of P-type material, the transistor is said to be an **N-P-N** (NPN) type. The differences lie in the polarity of power supplies and signals rather than in the way that the transistors act. For most of this chapter, we shall concentrate on the NPN type of transistor, simply because it is more widely used. Figure 8.1 shows the symbols for the NPN transistor and its opposite, the PNP type.

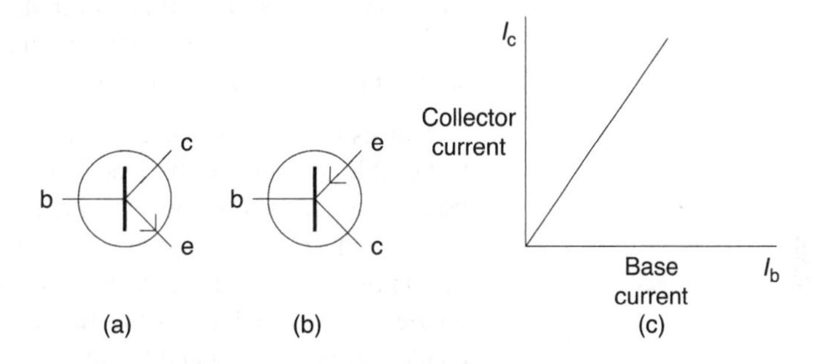


**Figure 8.1** (a) NPN BJT symbol, (b) PNP BJT symbol, and (c) $I_c/I_b$ graph for NPN BJT

The transistor has three connections, called **emitter**, **base** and **collector**, shown labelled as e, b and c in Figure 8.1. For the NPN type, the transistor is connected so that the collector is positive, with the emitter at a lower voltage, often earth (zero) voltage. With no connection to the base, this arrangement does not pass current. When a small positive voltage, about 0.6 V for a silicon transistor, is applied to the base, a base to emitter current, $I_b$, will flow, and when a base current flows, a current, $I_c$, also flows between the collector and the emitter. Figure 8.1(c) shows these quantities graphed. Note that the direction of the arrow on the emitter part of the symbol is used to distinguish NPN from PNP: it shows the (conventional) direction of current.

The collector current, $I_c$, is always proportional to the base current $I_b$; a typical value is that the current between emitter and collector is, on average, some 300–800 times the current between base and emitter. The exact size of this ratio depends on how thin the base layer is, and it varies considerably from one transistor to another, even when the transistors are mass-produced in the same batch. This value is variously known as **forward current gain** or $\mathbf{h_{fe}}$.

The action of the PNP transistor is similar, except that polarities are changed, with the collector negative, and a negative bias voltage applied to the base. In circuits that use both NPN and PNP transistors, the PNP symbols are often drawn upside down so that the emitter of the PNP transistor can be drawn connected to the more positive voltage.

Transistor types include small-signal transistors, radio frequency (RF) transistors, power transistors and switching transistors. The small-signal types are used for amplifiers that work at frequencies ranging from audio to the lower radio frequencies. The RF transistors are intended for the higher radio frequencies, and some specialized types can be used at super high frequency (SHF) (in satellite receiving dishes, for example). Power transistors are used for amplifier outputs, for driving solenoids and motors, for television use, and for transmitter outputs. Switching transistors are used with digital circuit to switch large amounts of current or high voltages in very short times, in the order of nanoseconds.

Transistors exist in a huge variety of sizes and shapes, and it is important to be able to identify the electrodes (emitter, base and collector). Data on transistors can be obtained from manufacturers' books, independent source-books and Internet sources. Two well-known source-books are Tower's International Transistor Selector and the International Transistor Equivalents Guide (by Michaels). On the Internet, try sites such as www.ajpotts.fsnet.co.uk/transistors.html and the more extensive www.datasheetarchive.com/

For older equipment, the website http://vintageradio.me.uk/info/transistorcode.htm is very useful in identifying transistors and suggesting replacements. It also contains an extensive range of data sheets, although these can be difficult to read on a small monitor screen. You should use the 'Save Target As' facility of the browser to make a copy that you can load into a graphics program and magnify.

If no data is obtainable, you can use a resistance meter to identify the connection, because for an NPN transistor the base will conduct to both other electrodes when the base is positive. If you connect the positive lead of the resistance meter to one lead, and test by holding the negative lead to the other two in turn, the positive lead is on the base if both tests show a low resistance. You may have to try three times until you clip the positive lead to the correct connection.

Transistor types are coded either by the US MIL specification or by the European (Pro-Electron) lettering system. The US system starts with 2N followed by a code number; the 2N portion identifies the device as a transistor (1N indicates a diode). The full description can be found only if you can identify the manufacturer by looking up the code number.

The European system, used for all types of semiconductors, uses two or three letters followed by a number. The letters indicate the type of device, with the numbers indicating how recent the design is. Table 8.1 shows the meanings of the letters.

**Table 8.1** European (Pro-Electron) coding letters

| Letter | Meaning |
|---|---|
| The first letter indicates the semiconductor material used: | |
| A | Germanium |
| B | Silicon |
| C | Gallium arsenide and similar compounds |
| D | Indium antimonide and similar compounds |
| R | Cadmium sulphide and similar compounds |
| The second letter indicates the application of the device: | |
| A | Detector diode, high speed diode, mixer diode |
| B | Variable capacitance (varicap) diode |
| C | AF (not power) transistor |
| D | AF power transistor |
| E | Tunnel diode |
| F | RF (not power) transistor |
| G | Miscellaneous |
| L | RF power transistor |
| N | Photocoupler |
| P | Radiation detector (photodiode, phototransistor, etc.) |
| Q | Radiation generator |
| R | Control and switching device (such as a thyristor) |
| S | Switching transistor, low power |
| T | Control and switching device (such as a triac) |
| U | Switching transistor, high power |
| X | Multiplier diode (varactor or step diode) |
| Y | Rectifier, booster or efficiency diode |
| Z | Voltage reference (Zener), regulator or transient suppressor diode |

The figures or letters following indicate the design. A three-figure serial number is used for 'consumer types', used in domestic radio, television, tape-recorders, audio, etc. A serial consisting of a letter (Z Y, X, W, etc.) followed by two figures means a device for professional use (transmission, etc.). The European system is much more informative, because you can tell the type of device immediately from the code. The 1N, 2N system tells you only whether the device is a diode or a transistor.
AF: audio frequency; RF: radio frequency.

The Japanese system is shown in Table 8.2.

**Table 8.2** Japanese transistor coding

| *Code* | *Device type* |
|---|---|
| 2SA | PNP transistor |
| 2SB | PNP Darlington |
| 2SC | NPN transistor |
| 2SD | NPN Darlington |
| 2SJ | P-channel MOSFET or JFET |
| 2SK | N-channel MOSFET or JFET |
| 3SK | Dual-gate N-channel FETs |

### Practical 8.1

Test both PNP and NPN transistors using a multimeter. For an NPN transistor you should find a low resistance when the meter is connected with base positive and emitter or collector negative. For the PNP transistor, the low resistance reading will occur for base negative and emitter or collector positive. All other readings should show high resistance. Make also a set of readings on transistors that are known to be faulty with either short-circuit (s/c) or open-circuit (o/c) connections.

The common BJT faults are base-emitter s/c or o/c and base-collector s/c or o/c. The o/c faults will prevent any collector current from flowing, revealed by the collector voltage rising to the same level of power supply voltage. The s/c faults will result in excessive collector current flowing so that the collector voltage is very low. Note that faults in the bias components can also cause these symptoms.

## BJT amplifier

Figure 8.2 shows a simple voltage amplifier constructed using a single bipolar junction transistor, in this example, an NPN type. The input signal is taken to the base, whose steady voltage level is fixed by two resistors. The resistor in the emitter circuit controls the steady (bias) current. The load resistor is placed in the collector circuit, and the output signal is taken from the collector. The same principles are used when the load is an inductor or a tuned circuit, or when the input is taken from a transformer. At this stage, no component values are shown, but usually the capacitors are of high values, several microfarads, and the resistors have low values, a few kilohms.

Although this form of simple amplifier is seldom seen now because of the extensive use of ICs, you should be able to identify the input and output points, and the load. The resistor in the emitter circuit is part of the bias arrangement, along with the resistor divider chain connected to the base. The emitter resistor is usually in parallel with a capacitor, and a s/c fault in this capacitor will result in the transistor passing as much current as the

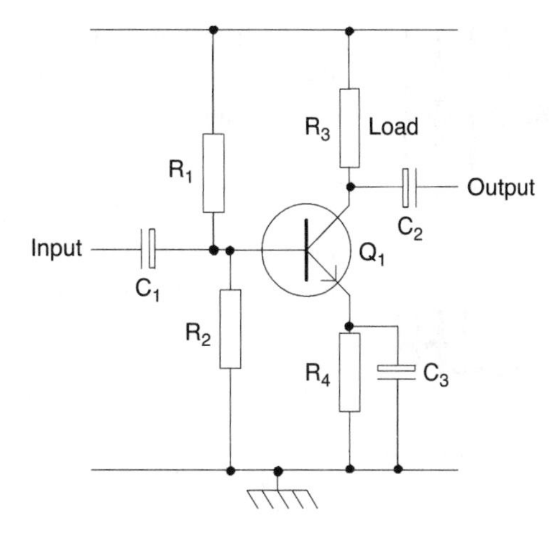


**Figure 8.2** A simple BJT voltage amplifier. Note the use of Q to identify a transistor, because Tr can be used for a transformer

load permits, with a very low voltage at the collector, and no amplifying action. Similarly, s/c faults in the base bias resistors will cause the transistor to be biased either fully on or fully off. The simple amplifier will be more fully explained in Chapter 12, with a practical exercise in construction and testing.

Note that many ICs are constructed using a different type of transistor, the MOSFET (metal oxide semiconductor field-effect transistor). Because you will seldom encounter this type of transistor other than as part of an IC, its construction and action will not be covered here. More details of MOS transistors are provided in the L3 section (Chapter 28).

## Practical 8.2

Assemble the simple BJT amplifier circuit of Figure 8.2, using the component values: $R_1$ = 47k, $R_2$ = 10k, $R_3$ = 2k2, $R_4$ = 560R, $C_1$ = 10 μF, $C_2$ = 10 μF, $C_3$ = 100 μF, $Q_1$ = BC107 or equivalent.

## Heat sinks

When a transistor conducts, the current that flows between the collector and the emitter causes heat to be dissipated. Most of this heat comes from the collector-to-base junction, and because the heat is concentrated in a small area, it can cause damage, eventually melting the junction and destroying the transistor instantly.

When transistors are used for circuits in which large amounts of current flow, some way of removing this heat is needed, and is provided by using

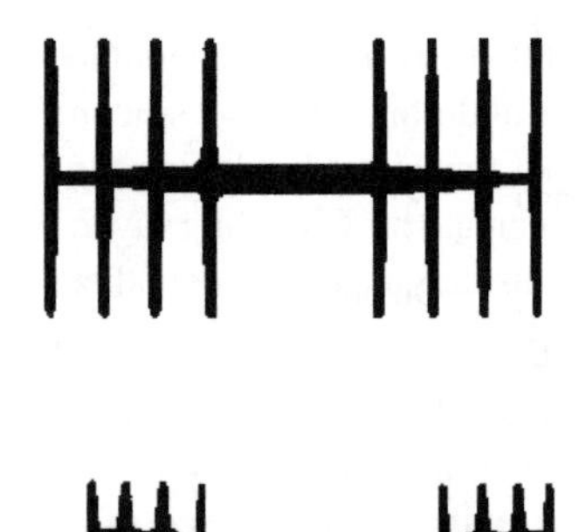

**Figure 8.3** Typical heat-sink shapes, seen side-on

a heat sink. For small transistors this can be a simple metal fin clipped to the metal body of the transistor. A heat sink for power transistors (Figure 8.3) is a metal plate, often quite heavy and fitted with fins. The power transistor is bolted to this plate, usually with a thin layer of heat-sink grease between the transistor and the plate. This grease is a silicone grease that conducts heat well, and is also a good electrical insulator. The heat from the junction spreads to the plate then so to the air, preventing the transistor from becoming too hot. The use of a heat sink allows more power to be dissipated, but the power must be kept within the limits that the heat sink has been designed for, otherwise damage can still be done. Heat sinks for computer ICs often incorporate fans.

> Note that modern power transistors can run so hot that they will cause burns if you touch them. A fuse cannot protect a transistor. A fuse takes several milliseconds to blow, but a transistor can have its junctions vaporized in less than a microsecond.

## Integrated resistors

Many modern circuits make use of sets of resistors with identical values, and in these circuits it is usual to use an integrated resistor set. The aim is to provide a single component for a circuit that requires a large number of identical resistors, and these integrated resistors can be packaged as SIL (single in line) or DIL (dual in line), referring to the line(s) of pins that are used for connections. Typically, these units will contain anything between five and 15 resistors; one very common value is eight. These resistor units are often called thick-film circuits because of their construction from comparatively thick metal films. Figure 8.4 shows typical packaging used for integrated resistors. Testing a resistor in such an array is no more difficult than testing a single resistor on a board, provided that you can locate the connections to each end of the resistor you want to test.

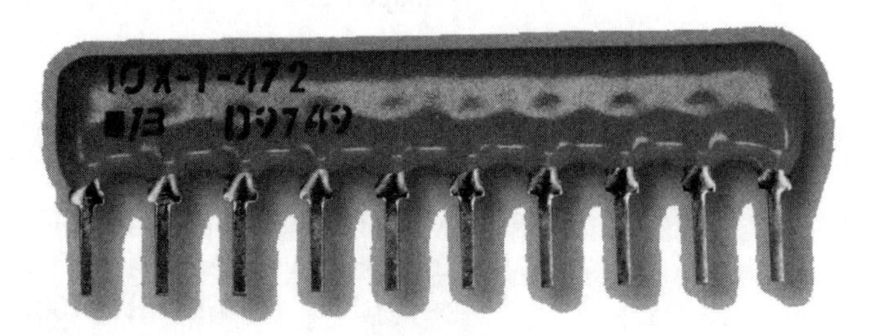


**Figure 8.4** A typical SIL integrated resistor package [photograph: Alan Winstanley]

## Integrated semiconductor circuits

A transistor of any type, bipolar or field-effect (FET), is manufactured by a set of operations on a thin slice, or wafer, of silicon crystal. These operations include selective doping and oxidation (for insulation), and the areas that are affected can be controlled by the use of metal masks placed over the semiconductor. By using different amounts of doping of a strip of semiconductor material, the resistance of the strip can be controlled, so that it is possible to create a resistor on a semiconductor chip. In addition, because silicon oxide is an excellent insulator, it is possible to make capacitors by doping the semiconductor to make a connection, oxidizing to create an insulating layer, and depositing metal or semiconductor over the insulation to form another connection.

Since transistors, resistors and capacitors can be formed on a silicon crystal chip, then, and connections between these components can also be formed, it is possible to manufacture complete circuits on the surface of a silicon chip. The transistor types can be bipolar or FETs, or a mixture of both, but FETs are preferred because they are easier to fabricate in very small sizes. Virtually all digital ICs are of the FET type, but some analogue signal devices use integrated BJTs.

The advantages of forming a complete circuit, compared to the discrete circuit in which separate components are connected to a printed circuit board (PCB), are very great. The obvious advantages are that very small complete circuits can be made. It is possible to pack a huge number of components on to one chip simply by making the components very small, and the technology of creating very small components has advanced spectacularly. The idea of packing three million FET or more transistors on to one chip might have seemed unbelievable only a few years ago (and it is still hard to imagine even now), but this is the extent to which we have come. The advantage that this provides is that very complicated circuits can be manufactured that simply would not be economic to make in discrete form.

Another advantage of integrated construction is of cost. The cost of making an IC that contains 50 000 transistors is, after the tooling has been paid for, much the same as the cost of making a single transistor. This is the basis of the £2.50 calculator, the television remote control, the electronic toys and the small computer. The cost of ICs has been steadily reducing with each improvement in the technology, so that prices of computers go down even faster than the prices of houses go up. The greatest advantage from the use of integration, however, has been reliability, and it was the demand for 100% reliability in the electronics for the space missions that fired the demand for ICs.

> Modern ICs can be formed with several million transistors on one chip, and with very elaborate connections. Such ICs are used in computing and communications, and for the decoders for digital radio and television.

Consider a conventional circuit that uses 20 transistors, some 50 resistors and a few other components. Each transistor has three terminals and each resistor has two terminals, so that the circuit contains some 160 soldered connections. One faulty connection, one faulty component, will create a fault condition, and the greater the number of components and connections there are the more likely it is that a fault will develop.

By contrast, any IC that will do the same work might have only four terminals, so that it is connected to a PCB at four points only. The IC is a single component which can be tested: if it works satisfactorily it will be used. Its reliability should be at least as good as that of a single transistor, with the bonus that each internal part will always be working under ideal conditions if the IC has been correctly designed.

The reliability of a single component is 20 times better than the reliability of 20 components that depend on each other. If the IC does the work of 1000 components, its reliability will be about 1000 times better than that of a single component, and so on. This enormous improvement in reliability, many times greater than any other advance in reliability ever made, is the main reason for the overwhelming use of ICs. When electronic equipment fails, the first thing to check is the connections to the mains plug, not the state of the ICs.

For these reasons, circuits that use separate (discrete) transistors are becoming a rarity in modern electronics. When we work with ICs, circuits become very different. We may still be concerned with d.c. bias voltages, but usually at just one terminal. We are still concerned with signal amplitudes and waveshapes, but only at the input and the output of the IC. What goes on inside the IC is a matter for the manufacturer, and we simply use it as advised. That said, some of the equipment that comes in for servicing may use quite a large number of separate (*discrete*) transistors along with some early types of ICs. This means that we cannot ignore the older components, and we have to know about the later technology even though its reliability is so much better.

The IC is a complete circuit, and servicing amounts to deciding whether the IC is correctly dealing with the signal that is applied to its input. If it is not, then it has to be replaced. If the input and output signals are correct, the fault lies elsewhere. The use of ICs has not complicated servicing, nor has it made designers redundant. On the contrary, like all technical advances, it has resulted in a huge increase in the amount of items to service and the variety of goods that can be designed.

IC packages vary enormously, but many common types use the DIL type of packages illustrated in Figure 8.5. Specialized ICs such as are used for

**Figure 8.5** Typical IC packages: (a) DIL, and (b) 80 lead quad flat pack (QFP) [photographs: Alan Winstanley]

computers use the square package, and some have a very large pin count, approaching 1000 in 64-bit microprocessors.

In addition to multipurpose ICs, manufacturers use **application-specific integrated circuits (ASICs)**. These are designed and built to order for some specific part of a hi-fi amplifier, television receiver, computer, etc. Some of these use large packages, sometimes of the square type that is favoured for microprocessors. Such circuits are also likely to use a large number of pins.

The largest class of ICs is the **digital** type, including the microprocessor and memory chips that are produced in such enormous numbers. These types of IC were in use before ICs started to be available for other purposes, so that the digital types are in many ways more advanced in design and construction. They are almost all of FET construction, and some use 420 or more pins in the package. The thermal dissipation of computer microprocessor chips is such that a fan and heat-sink assembly must be clamped over the chip.

Since ICs are much too small to be used directly, **packages** are used. The tiny IC chip is embedded in plastic, and metal wires connect it to pins that are rugged enough for insertion into printed circuits. The most common packaging system is the DIL type, with two rows of pins. On this type of packaging, the pins are numbered starting from pin 1, which is located by a punchmark, or from its position, and the numbering continues down one side and up the other side. The surface mounted component (SMC) ICs are designed for surface mounting, and are very small, often no larger than a blob of solder. Note that pin numbering is always shown as seen looking down on the mounted IC, not from the pin side.

Data on ICs can be obtained from manufacturers' booklets, on computer data disks, or by way of other manuals. Coding of ICs will be by manufacturers' type-lettering and numbering, by Pro-Electron coding letters and numbers, or for digital ICs following the 7400 or 4000 family series that is noted in Chapter 13.

## Operational amplifier (op-amp)

A large number of ICs are intended for digital circuits, but there are others that are of the analogue type, intended for purposes such as amplification and filtering. For most purposes an IC of the type called the **operational amplifier** (usually shortened to **op-amp**) will be used in place of a multistage amplifier constructed from individual transistors. You do not usually know (nor need to know) the internal circuit of such an IC, and the important point to recognize is that the gain and other features of the op-amp will be set by external components such as resistors. The voltage gain of the circuit alone is typically very high, 100 000 or more. Figure 8.6 shows a typical op-amp symbol.

The point about performance being controlled by resistors and other external components (in a feedback loop, see Chapter 12) is true also of most amplifiers that have been constructed using individual transistors, but since the op-amp consists of one single component, it is very much easier

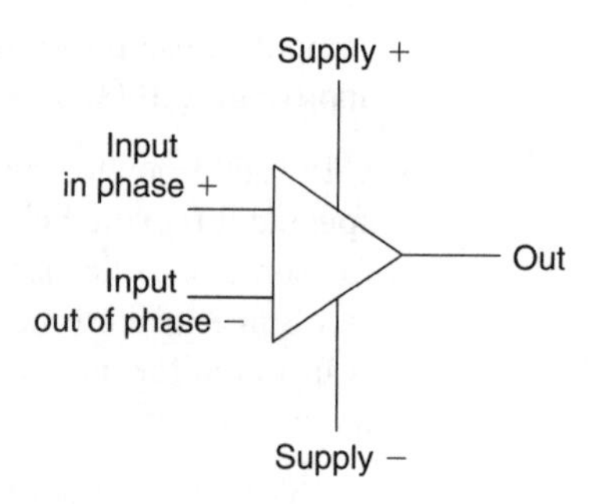


**Figure 8.6** A typical op-amp symbol

to see which components control its action, since only these components are visible on the circuit board.

Operational amplifiers can be packaged as a single unit, or as a set. Single units are usually packaged as eight-pin DIL, but the multiple unit will be placed in larger packages.

### Practical 8.3

Test a ready-assembled op-amp amplifier. Apply a 1 kHz 20 mV signal to the input and measure the output signal using the oscilloscope. Remembering that the output measurement will be peak-to-peak, calculate the gain of the amplifier at this frequency.

## IC handling

MOSFETs and ICs that are based on MOSFETs require careful handling because of the risk of electrostatic damage. Electrostatic voltages arise when materials are rubbed together, and synthetic fibres such as nylon can rub on other materials to develop voltages of several kilovolts. These voltages can cause sparks, and although the currents that flow are usually very small, the gate insulation of a MOSFET will not survive having such a voltage between gate and channel connections.

Although most devices incorporate over-voltage diodes that will conduct at a lower voltage than would damage the gate connections, there is always some risk of damage when devices are being inserted into boards. MOS devices that are soldered to boards are almost immune from accidental damage from electrostatic effects, because most boards will have resistors wired between the gate and the source or emitter terminals. Even a resistor of several thousands of megaohms is enough to reduce electrostatic voltages to a fraction of a volt because of the current that passes through the resistor.

Manufacturers of MOSFET devices need to use precautions such as using earthed metal benches, with staff wearing metal wrist clamps that are also earthed, but such precautions are seldom necessary for servicing purposes. The main risk is in opening the wrapping and taking out the IC, and you should try to avoid touching the pins. Sensitive ICs are often sold with the pins embedded in conducting plastic foam, and one useful precaution is to short the pins together (by using a metal clip or wrapping soft wire around them) until the device is soldered into place. If special precautions are specified for a device, these should be used.

## Practical 8.4

Given a printed circuit board with a known faulty component, desolder the faulty component and replace with one known to be good.

## Multiple-choice revision questions

8.1 A BJT:
(a) can be positive or negative
(b) conducts using electrons only
(c) is unaffected by temperature changes
(d) can be PNP or NPN.

8.2 A transistor is coded as BD followed by a number. It must be:
(a) Japanese
(b) an AF power transistor
(c) intended for RF amplification
(d) intended for US military applications.

8.3 An NPN transistor will conduct when:
(a) the collector and base are both positive relative to the emitter
(b) the collector and base are both negative to the emitter
(c) the collector is positive and the base negative relative to the emitter
(d) the collector is negative and the base positive relative to the emitter.

8.4 For any transistor, a plot of the following quantities is fairly linear:
(a) collector voltage and base current
(b) collector current and base voltage
(c) base current and collector current
(d) base voltage and base current.

8.5 An IC is described as DIL. Does this mean:
(a) it contains diodes in line
(b) it uses a package with two parallel rows of pins
(c) it must be constructed from MOS components
(d) it must be constructed from surface mounting components?

8.6 A heat sink:
(a) totally removes heat
(b) changes heat to electrical energy
(c) changes electrical energy to heat
(d) passes heat from a semiconductor to the air.

# Unit 4
## Electronic systems

### Outcomes

1. Demonstrate an understanding of waves and waveforms and apply this knowledge in a practical situation
2. Demonstrate an understanding of input and output transducers and apply this knowledge safely in a practical situation
3. Demonstrate an understanding of electronic modules and apply this knowledge safely in a practical situation.

# 9 Other waveforms

## Rectangular waves

You should at this point revise the section on waveforms in Chapter 6, which deals mainly with sine waves. A sine wave can be produced by rotating a coil of wire in the field of a magnet and is therefore the waveform that is produced by alternators. It can also be produced by oscillator circuits and is the type of wave used for carrying radio transmissions (see later). Another important waveform is the rectangular wave, shown in Figure 9.1, whose amplitude and frequency can be measured and noted in the same way as for a sine wave. In addition, rectangular waves require an additional measurement, called their **mark-to-space ratio**. It is possible for two rectangular waves that have the same period (and therefore the same frequency) and the same peak voltage levels (amplitude) to have quite different shapes and also quite different values of average voltage. For a rectangular wave, the **mark** is the length of time for which the voltage is positive and the **space** is the length of time for which the wave is either at zero or at a negative voltage.

A wave with a large mark-to-space ratio will thus have an average value nearly equal to its positive peak value, while a wave with a small mark-to-space ratio will have an average value nearly equal to its negative peak value (in the example shown in Figure 9.1, almost zero). Varying the mark-to-space ratio of a rectangular wave of voltage is often used as a method of d.c. motor speed control because, using electronic methods, this is simpler than varying the amplitude and much less likely to allow the motor to stall.

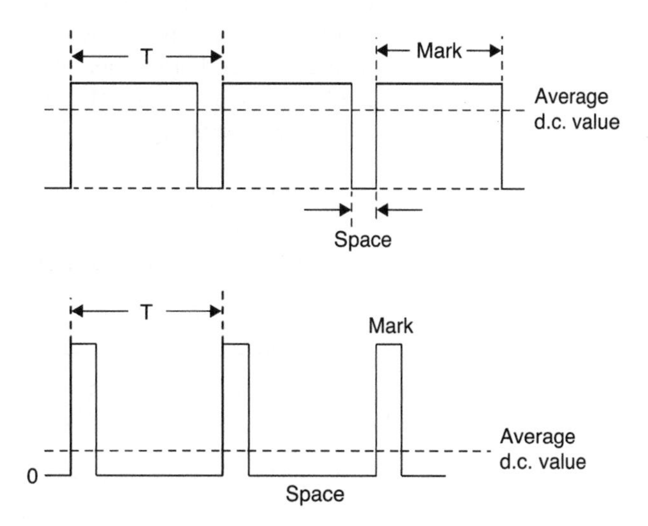


**Figure 9.1** Rectangular waveforms or pulses showing mark and space and average values

## Fundamental and harmonics

A perfect sine wave has just one single value of frequency, but waves that are not of sine wave shape (**non-sinusoidal waves**) contain other frequencies, called **harmonics**. An instrument called a **spectrum analyser** can detect what harmonic frequencies are present in such waves.

One frequency, called the **fundamental frequency**, is always present. It is the frequency equal to 1/period for the waveform. No frequencies of less than this fundamental value can exist for a wave, but higher frequencies which are whole number multiples of the fundamental frequency (i.e. twice, three times its value and so on) are often present. A pure sine wave consists of a fundamental only.

When more than one frequency is present the amplitudes of the waves add to give an overall voltage at any particular time, and this resulting voltage, plotted against time, provides another waveform, the **resultant**. This resultant waveform will have a waveshape that is different from the waveforms that were added and the idea of adding two waves can be extended to the addition of the instantaneous values of many waves. The principle is illustrated here, as follows.

It is possible to show that any square or triangular waveform consists of the addition to a fundamental sine wave of an infinite number of its own harmonics. The distortions produced by non-linearity in amplifiers, in particular, can be analysed in this way. The precise shape depends on the amplitude of the harmonics as well as the frequencies. If, however, a square or triangular wave is passed through a low-pass filter which removes all the higher frequencies, the process is reversed and the higher order harmonics are removed from the input wave. When a 1kHz square or triangular wave, for instance, is passed through a filter which attenuates all frequencies above 1.5kHz, the resulting output in both cases will be a sine wave of 1kHz frequency.

Figure 9.2 shows half a cycle of a fundamental with the effect of adding odd harmonics (3rd, 5th, 7th, etc.) to a fundamental, amplitude A, with the amplitude of each harmonic inversely proportional to the square of its harmonic number. For example, we use A/9 of the 3rd harmonic, A/25 of the 5th harmonic and so on. The result, if this is taken to an infinite number of harmonics, is a triangular wave.

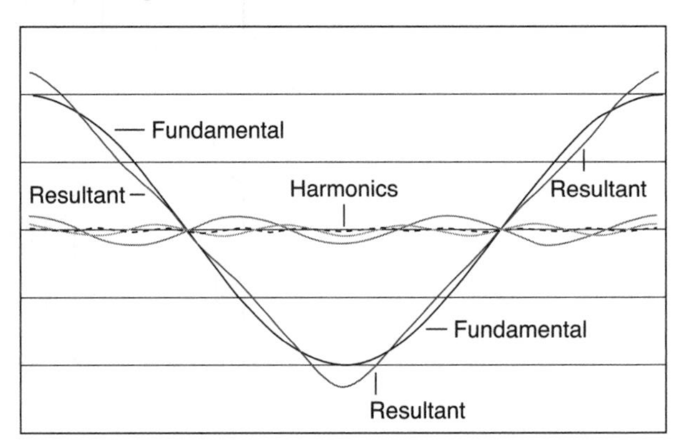


**Figure 9.2** Adding odd harmonics of constant amplitude to produce a near-triangular wave

Figure 9.3 shows the waveshapes that result from adding the 3rd and 5th harmonics to a sine wave, and the odd harmonics up to the 15th. The higher amplitudes of the harmonics are in this case inversely proportional to the harmonic number (rather than to the square), so we use A/3 of the 3rd harmonic, A/5 of the 5th harmonic, etc., so that very high harmonics have more effect on the waveshape than for the example in Figure 9.2. Pulse waveforms contain a mixture of both even and odd harmonics.

The waveshape depends not simply on whether odd or even harmonics are added, but on the relative amplitude of each harmonic and the phase of each harmonic relative to the fundamental. This is why there are no hard and fast rules about using odd or even harmonics to produce a particular waveshape.

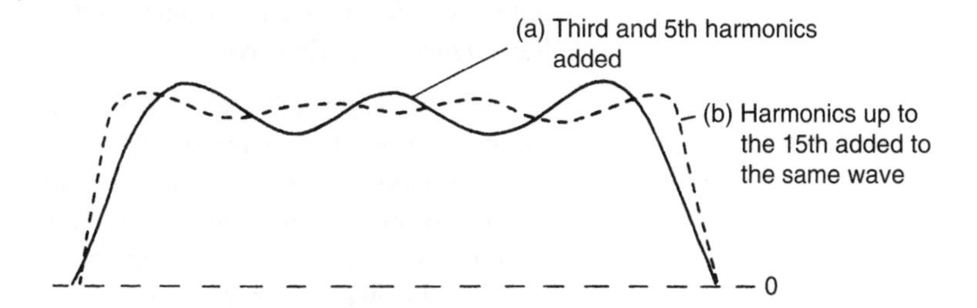


**Figure 9.3** Adding harmonics of different amplitudes to a fundamental to make a square wave

The principle of **wave analysis** is important as a method of measuring amplifier distortion. A perfect sine wave signal applied to the input of an amplifier will produce a perfect sine wave output only if the amplifier causes no distortion. Any alteration in the waveshape brought about by distortion will cause harmonics to be present in the output signal. These harmonics can be filtered out and measured to enable the amount of the distortion that has taken place to be calculated.

## Acoustic and electromagnetic waves

Although the audio frequencies in the range of around 20 Hz to 20 kHz are the same for both sound and for audio-frequency radio waves, these are entirely different types of waves. One important difference is **propagation** method. Sound waves are mechanical waves that spread (**propagate**) in solids, liquids or gases by vibrating the molecules. Each vibrating molecule passes the vibration to its neighbour, so that the wave propagates at a speed that depends on the density and elasticity of the material. Sound **cannot** travel in a vacuum; there are no sounds in space. The speed of sound is very much faster in liquids and dense solids than it is in air (speed around 300 m/s), so that sound waves travel more quickly in liquids and solids. In addition, the wave is a **longitudinal** one, meaning that the molecules vibrate in the same line as the line of motion of the wave (Figure 9.4).

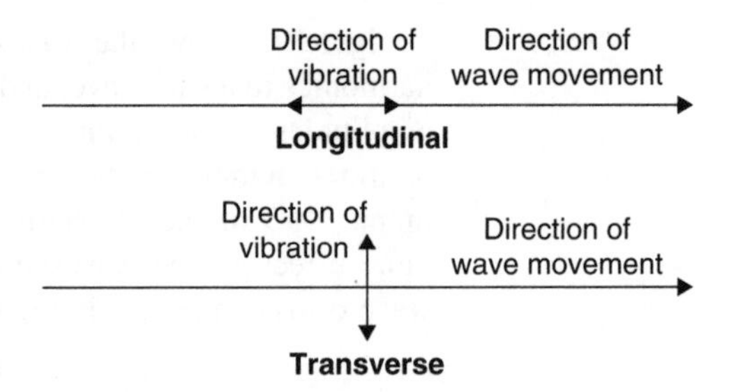


**Figure 9.4** Longitudinal and transverse waves

Acoustic waves are not necessarily of the frequency range that we can hear (the audible range). They can be **infrasonic**, meaning that the frequency is too low for the ear to detect, or **ultrasonic**, meaning that the frequency is too high for the ear to detect. Ultrasonic waves are used in electronics to provide power for ultrasonic cleaners and in a type of filter called **SAW**, meaning **surface acoustic wave**.

Radio waves, which are electromagnetic waves, of any frequency travel (propagate) because of the oscillation of electric and magnetic fields, and this can happen most easily in a vacuum (called **free space**). The waves are **transverse**, meaning that the vibration is directed at right angles to the direction of propagation, and their speed is typically 300 *million* m/s in free space, and **slower** in other materials.

Electromagnetic waves, unlike sound waves, can travel in a vacuum such as free space. The **wavelength** of a wave is a quantity that applies to any wave that is travelling in space or along a wire. It can be measured easily only when the wavelength is short, and is defined as the closest distance between neighbouring peaks of the wave. The idea is illustrated in Figure 9.5.

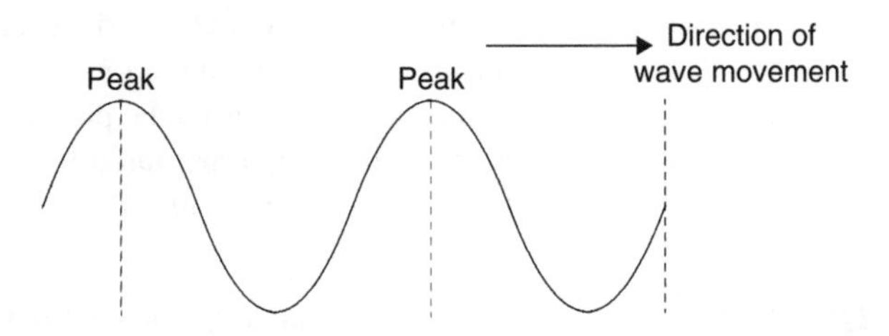


**Figure 9.5** Wavelength of a wave, the distance between peaks of the same type

In free space, the speed of all radio waves is identical at $3 \times 10^8$ m/s (remember that $10^8$ is the figure 1 followed by eight zeros, so $3 \times 10^8$ is 300 million). This is the same speed as that of light, because light is also an electromagnetic wave. By comparison, the speed of sound in still air is about 332 m/s, but this is affected by movement of the air or by movement of the source of the sound. The relationship can be written using the triangle $\frac{c}{\lambda f}$ as a reminder of the three possible forms of the equation:

$$c = \lambda f,\ \lambda = c/f \text{ and } f = c/\lambda$$

Wavelength calculations are of particular importance in the design and use of aerials for high frequency (mainly television) signals.

The peak points of the wave are called **antinodes**. They can sometimes be located on wires with the aid of lamp indicators (see Figure 9.6), but the wavelength of a wave is usually a quantity that is calculated. The formula used is $\lambda = c/f$, where $\lambda$ (Greek letter lambda) is the wavelength in metres, $c$ is the speed of wave travel in metres per second, and $f$ is the frequency of wave in Hz.

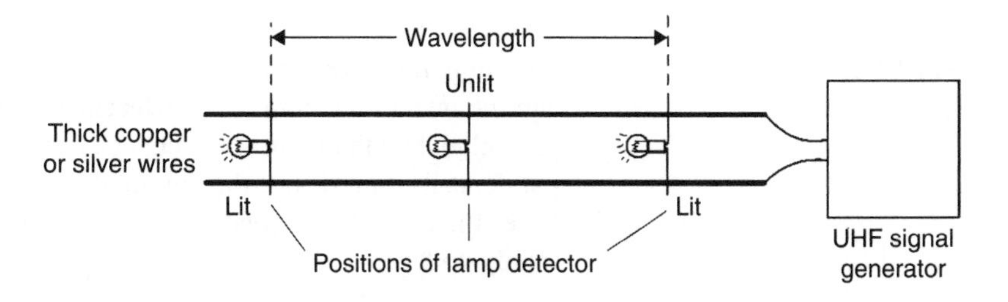


**Figure 9.6** Measuring the wavelength of UHF waves travelling along wires

**Example:** What are the wavelengths, in free space, of waves of the following frequencies: (a) 1 MHz, (b) 30 MHz, (c) 900 MHz?

**Solution:** Use the equation $\lambda = c/f$, with $c = 3 \times 10^8$ m/s. Then, (a) 1 MHz is $10^6$ Hz, so $\lambda = 3 \times 10^8/10^6 = 3 \times 10^2 = 300$ m, (b) 30 MHz is $3 \times 10^7$ Hz, so $\lambda = 3 \times 10^8/3 \times 10^7 = 10$ m, (c) 900 MHz is $9 \times 10^8$ Hz, so $\lambda = 3 \times 10^8/9 \times 10^8$ = one-third of a metre, or 33.3 cm.

A wave on a wire or in a circuit (but **not** in space) may be either pure a.c. or unidirectional. A pure a.c. wave has an average value of zero, so it can cause no deflection of a d.c. meter. A wave that is fed through a capacitor or taken from the secondary of a transformer (Figure 9.7) is always of this type, apart from certain short-lived (transient) effects that occur immediately after the circuit is switched on.

A unidirectional wave is one that has a steady average value that can be measured by a d.c. meter. Such a wave has a d.c. component, which means that its d.c. amplitude must be measured by a d.c. meter. The a.c. component of the wave can also be measured when the d.c. component is absent, by inserting a capacitor (which may have to be of large value) between

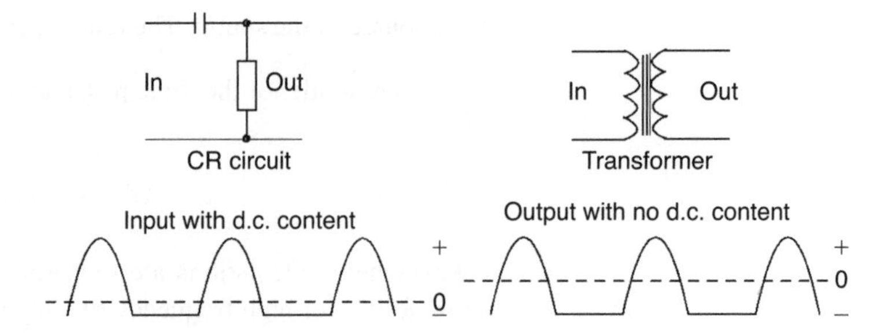


**Figure 9.7** Circuits that transmit no d.c. component

the circuit and the a.c. meter. This assumes that a suitable a.c. meter is obtainable. Most a.c. meters will give correct readings only for sine-wave waveforms.

## Waveshaping

Two important types of waveshaping circuit are the differentiating circuit and the integrating circuit. A **differentiating circuit** is one that has an output only when the input wave changes amplitude. The amplitude of its output depends on how fast the amplitude of the input wave itself changes: the faster the change, the greater the output. There is no output if the input is a steady voltage.

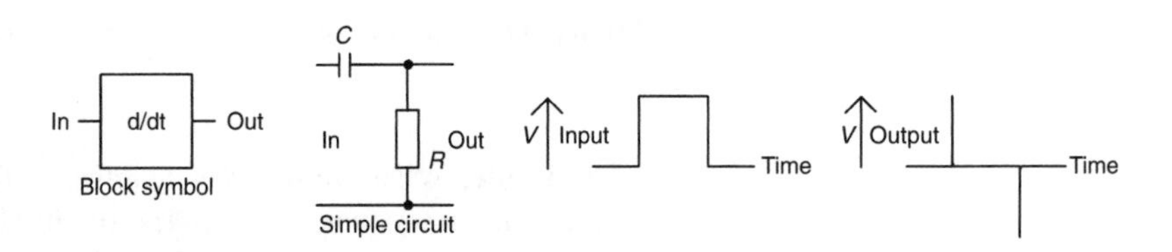


**Figure 9.8** A differentiating circuit

The normal differentiating circuit will act as a high-pass filter for sine waves. Its effect on a square wave is shown in Figure 9.8. As the illustration shows, the effect is to emphasize the changes in the wave so that the output consists of a spike of voltage for each sudden change in voltage level of the square wave. The direction of the spike is the same as the direction of voltage change of the square wave.

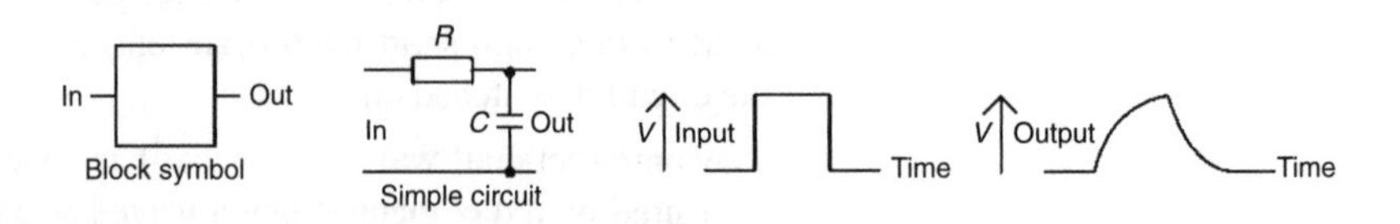


**Figure 9.9** An integrating circuit

An integrating circuit removes sudden amplitude changes from a waveform, its effect being thus the opposite of that of the differentiating circuit. Integrating action on a square wave is shown in Figure 9.9. The circuit's action on a sine wave is that of a low-pass filter, smoothing out any sudden changes, the opposite of the action of the differentiating circuit.

When simple RC circuits are used for differentiating and integrating, the circuit property called its **time constant** becomes important. It is measured by multiplying the value of *R* by the value of *C*. If we measure *C* in farads and *R* in ohms we get a time constant value in seconds, but it is more realistic to measure *C* in μF and *R* in kΩ, to give time constants in milliseconds. Another useful rule is that using *C* in nF and *R* in kΩ gives time constants in μs (microseconds).

A differentiating circuit requires a time constant that is **small** compared with the wave period. An integrating circuit requires a time constant that is **large** compared with the wave period. The time that is calculated from the time constant is the time to reach 63% of the final value. For an integrating circuit, this means 63% of the peak voltage of an input that has a sharp leading edge. For a differentiating circuit, this means that the voltage reduces from the peak input voltage (for a sharp rising edge) to 63% of the distance from the peak voltage to zero, meaning that it reaches 37% of the peak voltage measured from the zero line. These changes are easier to understand if we look at them graphically (Figure 9.10).

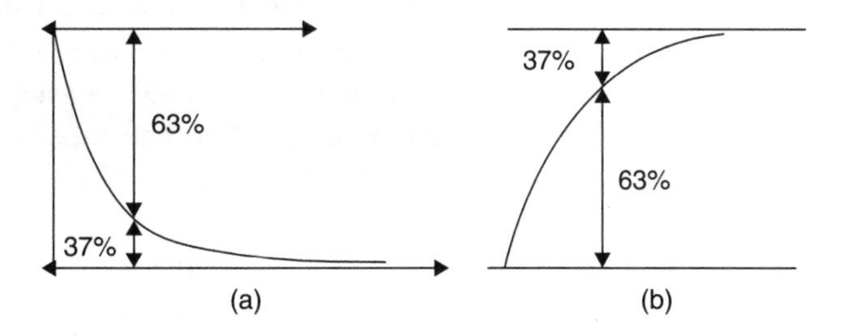


**Figure 9.10** Time constant and the 63% rule: (a) differentiating, and (b) integrating

One result of these requirements is that differentiating or integrating circuits that are designed to handle a given wave frequency will have a different effect when a wave of either higher or lower frequency reaches them as an input.

The units in which the time constant is calculated must be strictly observed. If *R* and *C* are given in ohms and farads, respectively, the time constant RC will appear in seconds. If (as is more likely) the units of *R* and *C* are given in kilohms and microfarads, respectively, the constant RC will appear in milliseconds (ms). If *R* is in kilohms and *C* in nanofarads, RC will appear in microseconds (μs).

**Examples:** Calculate the time constants of (a) 10K and 0.033 μF, (b) 470R and 0.1 λF, (c) 56K and 820 pF.

**Solutions:** (a) With $R = 10k$, $C$ must be converted into nanofarads, 0.033 μF = 33 nF. The time constant is then 33 × 10 = 330 μs. (b) $R$ must be converted into kilohms: $470R = 0.47k$. With $C = 0.1\,\mu F$, the time constant is 0.47 × 0.1 = 0.047 ms, or 47 μs. (c) With $R = 56k$, $C$ must again be converted into nanofarads: 820 pF = 0.82 nF. The time constant is then 56 × 0.82 = 45.9 μs, or about 46 μs.

Except for a few low-frequency circuits, the unit of microseconds is the most convenient for use in time constant calculations.

## Practical 9.1

Construct the waveshaping circuits shown in Figure 9.11. For each circuit in turn, connect the oscilloscope to the output of the circuit and a square wave signal generator to the input. Calculate the time constant for the circuit. Now adjust the square wave generator until it produces a wave of 1 V peak-to-peak (p-p) at 1 kHz. Sketch the output waveform from the waveshaping circuit.

Set the generator to a frequency of 100 Hz and repeat the measurement. Repeat the measurements using 100 Hz and 1 kHz sine waves and confirm that the only effect is alteration of phase shift.

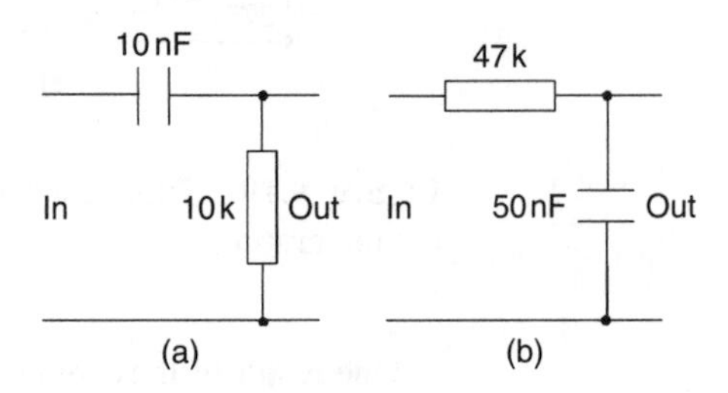


**Figure 9.11** Circuits for Practical 9.1: (a) differentiating, and (b) integrating

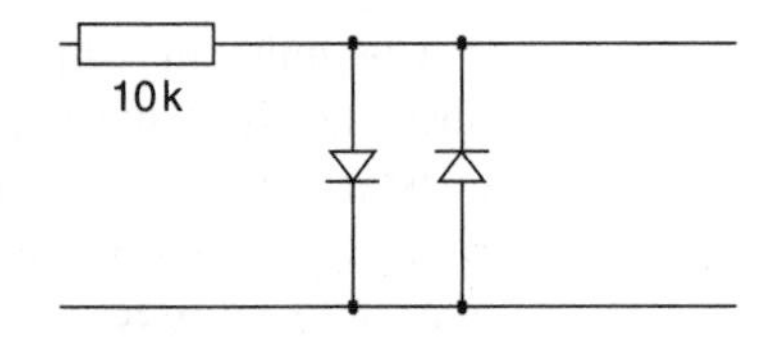


**Figure 9.12** Circuit for Practical 9.2

## Practical 9.2

Construct the circuit shown in Figure 9.12, using antiparallel diodes. Examine the output of the circuit when a 6 V p-p sine wave is applied to the input.

## Multiple-choice revision questions

9.1 A square wave has an amplitude of 10 V and a mark-to-space ratio of 1:3. It is used to drive an electric motor, and will produce the same speed as:
(a) a sine wave of 10 V peak
(b) a sine wave of 10 V r.m.s.
(c) d.c. of 2.5 V
(d) d.c. of 10 V.

9.2 A 1 kHz square wave generator is switched on and it causes interference on a radio. This is because of:
(a) the amplitude of the square waves
(b) the harmonics in the square wave
(c) the mark-to-space ratio of the square wave
(d) the frequency of the square wave.

9.3 Radio waves can travel in space but sound waves cannot. This is because:
(a) radio waves have higher frequencies
(b) sound waves have different waveshapes
(c) radio waves have shorter wavelengths
(d) sound waves need to have a material to vibrate.

9.4 The amplitude of a square wave can be reduced without seriously affecting its wave shape by using:
(a) a clipper
(b) an integrator
(c) a d.c. restorer
(d) a differentiator.

9.5 A d.c. signal component cannot be passed by a:
(a) capacitor
(b) resistor
(c) inductor
(d) Voltage dependent resister (VDR).

9.6 A 1kHz square wave is passed through a filter circuit with a cut-off frequency of 1.5 kHz. The output waveshape is:
(a) a square wave
(b) a sawtooth
(c) a sine wave
(d) a pulse train.

# 10 Transducers and sensors

## Definitions

A **transducer** will convert one form of power into another, and the efficiency of conversion is important. We cannot, for example, convert power in the form of heat into electrical power with efficiency above 45%, but we can convert electrical power into heat with an efficiency of well above 90%. Some transducers, however, have very low efficiency: a high-quality loudspeaker may have an efficiency (converting electrical power into acoustic power) of 1% or less.

A **sensor** also converts power from one form to another, but the efficiency is unimportant. The important factor for a sensor is *linearity*, because sensors are used in detection and measuring actions. Linearity means the extent to which the output is proportional to the input, so that a perfectly linear device would have a graph of output plotted against input that would be a straight line. A few devices (a microphone, for example) may be used as either a sensor or a transducer.

Transducers that are used along with electronics circuits can be either input or output transducers. An **input transducer** will convert power from some other form into electrical signals; an **output transducer** will convert electrical signals into some other form of power. The simplest input transducer is a switch, which converts mechanical movement into an on/off electrical signal.

## Switches

Switches have a low resistance between contacts in the **ON** setting, and a very high resistance in the **OFF** setting. The value of resistance when the switch is on (*made*) is called the **contact resistance**. The amount of the contact resistance depends on the area of contact, the contact material, the amount of force that presses the contacts together and the way that this force has been applied. A wiping action provides a lower resistance than a simple pressing action.

Switch contact configurations are primarily described in terms of the number of poles and number of throws or ways. A switch **pole** is a moving contact, and the **throws** or **ways** are the fixed contacts against which the moving pole can rest. The term *throw* is usually reserved for mains switches, mostly single- or double-throw, and *way* for signal-carrying switches. A single-pole, single-throw (SPST) switch will provide on/off action for a single line, and is also described as single-pole on/off. Such switches are seldom used for a.c. mains nowadays and are more likely to be encountered on d.c. supply lines. Figure 10.1 shows the most common switch arrangements in circuit diagram form. For a.c. use, safety requirements call for both live and neutral lines to be broken by a switch, so that double-pole single-throw (DPST) switches will be specified for this type of use.

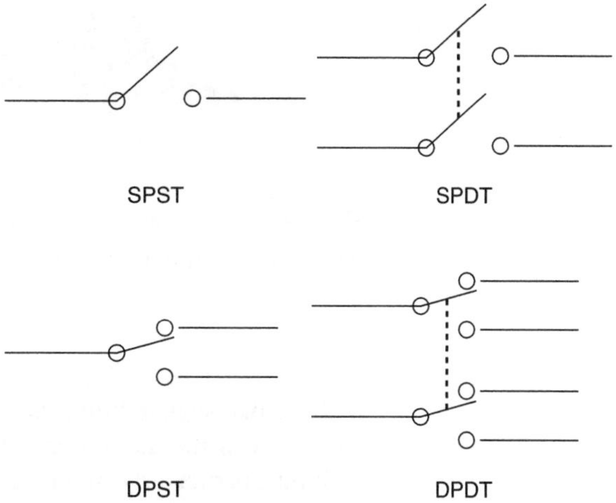

SPST SPDT

DPST DPDT

**Figure 10.1** Common switch arrangements and symbols

The configuration of a double-throw switch also takes account of the relative timing of contact. The normal requirement is for one contact to break before the other contact is made, and this type of break-before-make action is standard. The alternative is make-before-break (MBB), in which the moving pole is momentarily in contact with both fixed contacts during the changeover period. Such an action is permissible only if the voltages at the fixed contacts are approximately equal, or the resistance levels are such that very little current can flow between the fixed contacts. Once again, this type of switching action is more likely to be applicable in signal-carrying circuits. There may be a choice of fast or slow contact make or break for some switches.

A switch may be **biased**, meaning that one position is stable, and the other, off or on, is attained only for as long as the operator maintains pressure on the switch actuator. These switches are used where a supply is required to be on only momentarily, often in association with the use of a hold-on relay, or can be interrupted momentarily. A biased switch can be of the off–on–off type, in which the stable condition is off, or the on–off–on type, in which the stable condition is on. The off–on–off type is by far the more common. The ordinary type of push-button switch is by its nature off–on–off biased, although this is usually described for such switches as **momentary action** (see, however, *alternate action* below), and biased switches are also available in toggle form.

Switches may be operated in several different ways. The most common types are the toggle switches operated by a toggle lever or by a rocker action. Push-button switches are also commonly used, either alternate action, as a button that is pressed and released for ON and pressed and released again for OFF, or as a button that is pressed for ON and remains depressed until it is pressed again for OFF. The rotary switch is often used in electronic applications to select one out of several choices (Figure 10.2). A switch (or set of switches) can also be operated by a solenoid with a moving core, and this type of device is termed a **relay**.

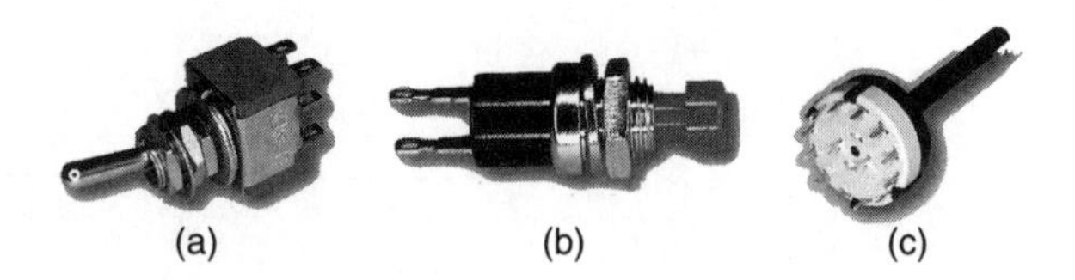


**Figure 10.2** Typical switch types, not to scale: (a) toggle, (b) push-button, and (c) rotary [photographs: Alan Winstanley]

Note that switch ratings are always quoted separately for a.c. and for d.c., with the a.c. rating often allowing higher current and voltage limits, particularly for inductive circuits.

## Practical 10.1

Use a multimeter to test the continuity of switch and relay contacts. Check the effect of operating the switch/relay.

## Microphones

The most familiar transducer for sound energy into electrical energy is the microphone, and microphone types are classified by the type of transducer they use. The overall sensitivity is expressed as millivolts or, more usually, microvolts of electrical output per unit intensity of sound wave. In addition, though, the **impedance** of the microphone is also important.

A microphone with high impedance usually has a fairly high electrical output, but the high impedance makes it very susceptible to hum pickup. A low impedance is usually associated with very low output, but hum pickup is almost negligible.

Another important factor is whether the microphone is **directional** or **omnidirectional**. If the microphone operates by sensing the pressure of the air in the sound wave, then the microphone will be omnidirectional, picking up sound arriving from any direction. If the microphone detects the velocity (speed and direction) of the air in the sound wave, then it is a directional microphone, and the sensitivity has to be measured in terms of direction as well as amplitude of the sound wave. The microphone types are classed as pressure or velocity operated, omnidirectional or in some form of directional response (such as cardioid) (Figure 10.3).

The **frequency response** of a microphone is usually very complicated because the housing of the transducer unit will act as a filter for the sound waves, causing the response to show peaks and troughs unless the housing is very carefully designed.

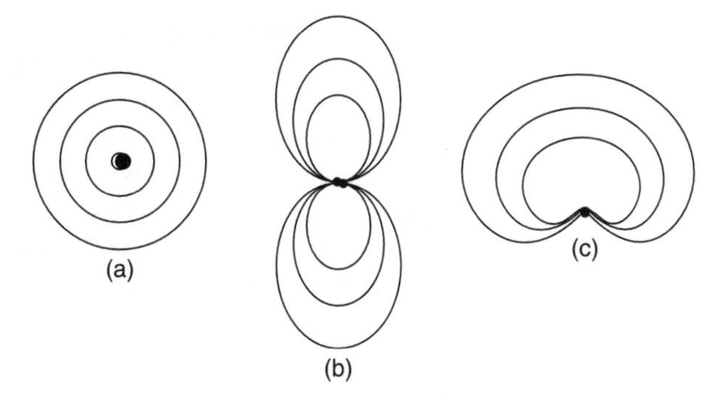


**Figure 10.3** Simplified response curves for microphone types: (a) omni-directional, (b) velocity-operated, and (c) cardioid (heart-shaped)

## Microphone types

Microphone types in common use include moving iron (dynamic), moving-coil, crystal and electret capacitor types. Of these, we shall pick out the dynamic and the electret types only.

The principle of the moving iron (dynamic, or variable reluctance) microphone is illustrated in the cross-section view of Figure 10.4. A powerful magnet contains a soft-iron armature in its magnetic circuit, and this armature is attached to a diaphragm with a flexible suspension so that sound waves will cause the diaphragm to vibrate. The magnetic flux in the magnetic circuit alters as the armature moves. A coil wound around the magnetic circuit at any point will give a voltage which is proportional to each change of magnetic flux, so that the electrical wave from the microphone is proportional to the acceleration of the diaphragm.

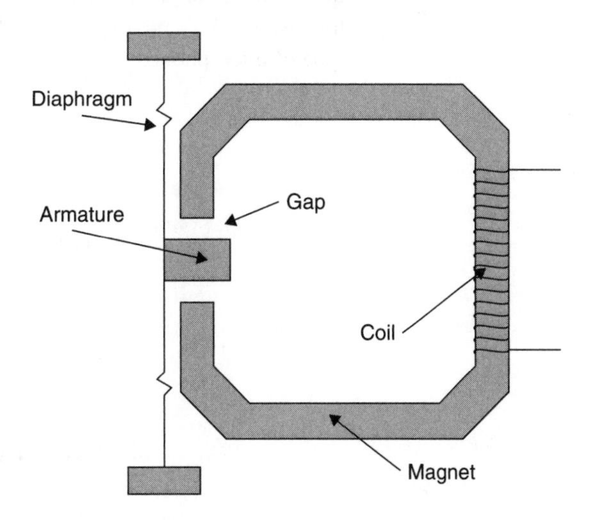


**Figure 10.4** Principle of the dynamic microphone

The linearity of the conversion can be reasonable for small amplitudes of movement of the armature, but very poor for large amplitudes. Both the linearity, and the amplitude over which linearity remains acceptable can be improved by appropriate shaping of the armature and careful attention to its path of vibration. These features depend on the maintenance of close tolerances in the course of manufacturing the microphones, so that there will inevitably be differences in linearity between samples of microphones of this type from the same production line. The output level from a moving-iron microphone can be high, in the order of 50 mV, and the output impedance is fairly high, typically several hundred ohms. Magnetic shielding is always needed to reduce mains hum pickup in the magnetic circuit.

Moving-coil and ribbon microphones are used professionally, but are not used with domestic electronic equipment. They have been omitted from this book.

The capacitor microphone is now used almost exclusively in electret form (Figure 10.5). An electret is the electrostatic equivalent of a magnet, a piece of insulating material which is permanently charged. A slab of electret is therefore the perfect basis for a capacitor microphone in which the vibration of a metal diaphragm near a fixed charge causes voltage changes between the diaphragm and a metal backplate. This allows very simple construction for a microphone, consisting only of a slab of electret metallized on the back, a metal (or metallized plastic) diaphragm, and a spacer ring, with the connections taken to the conducting surface of the diaphragm and of the electret. This is now the type of microphone that is built into cassette recorders, and even in its simplest and cheapest versions is of considerably better audio quality than the piezoelectric types that it displaced. The use of a field-effect transistor (FET) preamp solves the problem of the high impedance of this type of microphone.

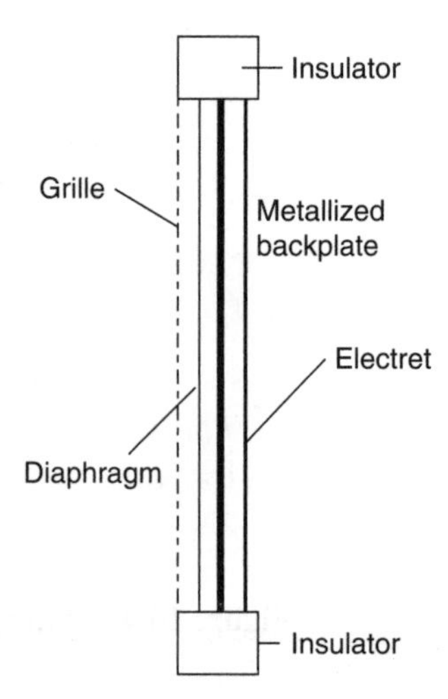


**Figure 10.5** Principle of electret capacitor microphone

## Tape playback

Magnetic tape uses the principle of *hard* magnetic material. Early tape recorders in 1898 used steel tape, but it was not until 1941 that the BASF company hit on the idea of coating a thin plastic tape with magnetic powder, using iron oxide. The recording head is made from a soft magnetic material with low hysteresis so that its magnetic flux will follow the fluctuations of a signal current through the winding (Figure 10.6a). The same type of head is also used for replaying tape, using the varying flux from the moving tape to induce currents in the head. The symbol for a tapehead is illustrated in Figure 10.6(b). The replay tapehead is another form of input transducer, converting the varying magnetic signal on the moving tape to an audio electrical signal in the windings of the head.

Domestic cassette recorders normally use just one head for both recording and replaying, with switched connections either to the output of an amplifier or to the input of the amplifier. A separate head is fed with the output from a high-frequency (typically 100 kHz) oscillator to act as an erase head only during recording.

See Chapter 16 for a more complete description of tape-recorder principles.

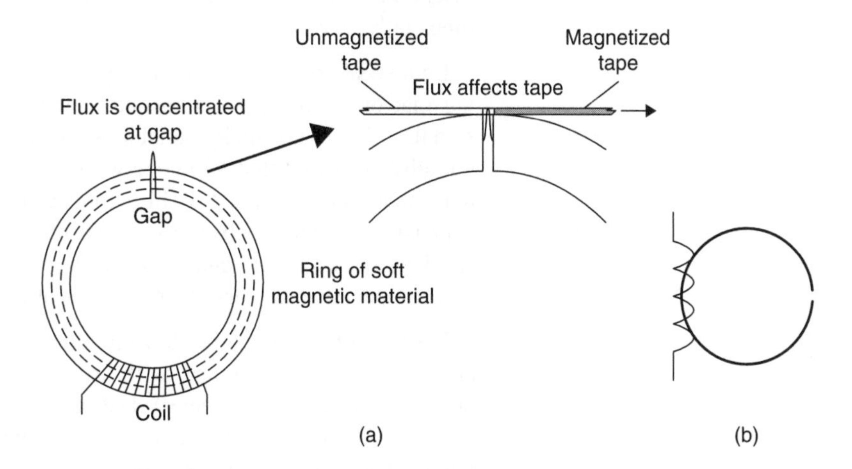


**Figure 10.6** (a) Principle of tapehead, and (b) symbol

### Practical 10.2

Test the continuity of the recording head of a tape/cassette recorder, using a multimeter. Note that the head will need to be demagnetized after this measurement: use a head demagnetizer if available.

## Output transducers

Typical output transducers used in electronics equipment include headphones and loudspeakers. Figure 10.7 shows the symbols that are used. Headphones (or earphones) have been in use for considerably longer than microphones or loudspeakers, since they were originally used for electric telegraphs. The power that is required is in the low milliwatt level, and even a few milliwatts can produce considerable pressure amplitude at the eardrum, often more than is safe for your hearing when the earphone is a snug fit into the ear.

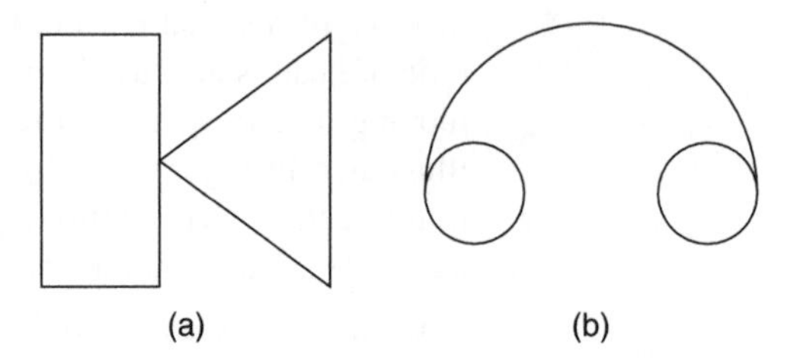


**Figure 10.7** Symbols for (a) loudspeaker, and (b) headphones

The methods that are used for loudspeakers and for headphones are essentially the same as are used for microphones, because the action is (unusually) reversible.

Loudspeakers for domestic equipment are often **full-range** units, meaning that one loudspeaker transducer deals with the whole range of audio signals. This is not really satisfactory because it is very difficult to make a loudspeaker unit that will cope with such a range, and the few full-range units that can be classed as hi-fi loudspeakers are usually of exotic (and expensive) design, such as the Quad electrostatic units. Small full-range speakers of acceptable quality are widely used in radios, cassette recorders and compact disc players at the low end of the price range.

For anything that can be (loosely) classed as hi-fi, separate loudspeakers are used for the audio range, with one (woofer) dealing with the lower frequencies and another (tweeter) dealing with the high frequencies. The two units are normally enclosed in one cabinet, but on speaker sets such as those intended for home cinema equipment, the tweeters are arranged around the room and a single subwoofer (for the frequencies in the range 30–150 Hz) is located centrally. The stereo effect depends almost completely on the higher frequencies, so the positioning of a woofer/subwoofer is not critical. A filter, called a **cross-over unit**, separates the signal into lower and upper frequencies so that the higher frequencies are fed to the tweeters and the lower frequencies to the woofers.

Most loudspeakers use the moving-coil principle (Figure 10.8). The coil is wound on a former at the apex of a stiff diaphragm which is held in flexible supports so that the coil is the intense field of a permanent magnet. A signal current through the coil will then vibrate the diaphragm, producing sound waves at the same frequency and with the same waveshape as the signal. To be really effective, a loudspeaker has to be housed in a cabinet (**enclosure**) whose design assists the loudspeaker to move a large volume of air, acting like a transformer for air waves. Even in a fairly good enclosure, the efficiency of a typical hi-fi loudspeaker is very low, around 1%.

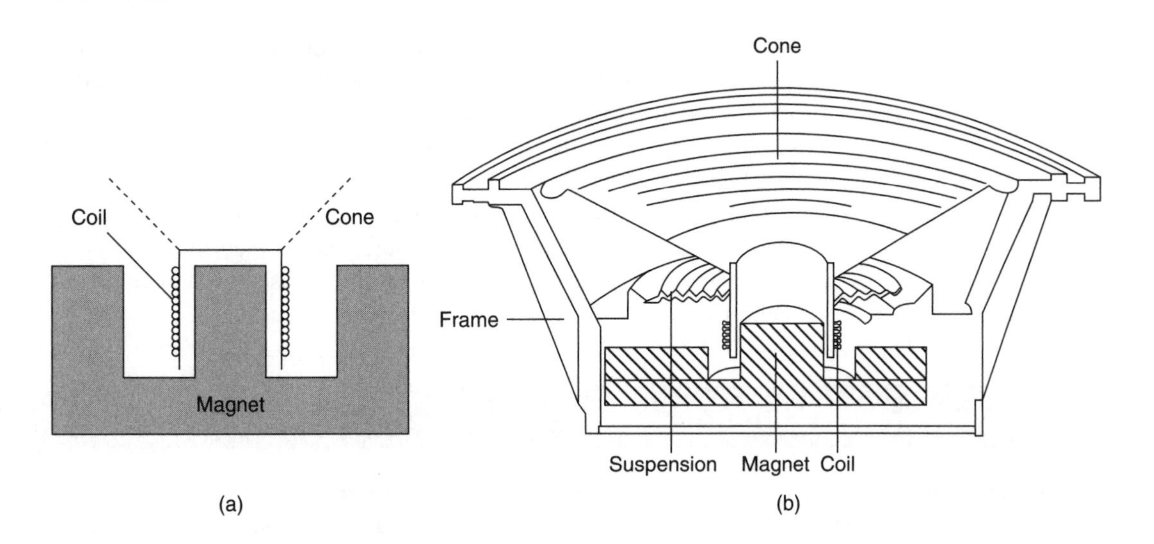


**Figure 10.8** Moving-coil loudspeaker: (a) cross-section of coil and magnet, and (b) cutaway view of complete unit

Miniature loudspeakers, typically of diameter 70 mm or less, are used as warning transducers in cars, signal devices on electronic equipment, and in toys. These small units can be used also as microphones, although their efficiency as a microphone is low and the frequency response is uneven. Confusingly, the same term is also used for hi-fi loudspeakers that are noticeably smaller than average.

## Headphones

Headphones can be constructed using any of the principles used for loudspeakers, but the most widespread and successful type is the **electrodynamic** (or **orthodynamic**) principle. This uses a diaphragm which has a coil built in, using printed circuit board techniques. The coil can be a simple spiral design, or a more complicated shape (for better linearity), and the advantage of the method is that the driving force is more evenly distributed over the surface of the diaphragm.

The **piezoelectric** principle has also been used for headphones in the form of piezoelectric sheets that can be formed into very flexible diaphragms. A piezoelectric material is one whose dimensions can be changed by applying an electric field, and which will generate an electric field if the material is compressed. Quartz crystals are piezoelectric, and another crystalline material that is extensively used is barium titanate.

Moving-coil headphones have also been manufactured and the quality that can be attained justifies the high prices that are attached to these units.

Subminiature earphones, often referred to as *earbuds*, are designed for use with portable CD or MP3 players, and run at a typical rated power of 3 mW. These use the same dynamic mechanism as larger earphones. The impedance is typically 32 Ω.

Moving-coil loudspeakers and earphones have a nominal value of impedance, typically 4 ohms, but a graph of impedance plotted against frequency is very complicated in shape, with peaks and dips caused by the mechanical resonances of the loudspeaker components and the enclosure. Small headphones usually operate with higher impedances, partly to prevent overdriving if they are connected to an amplifier output designed to pass power to a low-impedance loudspeaker.

Miniature loudspeakers can generally handle only the higher frequencies, but hi-fi headphones are miniature loudspeakers in enclosures that are designed to achieve a reasonable level response over a large part of the audio-frequency range.

## Multiple-choice revision questions

10.1 A transducer:
(a) measures power
(b) converts power to energy
(c) converts energy from one form to another
(d) converts high power to low power.

10.2 A switch is described as DPDT. It could be used particularly for:
(a) switching a single light on or off
(b) switching two lights on or off
(c) switching a tape recorder on or off
(d) switching both leads to a tapehead to either the input or the output of an amplifier.

10.3 A cardioid microphone:
(a) is equally sensitive in all directions
(b) has a heart-shaped response characteristic
(c) has a straight-line response characteristic
(d) has no response to the side.

10.4 A tape recorder:
(a) creates a magnetic pattern on tape
(b) creates an electric pattern on tape
(c) uses a capacitor for recording
(d) can record any frequency of sound.

10.5 Most loudspeakers are of the type known as:
(a) moving iron
(b) moving field
(c) moving coil
(d) moving flux.

10.6 Many makes of headphones use:
(a) the binary principle
(b) the photovoltaic effect
(c) the induction effect
(d) the electrodynamic principle.

# 11 Transducers (2)

## Permanent-magnet motor

The d.c. permanent-magnet motor, whose symbol and typical connections are illustrated in Figure 11.1(b), is used extensively in equipment that needs mechanical actions, such as video recorders and compact disc (CD) players. As the name implies, the motor uses a rotating armature with a winding connected through a commutator, with a field supplied by a permanent magnet. Figure 11.1(a) shows a simple example with a one-turn rotor. We can think of the motor as a transducer whose input is electrical and whose output is mechanical. Miniature general-purpose d.c. motors, using permanent magnet fields, can be controlled by the output from a comparatively low-output semiconductor amplifier. Typically they have voltage requirements of 6–12 V and currents of 200–400 mA. Small electric motors of this type can be used in models, rotating signs, display systems and to some extent in robotics.

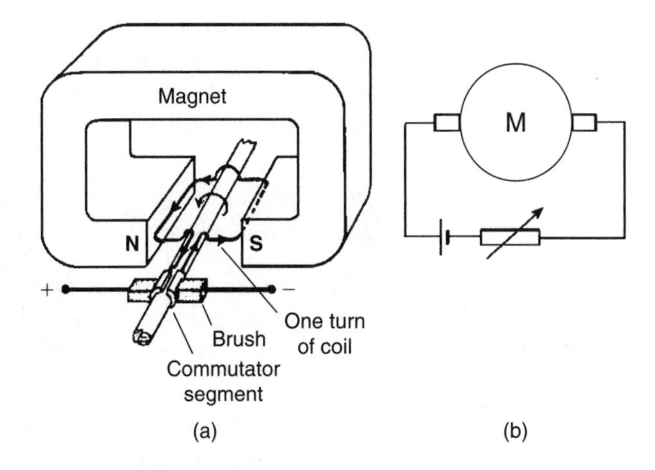


**Figure 11.1** (a) Permanent magnet motor principle, and (b) symbol and motor circuit with current controller

The permanent-magnet type of motor is favoured because its speed can be controlled by controlling the current through the armature winding, and the direction of rotation can be reversed by reversing the direction of current through the armature winding. The armature current can be supplied from a circuit incorporating a power transistor, so that the current can be closely regulated to provide precise speed control.

For closer control and higher power, the permanent magnet can be replaced by an electromagnet, energized by a **field** winding. Using separate control of current through the field, the speed of the motor can be controlled over a wide range, so that this type of motor is used where precise speed control is important. When the armature and the field windings are connected in parallel, the motor is said to be **shunt wound**, but if the field and armature windings are connected in series the motor is **series wound**. Series-wound motors have very high starting torque and can spin at very high speeds on a light load, although their rotational speed drops considerably under load. They can deliver high torque at low speeds under load. Shunt-wound motors have a very low starting torque (and may have to be assisted), but when running their speed is relatively constant as the load changes. For reversing a motor with a wound field, the direction of current through either the armature or the field (but not both) can be changed.

## Tape-recording and erase heads

These are transducers with an electronic signal input and a magnetic field output. The tape-record head is a transducer with electrical input and magnetic output, and one head can be used for both purposes, although this is not an ideal arrangement. The head gap is usually of around the same size as the thickness of the magnetic coating on the tape, and for a good-quality cassette recorder can be as small as $1\,\mu m$ ($10^{-6}\,m$). Ideally, the head gap for a recording head should be larger than for the reading head, and figures of around $6–10\,\mu m$ are common. The gap for the replay head should be as small as head technology can provide (allowing for the price of the recorder). The erase head has a comparatively large gap. It is fed from an oscillator, usually the same oscillator as provides bias (see Chapter 10), and its purpose is to erase the tape by applying an alternating field that is gradually reduced to zero. Revise the topic of tape replay head (Chapter 10).

### Practical 11.1

Measure the resistance of a tape-recording head. Use a demagnetizer on the head after making the measurement.

## Relays

A relay is a form of switch, with the mechanical action of moving the switch contacts being carried out by a solenoid and armature. The classical form of relay (Post Office type 8000) is illustrated in Figure 11.2. The solenoid is wound on a former around a core, and a moving armature forms part of the magnetic circuit. When the coil is energized, the armature moves

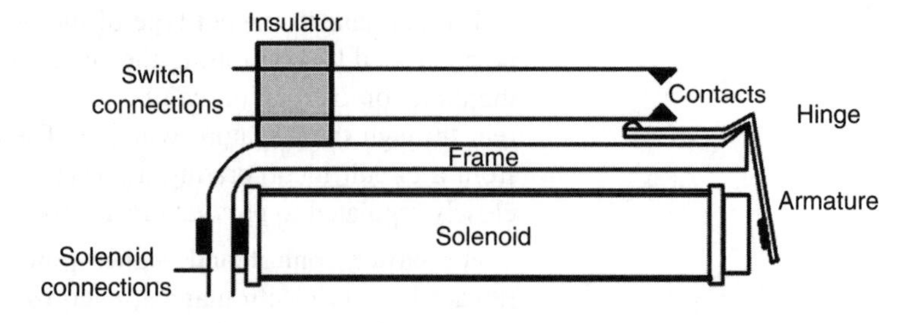


**Figure 11.2** The classic type 8000 relay construction. Modern relays are smaller, but the working principles are unchanged

against the opposition of a spring (or the elasticity of a set of leaves carrying the electrical contacts) so as to complete the magnetic circuit. This movement is transmitted to the switch contacts through a non-conducting bar or rod, so that the contacts close, open or change over depending on the design of the relay.

The contacts are subject to the same limits as those of mechanical switches, but the magnetic operation imposes its own problems of make and break time, contact force and power dissipation. Like so many other electronics components, relays for electronics use have been manufactured in decreasing sizes, and it is even possible to buy relays that are packaged inside a standard TO-5 transistor can.

The relay has to be specified by the operating voltage and current, and also by the switching arrangements. These are commonly classed as normally open (NO), meaning that applying power to the relay will close the contact(s), or change-over (CO), meaning that applying power will change over the switch connections. Normally closed connections are less usual, and applying power to such a relay will open the contacts. Contacts are usually single or double pole, so that typical arrangements are SPNO, SPCO, DPNO or SPCO (also known as DPDT). At one time, separate symbols were used for these contact arrangements, but this has changed.

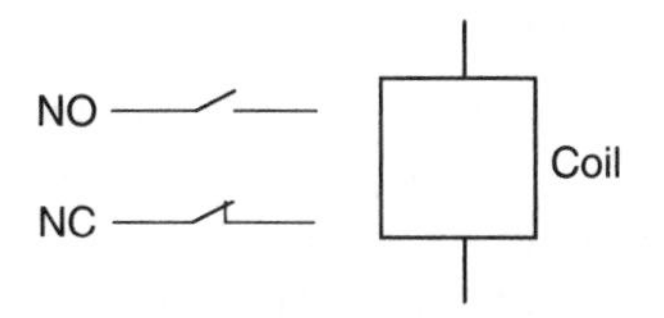


**Figure 11.3** Contact symbols for diagrams including relays

Circuit diagrams now use the **detached contact method**. The coil of the relay is shown in one part of the diagram (in the collector circuit of a transistor, perhaps), and the switching contacts are in another part of the diagram. This avoids the need to show long and confusing connecting lines on a diagram. It also allows the use of two standard contact symbols, one for normally open and the other for normally closed. These symbols are illustrated in Figure 11.3.

### Practical 11.2

Measure the resistance of a relay operating coil, using a multimeter. Measure also the resistance of closed contacts.

## Monochrome cathode-ray tube

The cathode-ray tube (CRT) is a transducer component which nowadays is just about the only device you are likely to come across that uses the same principles as the old-style radio valves. The CRT is used to convert variations of voltage into visible patterns, and is applied in instruments (oscilloscopes), for television, and for radar. Colour tubes are used for television and monitor purposes, but monochrome (black and white) tubes are still used in considerable numbers for instruments and for low-cost closed-circuit television (CCTV). For computer and (gradually) television receivers they are being replaced by LCD and plasma devices. They, in turn, are likely to be replaced by other types of displays in the near future, such as SED (or FED), OLED and PDP.

The three basic principles of the CRT are:

- Electrons can be released into a vacuum from a red-hot cathode (a metal coated with metal oxides).
- These electrons can be accelerated and their direction of movement controlled by using either a voltage between metal plates or a magnetic field from a coil that is carrying an electric current.
- A beam of electrons striking some materials such as zinc sulphide will cause the material (called a phosphor) to glow, giving a spot of light as wide as the beam.

In the monochrome tube electrons from the **cathode** are attracted to the positive **anode** by an accelerating voltage of several kilovolts. On their way, they have to pass through a pinhole in a metal plate, the **control grid**. The movement of the electrons through this hole can be controlled by altering the voltage of the grid, and a typical voltage would be some 50 V negative compared to the cathode. At some value of negative grid voltage, the repelling effect of a negative voltage on electrons will be greater than the attraction of the large positive voltage at the far end of the tube, and no electrons will pass the grid: this is the condition called **cut-off**.

Electrons that pass through the hole of the grid can be formed into a beam by using metal cylinders at a suitable voltage. By adjusting the voltage on one of these cylinders, the **focus electrode**, the beam can be made to come to a small point at the far end of the tube. This end is the **screen**, and it is coated with a material (a **phosphor**) that will glow when it is struck by electrons. The phosphor (which does **not** contain phosphorus) is usually coated with a thin film of aluminium so that it can be connected to the final accelerating (anode) voltage. The whole tube is pumped to as good a vacuum as is possible; less than a millionth of the normal atmospheric pressure.

This arrangement will produce a point of light on the centre of the screen, and any useful CRT must use some method of moving the beam of electrons. For small instrument CRTs a set of four metal plates can be manufactured as part of the tube and these deflection plates will cause the beam to move if voltages are applied to them. The usual system is to arrange the plates at right angles, and use the plates in pairs with one plate at a higher voltage and the other at a lower voltage compared to the voltage at the face of the tube. This system is called **electrostatic deflection**.

A typical electrostatically deflected CRT is intended for oscilloscope use, and will use a screen size of, typically, 5–7 inches diameter. Rather than earthing the cathode and connecting the anode to several kilovolts voltage, the tube is used with the anode earthed, and the cathode at a high negative voltage. This makes it possible to connect the deflection plates directly to a modest voltage that can be supplied from transistor circuits. The other voltages for focus and accelerating electrodes can be supplied from a resistor network connected between the earth and the most negative voltage (applied to the grid). Figure 11.4 shows the symbols for the electrodes and the typical voltage in a small instrument CRT. Note that high voltage levels exist particularly around the cathode.

In oscilloscope use, steady voltages are applied to the anodes, although the supply to the second anode can be varied to allow for focusing the electron beam on the screen. Alternating current is applied to the heater coil, and a modulating pulse is applied to the cathode (as well as a d.c. bias) to switch the electron beam on and off. The vertical deflection plates are supplied from an amplifier which provides a d.c. bias, and is used to place an alternating signal voltage. This is an amplified version of the signal input to the oscilloscope.

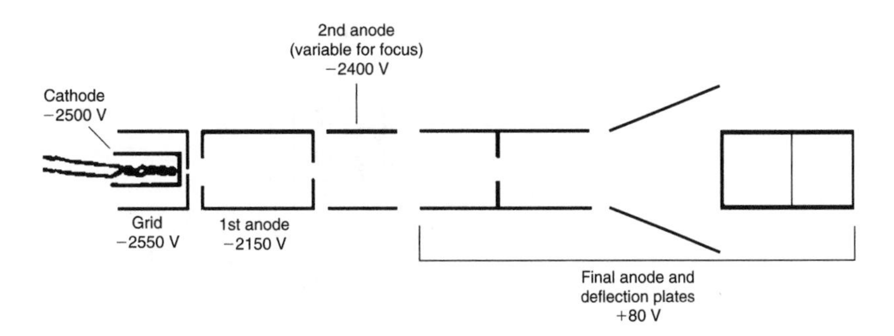


**Figure 11.4** Electrode symbols and typical voltages used for a small instrument CRT

The form of deflection that is most common for CRTs is a linear sweep. This means that the beam is taken across the screen at a steady rate from one edge, and is returned very rapidly (an action called **flyback**) when it reaches the other edge. A sawtooth waveform is needed to generate such a linear sweep. An electrostatic tube can use a sawtooth voltage waveform applied to its horizontal deflection plates.

There is an alternative method for deflecting the electron beam which is used for larger tubes, particularly for computer monitors, radar and television. A beam of electrons is a current flowing through a vacuum, and magnets will act on this current, deflecting the beam. A tube of this type (Figure 11.5) can have a much shorter neck and a much wider angle of deflection than an instrument tube with electrostatic deflection. The easiest way to carry out beam deflection is to place coils around the neck of the tube and pass current through these coils to control the beam position on the face of the tube. This **magnetic deflection** method is better suited for large CRTs such as are used for simple radar, CCTV or computer monitor applications. The coil windings are usually shaped so that they can be placed over the curved portion of the neck of the tube, as illustrated in Figure 11.6.

Figure 11.7 shows typical connections to a small monochrome monitor or television tube. These are fewer than for the electrostatic type because there are fewer electrostatic components (only focus is achieved by voltage on a focus cylinder) and the magnetic components are external. A magnetic-deflection system must use a sawtooth current applied to its deflection coils. The difference is important, because the electrostatic deflection requires only a sawtooth voltage with negligible current flowing, but the magnetically

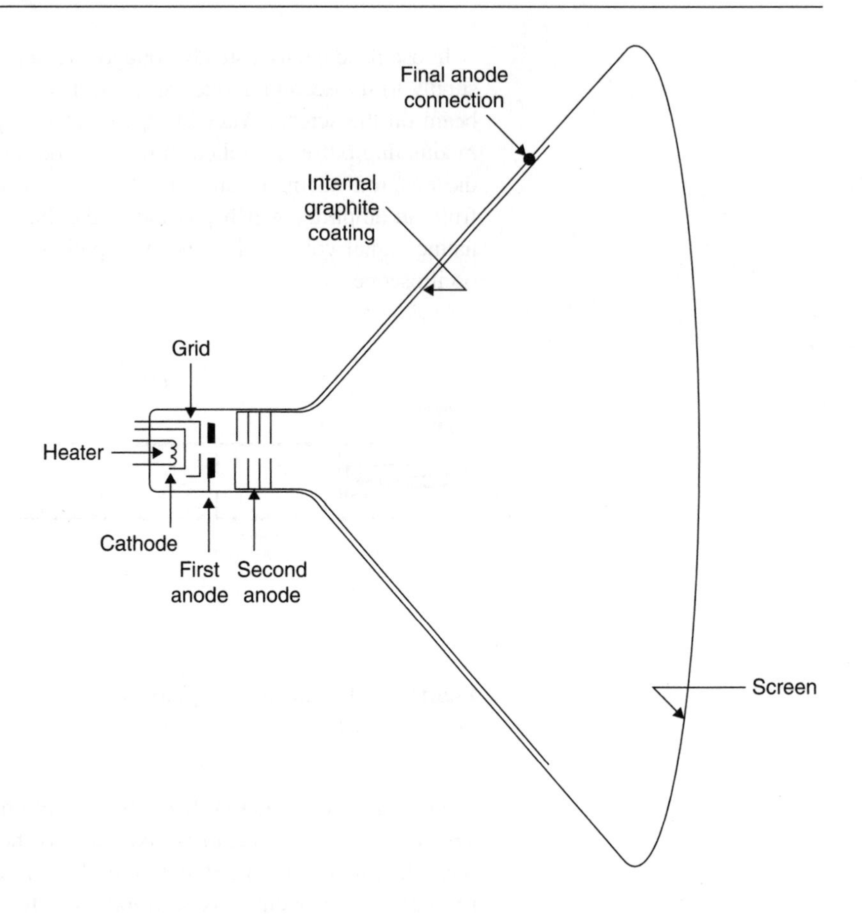


**Figure 11.5** Cross-section of a magnetically deflected tube

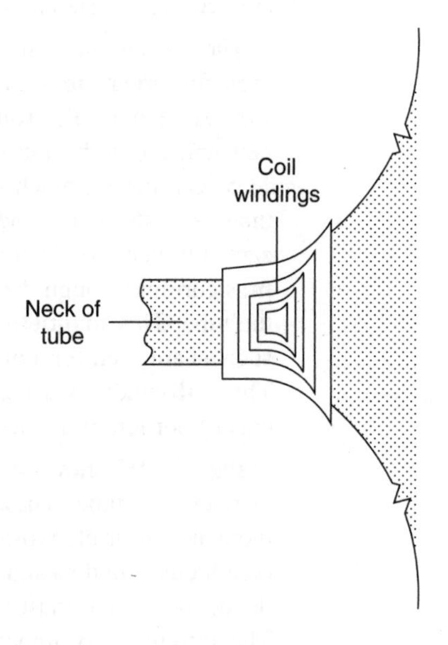


**Figure 11.6** Position of the deflection coils

deflected tube requires a sawtooth current, and the voltage across the deflection coils will not be a sawtooth, because the coils act as a differentiating circuit converting a sudden change of voltage into a steadily rising or falling current. In fact, the voltage waveform is a pulse, and this is used in television receivers to generate a very high voltage for the CRT. Chapter 18 deals with colour CRTs and their use in television receivers, and Chapter 22 deals with computer monitors.

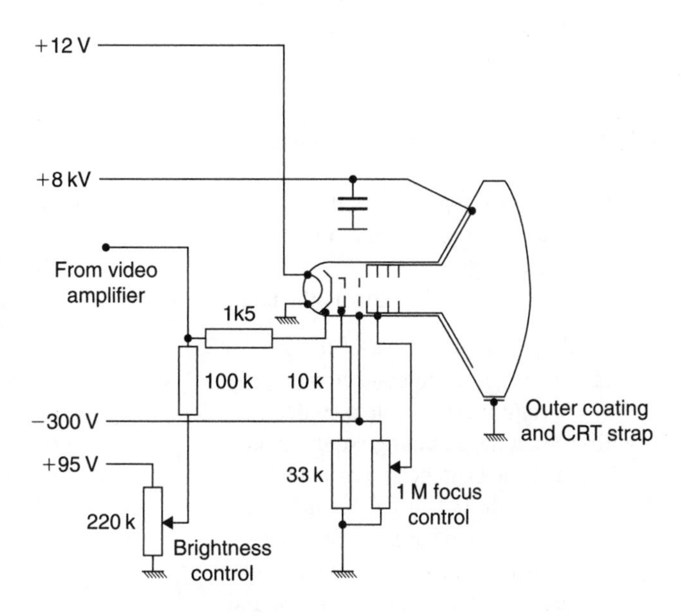


**Figure 11.7** Typical connections to a monochrome monitor tube

## CRT faults

Experience with CRTs has proved them to be remarkably reliable components, and it is not unusual for a television colour CRT to have a life of 25 years or more. When a tube becomes faulty, the usual fault is loss of emission from the cathode, so that screen brightness drops and the brightness (grid voltage) control has to be taken to its limit to provide a visible trace. At one time it was a normal servicing procedure to revive failing emission by running the tube heater at a higher current than normal, and applying a small positive bias to the grid. This often provided a few years of extra life, but with the improvement in materials used in cathodes, loss of emission is now quite rare.

Other faults, also quite rare nowadays, are grid-cathode shorts and open circuits. A grid-cathode short will make the tube show a permanent bright trace, and for a monitor tube no picture information. Open circuits will result in no form of trace or picture on the screen. In either case, the tube will have to be replaced, and because of the hazard of implosion (the vacuum inside causing the glass to fly inwards when it shatters) you should take care to ensure that the tubes (old and replacement) are protected and your eyes shielded when you replace a tube. Large television tubes are a much greater hazard because of their weight and glass content, but at the time of writing they are being replaced by plasma and LCD screens.

## Multiple-choice revision questions

11.1 A permanent-magnet motor:
(a) uses a wound armature and field
(b) uses a wound field only
(c) uses a wound armature
(d) uses no armature.

11.2 The gap in the record head of a tape recorder:
(a) is for the tape to pass through
(b) allows for expansion of the metal
(c) must be large
(d) is where the flux affects the tape.

11.3 A relay is particularly useful:
(a) when more than one actuating circuit must operate a switch
(b) when the actuating circuit is d.c. and the switched circuit is a.c.
(c) when the actuating circuit is a.c. and the switched circuit is d.c.
(d) when complete isolation is needed between the actuating circuit and the switched circuit.

11.4 The cathode of a cathode-ray tube:
(a) provides a source of electrons at all times
(b) provides a source of electrons when heated
(c) gathers electrons from the anode
(d) deflects electrons from the grid.

11.5 In an instrument CRT:
(a) the beam is magnetically deflected
(b) the heater must be fed with d.c.
(c) deflection is electrostatic
(d) the phosphor must be white.

11.6 A monitor or small television tube:
(a) is usually magnetically deflected
(b) has a signal applied to its anode
(c) needs electrostatic deflection plates
(d) is longer than an instrument tube of the same screen size.

# 12 Electronic modules

The complexity of electronic systems means that it is often impractical to draw a full circuit diagram on one or two sheets of paper. In order to communicate the operation of the whole system and the details of its individual parts a hierarchical drawing structure is used, with block diagrams at the higher levels and detailed component level schematics at the lowest level (Figure 12.1).

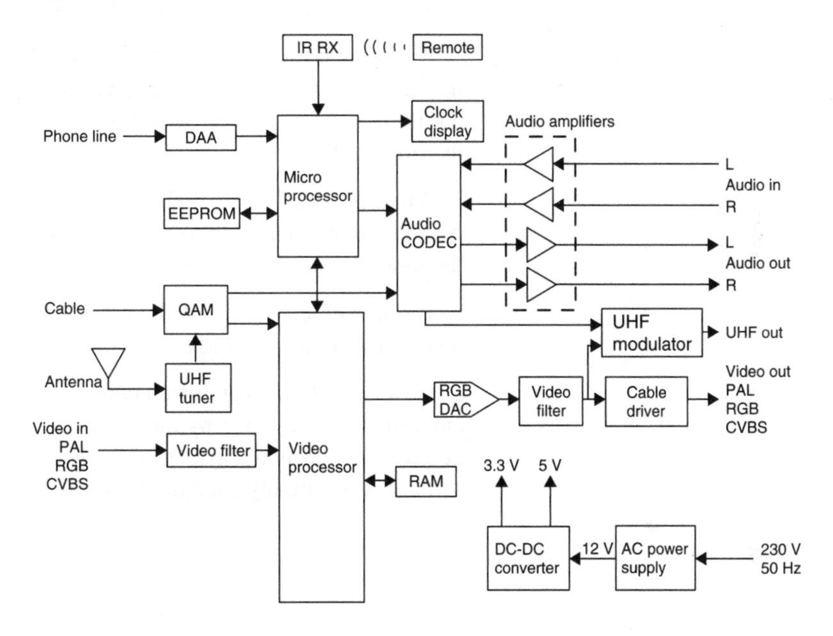


**Figure 12.1** A simplified block diagram of a digital television set-top box

The use of such a hierarchical approach aids understanding of systems and makes fault finding and repair of complex systems possible without the need to interpret design details within every block of the system. The complexity of integrated circuits (ICs) that are used in modern systems, and the increasing complexity of systems themselves mean that most electronic devices and equipment cannot usefully be treated as an arrangement of simple circuit components. Breaking the functions of a system down into functional system modules, with defined inputs and outputs, typically makes description and understanding easier and quicker. Although a module is an item that has clearly defined inputs and outputs, at block diagram level, it may cut across physical boundaries of printed circuit boards or complex chips in an actual device. A module could be a small part of a single IC, or an entire printed circuit board. The advantage of considering any electronic device as a set of modules in this way is that you do not require a detailed knowledge of how a module works in order to test its input and output signals and determine whether it is faulty. Normally, in consumer electronic devices a faulty module requires the replacement of the whole circuit board on which it is located, and any form of servicing work at the component level is now rare (and may be too expensive in labour costs).

## Amplifiers

A typical example of a module is an analogue **amplifier**, which has input and output terminals and a power supply. A signal at the input terminals will appear with greater amplitude at the output terminals. If this output signal has an identical waveshape in gain scale with the signal at the input terminals, the amplifier is termed a **linear amplifier**. Circuits that operate on signals of varying amplitude are called analogue circuits; compare this with digital circuits, which have defined states.

When the output signal has greater voltage amplitude than the input signal, the amplifier is said to have **voltage gain** ($G_v$). This gain is defined as the ratio $G_v = V_{OUT}/V_{IN}$, with both amplitudes measured in the same way, therefore either both r.m.s. or both peak measurements.

When the output signal has greater current amplitude than the input signal, the amplifier is said to have **current gain** ($G_i$). This gain is defined as the ratio $G_i = I_{OUT}/I_{IN}$, again with both measured in the same way.

The **power gain** ($G_p$) of the amplifier is defined as $G_p = P_{OUT}/P_{IN}$, and this is also equal to voltage gain × current gain, so that $G_p = G_v \times G_i$.

These gain figures are often converted into decibel form (see below). An amplifier is represented by the triangle symbol shown in (Figure 12.2). The output terminal is always assumed to be at the point of the triangle. This symbol is commonly used in block diagrams for the amplifier module.

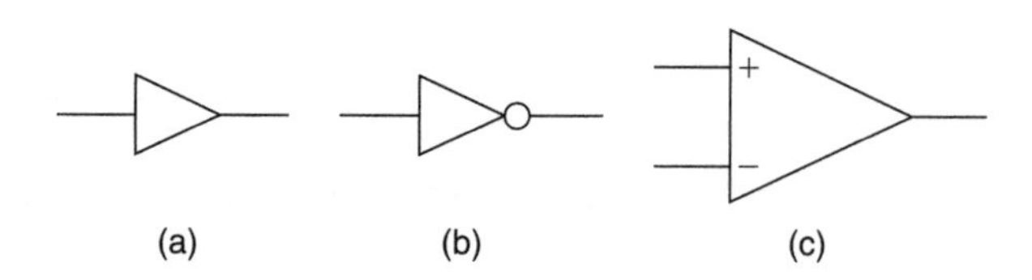


**Figure 12.2** Symbols for an amplifier: (a) non-inverting, (b) inverting, and (c) an op-amp

An **inverting amplifier** has an output that is the inverse of the input; this means that there is a phase shift of 180° between input and output. The **non-inverting** amplifier has an output that is in phase with its input.

Electronic systems can process signals over a very wide range of amplitudes from microvolts to many tens of volts. This large range makes it useful to use a logarithmic scale in comparing some measurements; we use the decibel (dB) scale, which makes use of logarithms. The logarithm of a number is the power to which the base must be raised to equal the number, so for example the base 10 logarithm of 100 is 2, that is $10^2 = 100$. The decibel is defined as: $dB = 10 \log [P_{OUT}/P_{IN}]$.

For example, when the output power of an amplifier is 100 times the input power, this is expressed as 20 dB, and if the output power is 1000 times the input power, the decibel ratio is 30 dB.

It is often more convenient to use voltage or current ratios than power, so we can recast the formula in terms of signal voltage measurements to become: dB = 20 log $[V_{OUT}/V_{IN}]$. This is true only when the output resistance equals the input resistance, but we often use it when the resistances are similar in size rather than perfectly equal.

A doubling of power can be expressed as $10\log 2 = 10 \times 0.301 = 3.01$ dB. Because fractional decibels are not normally audible, this is usually rounded down to 3 dB. However, in complex systems where the effects of noise, etc., are additive, it may be useful to work to values as low as 0.1 dB. Losses in a system are usually represented by negative values of decibels, so a halving of power is shown as −3 dB. The doubling of voltage or current levels is expressed as $20 \log 2 = 6.02$ dB or, when rounded down, 6 dB.

You can calculate the power ratio from the decibel figure using the antilog of the value, so the power corresponding to 1 dB = antilog(1/10) = $10^{1/10} = 10^{0.1} = 1.259$. (The antilog function raises 10 to the power of the argument supplied.) Therefore, an increase or decrease of 1 dB is equivalent to a change in power level of about 26%.

A decibel scale is normally used when the gain of an amplifier is plotted against frequency. The decibel corresponds to the smallest change that can be detected when listening to the amplifier. The **bandwidth** of an amplifier is defined as the frequency range between the two 3 dB or **half-power** points, which corresponds to the points where the voltage gain falls to 70.7% of its maximum value (and the power to half of its maximum value). Bandwidth can be expressed in one of two ways: by quoting either the two limiting frequencies or the difference between them. Thus, the bandwidth of an audio amplifier may be quoted as, for example, '40 Hz to 25 kHz', or the bandwidth of a radio frequency amplifier as '10 kHz centred on 465 kHz'. In either case, the ends of the quoted range are the frequencies at which the voltage gain is 3 dB down.

Bandwidth is assessed by plotting the actual voltage gain of the amplifier at a range of different frequencies. A graph of gain plotted against frequency (Figure 12.3) then shows the two frequencies (the 3 dB points) at which the gain has fallen to 70.7% of mid-band gain (3 dB down).

The fact that the power gain is half of the mid-band power gain at the 3 dB point is a convenient way of defining bandwidth for power amplifiers, so that you will see the phrase half-power bandwidth used in connection with power amplifiers.

Audio frequency (AF) amplifiers handle the range of frequencies lying within the approximate range 30 Hz to 20 kHz. This is about the range of frequencies that can be detected by the (younger) human ear. Such amplifiers are used in record players and tape recording, in the sound section of television receivers, in cinema sound circuits and in many industrial applications. Frequencies up to about 100 kHz, well beyond the range of the human ear, are often for convenience included in the AF range of a signal generator.

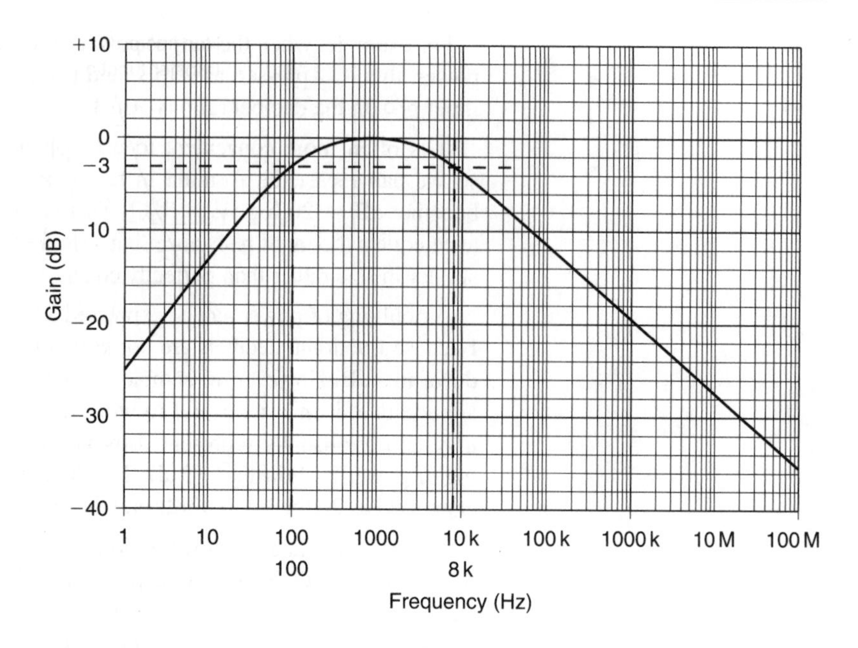


**Figure 12.3** Relative gain plotted against frequency, −3 dB points at 100 Hz and 8 kHz

Video amplifiers require a much larger bandwidth, typically d.c. to 6 MHz for television signals, and d.c. to 100 MHz or higher for instruments such as the oscilloscope. A d.c. amplifier, by contrast, will have a very small bandwidth, typically d.c. to around 10 Hz. Fault conditions on amplifiers of any type are indicated by low gain (or no gain at all), waveform distortion, reduced bandwidth, or changes in the input or output resistances.

## Distortion

The output of an analogue amplifier is never a perfect copy of the shape of the input signal. The difference is due to distortion that is caused by the imperfections in the amplifier, mainly because the plot of output against input is not a perfectly straight line (Figure 12.4). This **non-linear distortion** will result in a sine wave input giving an output that may look like a sine wave in an oscilloscope, but contains harmonics. This type of distortion can be reduced by careful design and by using a technique called negative feedback. For now, we are looking at the amplifier as a block rather than at details of design.

There are other forms of distortion that are much more noticeable than non-linear distortion, such as clipping and cross-over distortion. For a sine wave input, clipping means that the tips of a sine wave are not amplified, and this can affect one tip of a wave or both (Figure 12.5a). Cross-over distortion is created in amplifiers that use symmetrical positive and negative stages to amplify the positive and negative parts of the input waveform, hence the description cross-over distortion, meaning the region of low amplitude between the two parts of the amplifier where the input crosses from the positive half of the amplifier to the negative half. When the two halves are not well matched the whole amplifier does not amplify the lowest voltage levels

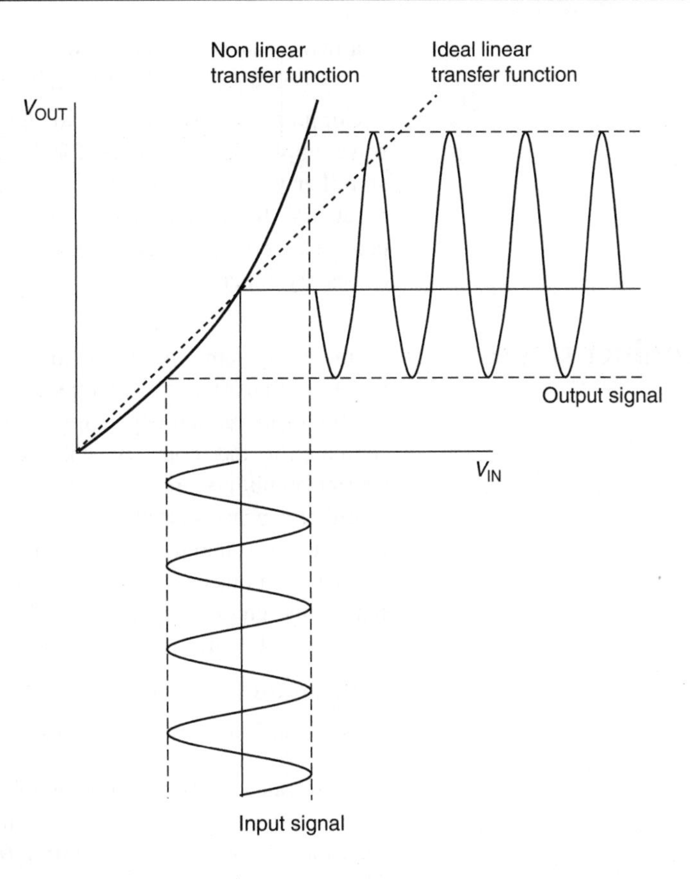


**Figure 12.4** Effect of non-linearity in amplification of a sine wave

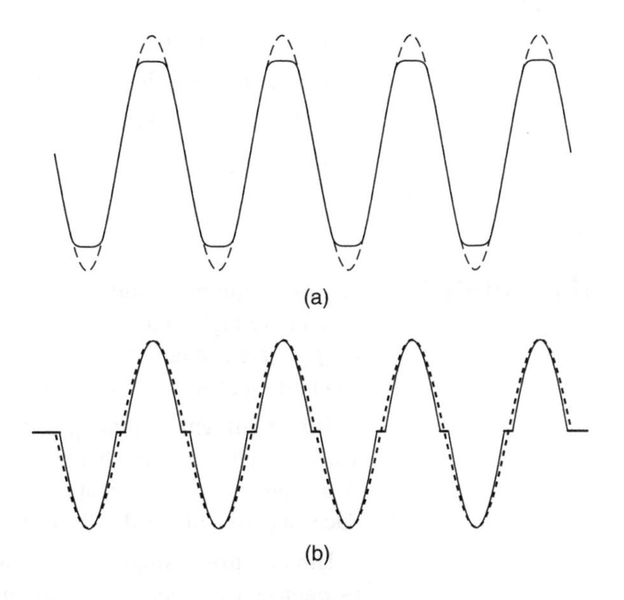


**Figure 12.5** Distortion of an amplifier output signal: (a) clipping, and (b) cross-over distortion

in a wave, causing a sine wave to form a small step just as it passes through the zero voltage level (Figure 12.5b).

Clipping is caused by overloading an amplifier, meaning that the input wave amplitude is too large for the amplifier to handle; for example, if an amplifier with a ±12 V supply and a gain of 100 has an input bigger than ±120 mV the output would need to exceed the supply rails, which is not possible. Cross-over distortion is usually a fault of incorrect transistor bias, caused by either faulty design or failure of feedback or bias stages.

## Amplifier stages

A single-stage amplifier is one that uses one amplifying device, usually some variety of transistor. An amplifier using one stage can have a high gain, 40 dB (×100) or more, and depending on the topology the output may be inverted compared to the input. Typically, voltage amplifiers are inverting, whereas current amplifiers are non-inverting. In sine-wave terms, the output of an amplifier is phase shifted by 0° or 180°. If the output of one such stage is used as the input to another identical stage, the gain will be much greater, 80 dB (×10 000) in this example, and the output signal will be phase shifted, by the sum of the phase shifts of the individual stages, from the input signal. The rules for finding the gain and phase of multistage amplifiers are therefore:

- gain in dB = gain of one stage × number of identical stages
- phase shift is phase shift of one stage × number of identical stages.

These rules apply to simple amplifiers without feedback; however, most amplifiers are designed to produce much more gain than is required in an open loop circuit and use **negative feedback** to close the loop, that is add a fraction of the output signal, inverted, to the input. This negative feedback has the effect, compared to the amplifier without feedback, of:

- decreasing gain
- decreasing non-linear distortion
- reducing cross-over distortion
- increasing stability.

Negative feedback is always used with IC amplifiers.

## Types of amplifier

**Audio frequency voltage** amplifiers have fairly high voltage gain, and medium to high values of output resistance. Audio frequency power amplifiers give large values of current gain. They have very low output resistance, so can pass large signal currents into low-resistance loads such as loudspeakers.

**Direct current** amplifiers, as their name suggests, are used to amplify d.c. voltages, although their bandwidth may extend up to quite high frequencies. A d.c. amplifier with a voltage gain of 100 (40 dB), for example, could produce an output of 1 V d.c. from an input of 10 mV d.c.

Direct current amplifiers are used in industrial electronic circuits such as photocell counters, in measuring instruments such as strain gauges, and in medical electronics (where they can be used to measure, for example, the electrical voltages in human muscles).

The main problem with d.c. amplifiers is their stability. It is difficult to ensure that their d.c. output does not vary, or drift, from one minute to the next, or change as ambient temperature changes. Voltage drift is thus a major difficulty with d.c. amplifiers, and instability which causes either low-frequency or high-frequency oscillations is often a source of operating trouble. Modern temperature-compensated integrated operational amplifier circuits have greatly improved the stability and drift of d.c. amplifiers. Until the mid-1980s a type of amplifier called the chopper stabilized amplifier was commonly used to amplify d.c. signals when very high voltage gains were required. Chopper amplifiers work by converting the input signal to a.c.; that is, chopping it with an oscillator signal and then amplifying the resulting narrow band a.c. signal before demodulating it to provide a d.c. output, this is much like an AM radio link. Chopper amplifiers are still found in some equipment.

**Wideband** amplifiers are amplifiers that provide the same value of gain over a large range of frequencies, typically from near d.c. to several megahertz. One important special application is as signal amplifiers in oscilloscopes, but it is true to say that most equipment used to make measurements on signal voltages will include a wideband amplifier.

No form of tuning is possible for such amplifiers, and their design is complicated by the requirement that they must provide the same values of voltage gain for d.c. as they do for high-frequency a.c. signals. Common faults include instability of bandwidth, variations in gain and the danger of oscillation.

**Video** amplifiers are a type of wideband amplifier originally used for the specific purpose of amplifying signals to be applied to cathode-ray tubes (CRTs). The video signal provided the picture information to be displayed. The term has developed to be used for any wideband amplifier that provides a signal to data-processing circuits, like analogue to digital converters. A minimum frequency range from d.c. to 5.5 MHz is necessary in video amplifiers used in television, but a wider range is required for computer monitors and signals such as radar.

If the waveshape at the output of the amplifier is not a true copy of the waveshape at the input, the amplifier is said to be producing distortion. Distortion can be caused by overloading, i.e. by putting in a signal input whose peak-to-peak voltage is greater than the amplifier can accept. If signal input is normal, distortion generally indicates a fault in the amplifier circuit itself.

Not all amplifiers are intended to be linear, however. Pulse amplifiers, for example, are not intended to preserve in every respect the shape of the wave that they amplify. A pulse is used for timing, and the leading edge of the pulse is the usual reference that is used. The one thing that a pulse amplifier must not do is alter the slope of the leading edges of the pulse inputs. Providing that the leading edge is sharp, the amplitude of the pulse is less important, and the shape of the portion of the pulse following the leading edge is also less important.

Voltage, current and power gain measurements can be made on any amplifier, and are often the key defining characteristics of an amplifier. To measure the gain of an amplifier, a small signal is injected into the input and the amplitude of the output signal is measured (see Figure 12.6). For most purposes, voltage gain is the figure of most interest, and it is the voltages of the input and output signals that have to be measured. Suitable input signals are obtained from a signal generator fitted with a calibrated attenuator, a circuit which reduces the signal by a given amount (e.g. ÷10, ÷100, ÷1000). The output signal is generally measured against an oscilloscope. Voltage gain is usually expressed in decibels.

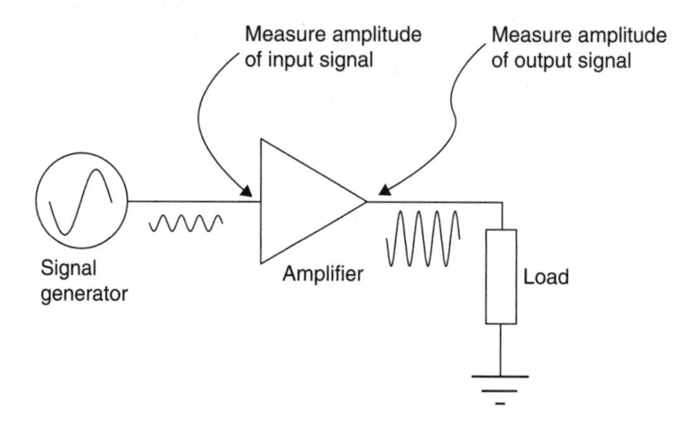


**Figure 12.6** Measuring voltage gain

A common problem is that the input signal from the signal generator may be too small to be measured by the oscilloscope, even when the oscilloscope is set at its maximum sensitivity. In practical work, for example, a signal as low as 30 mV p-p may have to be measured; but few of the oscilloscopes used in servicing can reliably measure such a quantity. To overcome this difficulty, the attenuator of the signal generator is switched so that the signal is made 10 times greater (in this example, to 300 mV), or even 100 times greater (to 3 V).

**Example:** An amplifier has an input of 5 mV and an output of 2.5 V, both measured peak to peak. What is its gain in decibels?

Quantities such as these can be readily measured with an oscilloscope; and when this has been done, the attenuator is reset to reduce the signal generator output to the value corresponding to a 30 mV input. Some signal generators provide a full-voltage output for the oscilloscope and an attenuated output for the signal to the equipment under test.

Similar methods may be used to measure the bandwidth of an amplifier. This is the range of frequencies over which the amplifier can be effectively used. The upper and lower limits of bandwidth are taken to be the frequencies at which the voltage gain of the amplifier is reduced to 70.7% of its value (−3 dB) in the middle of the frequency range.

As well as the change of gain, the phase of the output signal will change as the frequency is altered. At mid-frequency, the phase of the output will be either 0° or 180°, meaning that it is at the same phase as the input or inverted. Note that inverted and 180° phase have the same meaning only for sine waves. By the time that the gain has fallen to 70.7%, the phase will have changed also, by an angle of 45°.

Amplifier distortion is measured using distortion meters. A pure sine wave input is fed to the amplifier, and the output is applied to the distortion meter, which will detect any harmonics of the wave, indicating distortion. The amount of distortion is often quoted in terms of total harmonic distortion (THD), and for a hi-fi amplifier this figure would be expected to be less than 0.1%.

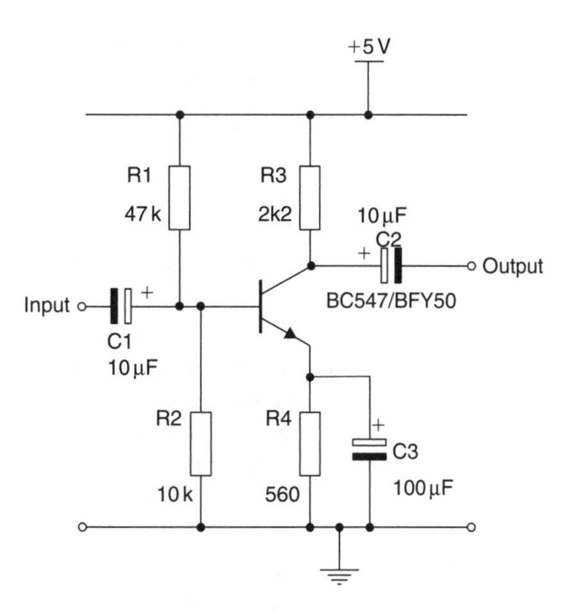


**Figure 12.7** Single transistor amplifier

## Practical 12.1

Construct the voltage amplifier shown in Figure 12.7, using a BC107 or BFY50 transistor. Connect the amplifier to a 9 V supply and check that the collector voltage falls within the range 4–5 V. Use the oscilloscope to set the sine wave of the signal generator to 30 mV p-p at 400 Hz.

Connect the output of the signal generator to the input of the amplifier, and the output of the amplifier to the oscilloscope. Measure the peak-to-peak amplitude of the signal from the output of the amplifier. Now calculate the voltage gain from the formula shown above.

Note that the *x*-axis of the graph along which frequency is measured is calibrated on a logarithmic scale in order to accommodate the wide range of frequencies covered by a typical amplifier bandwidth.

## Oscillators

An oscillator is an amplifier that provides its own input by using positive feedback. That is, a signal is taken from the output and fed back through some network of components into the input in phase, so as to maintain the output signal. The feedback network usually consists of frequency-determining components as shown in Figure 12.8. There are two conditions that have to be met to sustain oscillations:

- The amplifier gain must be greater than the attenuation of the feedback circuit.
- The feedback must be positive. That is, phase shift must be a multiple of 360°.

If the gain with feedback applied (the loop gain) is only just greater than 1, then the output wave shape will be a sine wave. If the gain is much greater than 1, oscillations will be violent and this results in pulse or square wave-shapes. The frequency of the oscillations depends on the values of the components in the feedback loop, and the output amplitude depends largely on circuit component and supply voltage values. The essential function of an oscillator is to convert the d.c. power supplied into it into a.c. power as the output signal and, like all energy converting devices, it dissipates heat owing to its inefficiencies.

The frequency-determining components may be part of the amplifier circuit or part of the feedback circuit, and they can be of various types. One very common type in the past has been the combination of an inductor and a capacitor (the LC circuit), which can use L and C in parallel (very high resistance for signals at the frequency of resonance) or in series (very low resistance for signals at the frequency of resonance). For lower frequencies, feedback circuits containing resistors and capacitors (RC filters) are used.

RC oscillators are frequently used in electronic systems to provide clocks and test frequencies, etc., since digital systems require square-wave clock signals. The most common type of RC oscillator is probably the relaxation oscillator (see Waveform generators, below). The type of RC oscillator that produces the best sine-wave output, that is with the least distortion, is the Wien bridge circuit, often used in audio signal generators. This circuit uses amplitude stabilization circuitry to obtain a pure sine wave from an RC oscillator. For oscillators whose frequency must be very precise and constant, a **crystal** is used to determine frequency. A typical crystal for this purpose consists of a thin, flat plate of quartz, ground and polished to great accuracy with metallized contacts deposited on both sides to form circuit connections. It is modelled by an equivalent LCR circuit.

## Waveform generators

Waveforms, other than sine waves, are generated by a class of oscillators described as waveform generators. These oscillators (also called aperiodic oscillators) typically contain at least one frequency-determining network which sets the period of oscillation, and an active device providing

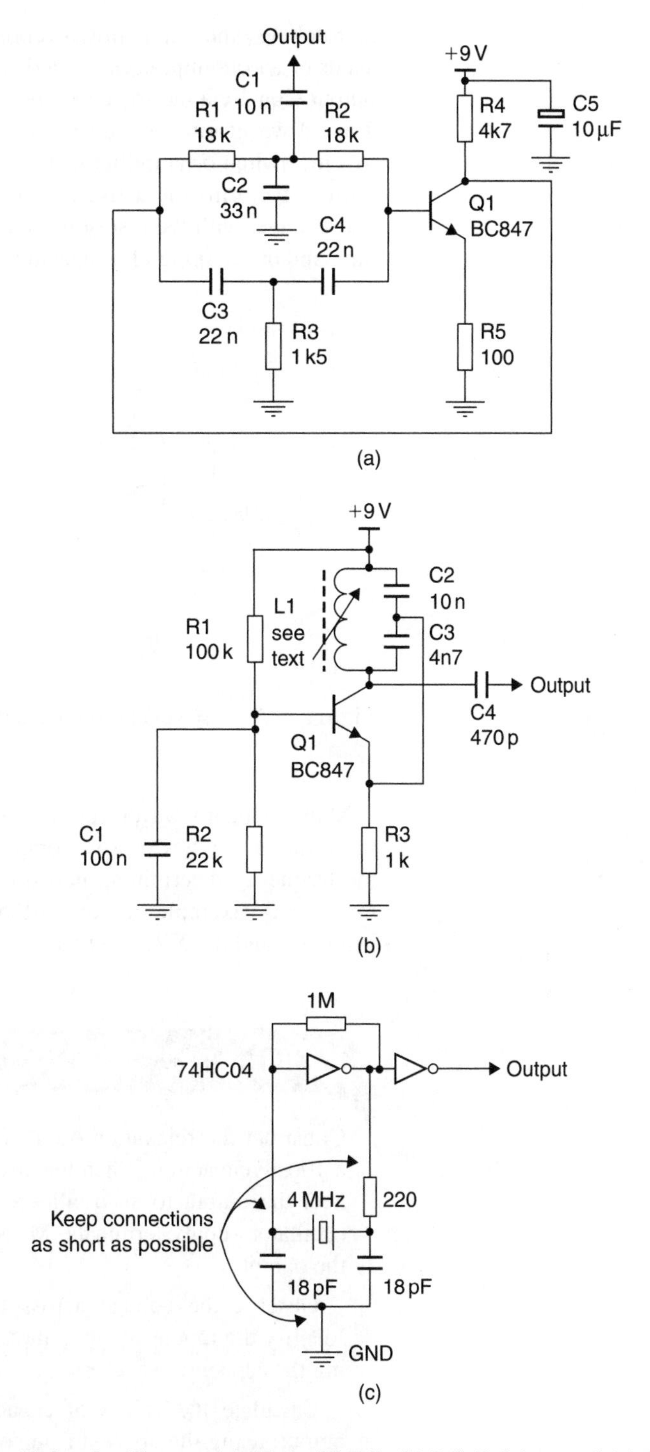


**Figure 12.8** Oscillators: (a) 'Twin T' RC, (b) LC Colpitts, and (c) CMOS crystal oscillator

a switching function or amplification. The active device, for example a transistor or an op-amp, is connected so that its input is provided from its own output signal via the frequency-determining network, by positive feedback. Typical waveforms are square, triangular or pulse, and in all such oscillators the switch or amplifier is not active for the entire time of the wave; for pulse waveforms the active device is used for only a short fraction of the entire wave, with the rest of the wave shape formed, typically, by the charging and discharging of a capacitor (Figure 12.9).

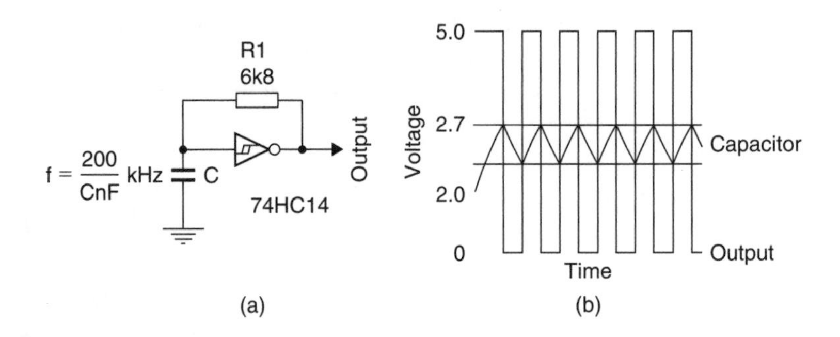


**Figure 12.9** (a) Relaxation oscillator, and (b) waveforms

Many standard waveform-generating circuits are available as ICs, with the addition of a few passive components (resistors and capacitors) to make the frequency-determining network produce stable and repeatable oscillator designs. Examples are the LM555 timer circuit, the ICL8038 waveform generator and the XR2206 function generator.

## Practical 12.2

Construct the relaxation oscillator circuit shown in Figure 12.8. Use a 100 nF capacitor. When the circuit has been built and checked, connect the output to an oscilloscope and connect a 5 V supply to the oscillator supply terminals. Measure the amplitude and frequency of the output.

Observe the voltage across the capacitor with an oscilloscope. Identify the input switching thresholds of the CMOS gate by measuring the capacitor waveform.

Calculate the values of capacitor required to produce a 50 kHz output, using the approximate equation shown in Figure 12.9. Select the nearest available capacitor value and retest the circuit; measure the frequency. Comment on the accuracy: how would a variable resistor help?

Note that an oscillator, operating correctly, should provide an output signal whenever a d.c. supply is connected to it. A faulty oscillator may be revealed by failure to produce an output signal (or at best an intermittent one) or by generating an output waveform of distorted waveshape, incorrect frequency or insufficient amplitude.

## Filters

Filters are circuits that act on waveforms to change their amplitude, their phase, or (often) both. Passive filter circuits that consist wholly of passive components will never increase the amplitude of a wave, but will either reduce it or leave it unchanged. The power gain of a passive filter is always less than unity; there are always losses, although they may be very small. The input and output amplitudes must be measured in the same way, so that if the input amplitude is measured as peak-to-peak, then the output amplitude must also be measured in this way.

Filter circuits do not change the *shape* of sine waves, but they do alter the shapes of other waveforms because the shape of a waveform depends on the frequency information it contains. Many filter circuits cause the *phase* of a sine wave to shift. The amount of change depends on the frequency of the signal.

Wave filters can be designed to cope with waves of any range of frequencies, from low (a few hundredths of a cycle per second, i.e. a cycle that lasts for many hundreds of seconds) to high radio frequencies of many hundreds or even thousands of megahertz. The components and circuit design of the filters vary considerably according to the frequency range they are designed to handle.

Whatever the range of frequencies handled by a filter, however, the main filter types are low-pass, high-pass, band-pass and band-stop. The symbols used in block and circuit diagrams to indicate these types of filter action are shown in Figure 12.10.

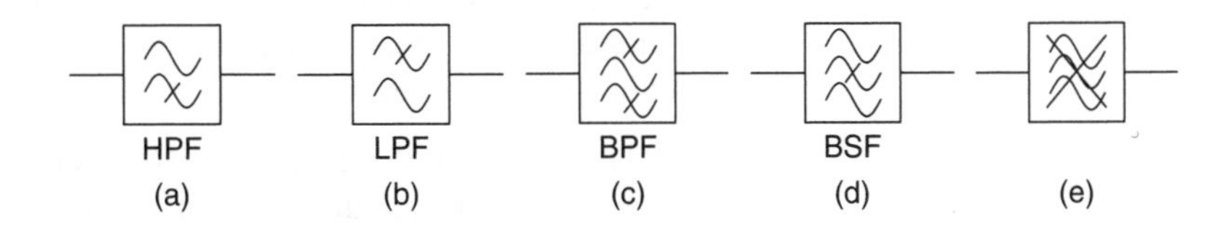


**Figure 12.10** Filter symbols: (a) high-pass, (b) low-pass, (c) band-pass, (d) band-stop, and (e) interference suppressor

A low-pass filter, as its name suggests, passes without attenuation the low frequencies of signals arriving at its input, but greatly reduces the amplitude of high-frequency signals, which are thereby heavily attenuated. In other words, a low-pass filter has a pass-band of low frequencies and a stop-band of high frequencies.

A simple form of low-pass filter is the RC filter shown in (Figure 12.11) together with typical performance graphs. These graphs show the variation in both gain and phase plotted against frequency which the filter is capable of achieving.

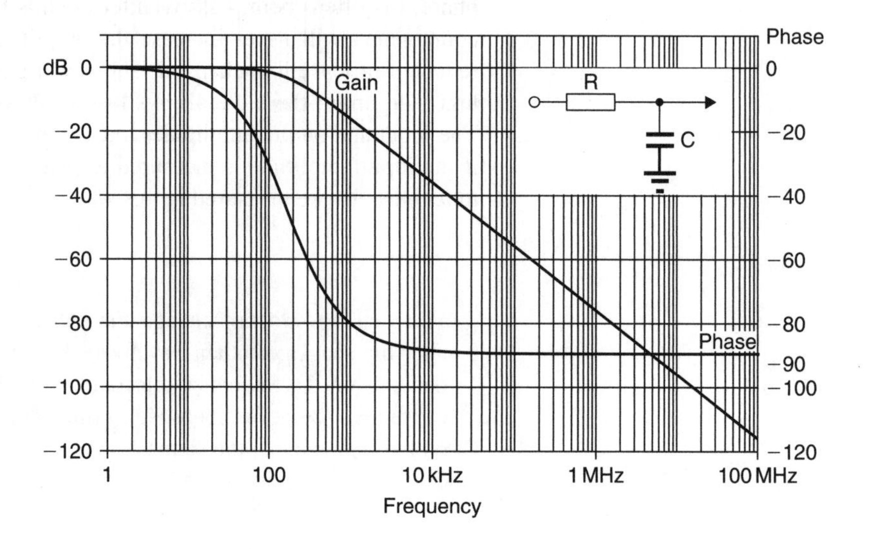


**Figure 12.11** RC low-pass filter and its performance graphs

Figure 12.12 shows a single CR high-pass filter with its measured gain/frequency performance graph. Simple RC filters do not give sufficient attenuation for most purposes, and filters which include inductors as well as capacitors and resistors are typically employed. Both the theory and the practical design of such filters present difficulties. Figure 12.13 illustrates examples of both low-pass and high-pass filters.

When inductors and capacitors are combined in a filter circuit, band-pass and band-stop filters can be constructed. Band-pass filters will pass a predetermined range of frequencies without attenuation, but will attenuate other frequencies both above and below this range. A band-stop filter has the opposite action, greatly reducing the amplitude of signals in a given band, but having little effect on signals outside it.

The basic action of filter circuits on sine waves is, as has been seen, to attenuate and change the phase of the waves, but to leave the shape of the wave unaltered. When the input wave to a circuit is not a sine wave, however, the action of even a passive circuit can change the shape of the wave (as well as its phase) considerably.

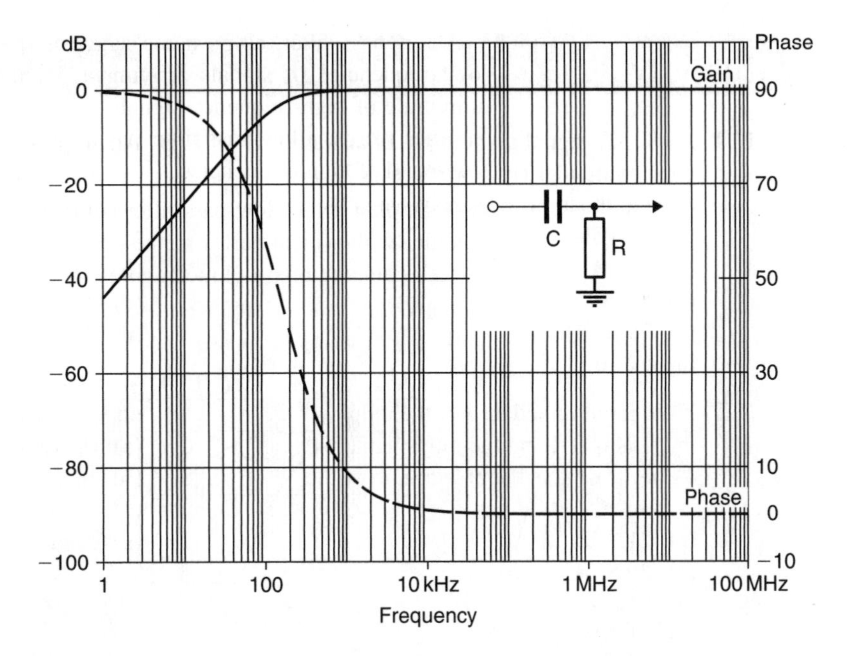


**Figure 12.12** Typical single-stage CR high-pass filter gain and phase versus frequency graph

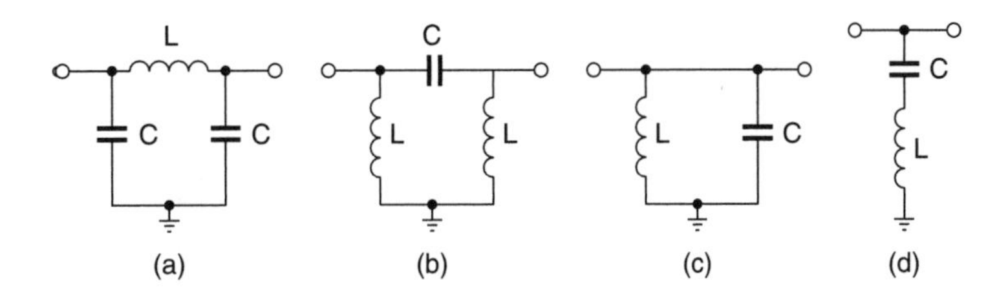


**Figure 12.13** LC filters: (a) low-pass, (b) high-pass filters, (c) band-pass, and (d) notch

## Multiple-choice revision questions

12.1 An 80 mV signal at the input of an audio power amplifier (PA) produces a 72 V drive to the loud speaker. What is the voltage gain?
(a) 90 dB
(b) 59 dB
(c) 29.5 dB
(d) 72 dB.

12.2 If the input impedance of the amplifier in Q1 is 600 Ω and the loudspeaker is a 4 Ω unit, what is the power gain of the amplifier?
(a) 21.7 dB
(b) 29.5 dB
(c) 51.3 dB
(d) 81 dB.

12.3 An amplifier with 50 Ω input and output impedance has a voltage gain of 30 dB. What is the power gain?
(a) 60 dB
(b) 30 dB
(c) 15 dB
(d) 10 dB.

12.4 Which one of the following is not an effect of using negative feedback?
(a) decreases gain
(b) increases stability
(c) increases gain
(d) reduces distortion.

# Unit 5
## Digital electronics

### Outcomes

1. Demonstrate an understanding of combinational logic circuits and apply this knowledge safely in a practical situation
2. Demonstrate an understanding of multivibrators, timers and logic systems and apply this knowledge safely in a practical situation
3. Demonstrate an understanding of input signals and output displays in digital systems and apply this knowledge safely in a practical situation.

# 13 Logic systems

## Numbers and counting

Ordinary counting is done using ones and tens: it is a **denary system**. In all number systems, the number that is raised to these different powers, like the denary 10, is called the **base**, and the numbers that are used for the powers are called the **exponents**. The number 476, for example, is six units, seven tens and four hundreds; remember that $10^1 = 10$ and $10^2 = 100$.

Ten, however, is not the only number that can be used as the base for a system of counting. Bases other than 10 can be used without any change in the method of writing the exponents. In the scale of 8, for example, the number 163 means three ones, plus six eights, plus one $8^2$ (=64), making a total of 64 + 48 + 3 = 115 in denary terms. Such a scale is used in some computing work, as is a scale of 16. In this latter hexadecimal scale, as it is called, we need to be able to express the denary numbers 10, 11, 12, 13, 14 and 15; and the first six letters of the alphabet (A, B, C, D, E and F, respectively) are used for the purpose.

Another counting system is the **binary scale**, whose base is 2. The distance to the left of the (suppressed) point is used to show what power of two is being used, and the only digits that are used are 0 and 1. Thus, the binary number 1101 is the equivalent of a denary number calculated as follows: the first place to the left of the suppressed binary point represents $2^0$, or 'ones'. There is a 1 in that place, so it must be counted. The second place to the left of the point represents $2^1$, or 'twos'. There is a zero in this place, so nothing is added to the count. The third place represents $2^2$, or 4; and a 1 in this position means that 4 must be added to the count, just as a 1 in the fourth place to the left ($2^3 = 8$) means that 8 must be added. The full denary equivalent is therefore 1 + 0 + 4 + 8 = 13.

Binary figures to the right of the binary point are used to express binary fractions, such as ($2^{-1}$, $2^{-2}$, $2^{-3}$, $2^{-4}$) and so on to further inverse powers of 2.

Table 13.1 shows powers (up to 9) for denary and binary systems. This illustrates that binary numbers must contain more digits than their denary equivalent. One digit in a hexadecimal number is the equivalent of four binary digits, so the hexadecimal scale is more compact than the binary scale, and conversions to and from hexadecimal are simpler.

Denary numbers are converted to binary numbers by the following process. Divide the denary number by 2, and write down any remainder at the head of a separate remainder column. Divide what is left by 2, and again write down any remainder in the remainder column, immediately under the first entry. Clearly, all entries in this remainder column must be either 1s (if there is a remainder after a division by 2) or 0s (if there is not).

**Table 13.1** Binary and denary powers

| *Power* | *Of ten* | *Of two* |
|---|---|---|
| 0 | 1 | 1 |
| 1 | 10 | 2 |
| 2 | 100 | 4 |
| 3 | 1000 | 8 |
| 4 | 10 000 | 16 |
| 5 | 100 000 | 32 |
| 6 | 1 000 000 | 64 |
| 7 | 10 000 000 | 128 |
| 8 | 100 000 000 | 256 |
| 9 | 1 000 000 000 | 512 |

Continue this process of division by two until the original denary number is reduced to 1. Now, since 'two into one doesn't go', put in a zero as the last entry in the result column, and the remainder 1 as the last entry in the remainders column. Then read off the remainder column starting from the bottom and moving upwards, and write down the ensuing string of 0s and ls, from left to right. The left-most bit is described as the most significant bit (MSB) and the right-most as the least significant bit (LSB).

**Example:** Convert the denary number 527 to binary.

**Solution**

| *Action* | *Result* | *Remainder* |
|---|---|---|
| Divide 527 by 2 | 263 | 1 |
| Divide by 2 | 131 | 1 |
| Divide by 2 | 65 | 1 |
| Divide by 2 | 32 | 1 |
| Divide by 2 | 16 | 0 |
| Divide by 2 | 8 | 0 |
| Divide by 2 | 4 | 0 |
| Divide by 2 | 2 | 0 |
| Divide by 2 | 1 | 0 |
| Divide by 2 | 0 | 1 |

Now start from the bottom of the remainder column and write out the digits in it from left to right. The binary number corresponding to denary 527 then appears as: 1000001111.

The conversion of binary numbers to denary is much simpler if Table 13.1 is used. Write out the binary figure, for example 100101. Ignore all the zeros in that figure, and concentrate on the correct power of 2 for each of the 1s according to its position in the binary number. Then add the denary figures on the bottom line, and the full denary equivalent of binary 100101 appears as $32 + 4 + 1 = 37$.

## Binary arithmetic

Adding binary numbers is carried out in the same way as for denary numbers, except that the rule is now that two 1s add to zero with 1 carried forward, while three 1s added together equal 1 and 1 carried forward. For example:

$$\begin{array}{rr} & 1011 \\ + & 1101 \\ = & 11000 \end{array}$$

In the units column the two 1s add up to binary 10 (spoken as 'one zero', not 'ten'). A zero is written down and 1 is carried forward. The next column then appears as 1 + 1 (carried) +0, which therefore again means that a zero is written down and 1 is carried forward. The same thing happens in the third column, but in the fourth the carry forward makes the total 1 + 1 + 1. So a 1 is now written down, leaving the carry forward 1 to appear alone in the fifth column on the left.

Binary subtraction can be done by the familiar 'borrow' method that is used in denary subtraction, but an easier method is available, called '2s complement'. Write down the larger number, and below it the smaller number 'complemented'. This means that every 0 in the smaller number becomes a 1, and every 1 a 0. Add this complemented binary number to the larger number and add another unit 1, to get the result. The figure on the extreme left of this sum is disregarded if it would make the number of digits larger, and what is left is the result of the binary subtraction.

Note that both the binary numbers in this method of subtraction must have the same number of digits. If they do not, the appropriate number of zeros must be added to the **left** of the smaller number.

**Example:** Subtract binary 110 from binary 1101.

**Solution**

| | |
|---|---:|
| Write down the larger figure | 1101 |
| The smaller number must be rewritten as | 0110 |
| Complement this binary number | 1001 |
| Add 1101 | 1101 |
| Now add 1 | 0001 |
| To get | 10111 |

Discard the left-hand digit, and the binary remainder is 0111, which is the result of the subtraction. You could, of course, convert the two binary numbers into denary, subtract, and then convert back to binary, but the 2s complement method is important because it allows us to make integrated circuits (ICs) that are able both to add and to subtract.

The peculiar advantage of the binary scale for electronic counting systems is that only the figures 0 and 1 need to be represented. It is much easier to design a circuit in which a transistor is biased either fully on or fully off than it is to design one in which varying values of collector voltage represent different figures in the denary scale. A binary system needs no carefully calculated bias voltages, and will not be greatly affected by changes in either supply voltage or component values. In addition, the recording and replaying of binary numbers is much less affected by faults in the recording process.

In a normal system of digital logic, '1' is represented by a comparatively high voltage, say 3 V, 5 V, 10 V, 12 V or whatever the supply voltage happens to be, and '0' by a much lower voltage, typically around 0–0.2 V. This system is called **positive logic**. It is also possible to represent 1 by a low (or negative) voltage and 0 by a higher voltage, a system called **negative logic**. These voltages are usually referred to only in terms of the digits 0 and 1.

The binary system described here is also known as 8-4-2-1 binary to distinguish it from other forms of binary code such as Gray code.

Another way of using binary code is in the form of binary-coded decimal (BCD). In a BCD number, four binary digits (or bits) are used to code one denary number, using the binary equivalents of denary 0–9. For example, the denary number 815 can be coded in BCD as 1000 0001 0101, the binary codes for the three separate denary digits. Arithmetic in BCD is very complicated, but the code is very useful when values have to be displayed.

## Combinational logic

A **gate** is a circuit that allows a signal to pass through when it is open, but blocks the signal when it is closed. The opening and closing of the gate is done by means of other electrical signals. A **logic gate** uses only digital signals of the 0 or 1 type, and the aim of all logic gate circuits is to ensure that the output becomes 1 only when some fixed combination of inputs is presented to the gate. A circuit using gates is called a **combinational circuit**. Figure 13.1 shows two logic actions carried out by switch circuits.

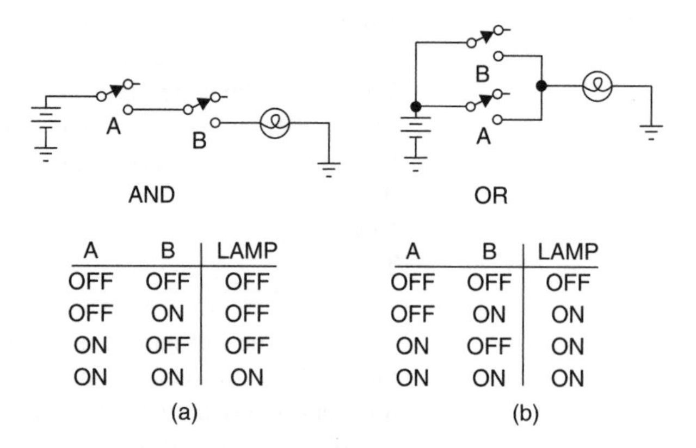


| A | B | LAMP |
|---|---|---|
| OFF | OFF | OFF |
| OFF | ON | OFF |
| ON | OFF | OFF |
| ON | ON | ON |

| A | B | LAMP |
|---|---|---|
| OFF | OFF | OFF |
| OFF | ON | ON |
| ON | OFF | ON |
| ON | ON | ON |

**Figure 13.1** Switch circuits for logic actions: (a) AND function: both switches must be on for the lamp to light, and (b) OR function: either switch can turn the lamp on

The simplest gate uses only one input and is the **NOT** gate or inverter. If the input is 0 then the output is 1. If the input is 1 then the output is 0. The simplest types of two input gates are called AND, OR, NAND and NOR gates, according to their action. The action of a gate is shown in its truth table, which is a list of the outputs that are obtained from every possible combination of inputs into the gate. A truth table is much easier to understand than the circuit diagram of a gate which, when an IC is used, may not even be available.

The truth tables for AND, OR, NAND, NOR and the NOT or INVERTER gate are shown in Figure 13.2. These show that the output of the AND gate is 1 only when both inputs are 1 (i.e. only when both input A and input B are at logic 1). The output of the OR gate is 1 when either input A or input B is at logic 1, or when both inputs are at logic 1.

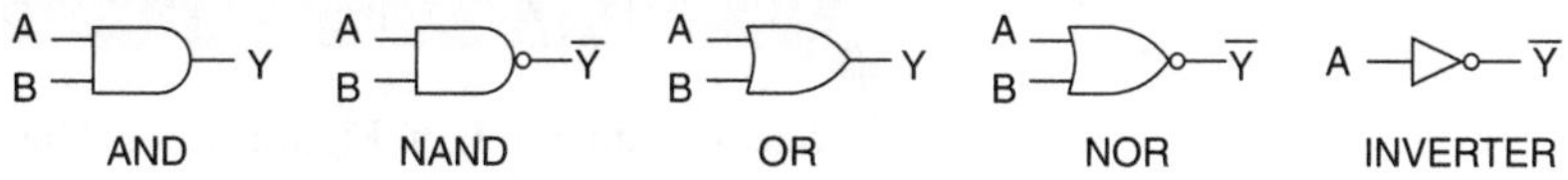

AND NAND OR NOR INVERTER

| A | B | Y | $\overline{Y}$ |
|---|---|---|---|
| 0 | 0 | 0 | 1 |
| 0 | 1 | 0 | 1 |
| 1 | 0 | 0 | 1 |
| 1 | 1 | 1 | 0 |

(a)

| A | B | Y | $\overline{Y}$ |
|---|---|---|---|
| 0 | 0 | 0 | 1 |
| 0 | 1 | 1 | 0 |
| 1 | 0 | 1 | 0 |
| 1 | 1 | 1 | 0 |

(b)

| A | $\overline{Y}$ |
|---|---|
| 0 | 1 |
| 1 | 0 |

(c)

**Figure 13.2** Truth tables for logic gates: (a) AND/NAND, (b) OR/NOR, and (c) INVERTER

The NAND and NOR gates are no more than the inverses of the AND and the OR gates, respectively. Check carefully through their inputs and outputs in the tables to see what this implies.

Conventionally, gate action is illustrated with reference to two inputs, although gates with larger numbers of inputs are manufactured and used.

The standard actions of the four types of gate mentioned are represented on diagrams by the symbols shown in Figure 13.3. Note that the British Standard (BS) symbols (now seldom used except in manufacturers' databooks) differ from the international symbols which will be found in most logic diagrams. Both sets of symbols should be known. The international symbols are sometimes described as United States Military Specification or Mil-spec.

By suitable connection of the appropriate standard gates, any type of truth table can be achieved. Conversely, since the truth tables of all the standard gates are known, a truth table for any combination of gates can be worked out. These design actions are outside the scope of this book.

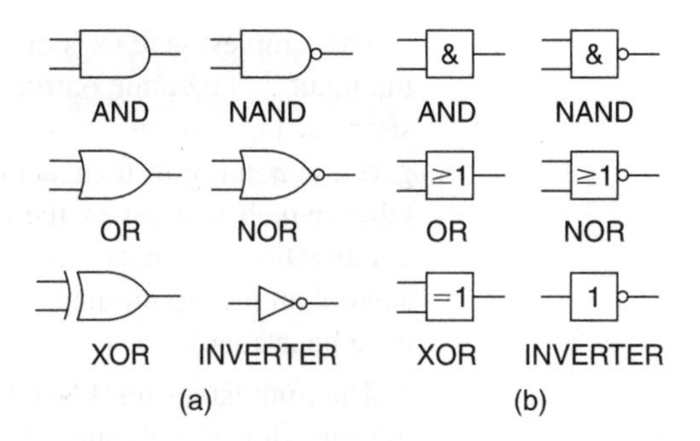


**Figure 13.3** Logic gate symbols: (a) international, and (b) BS symbols

## Practical 13.1

See the diagram of Figure 13.4. The SN74HC00 integrated circuit contains four NAND gates, each with two inputs, and is available as a 14-pin DIL pack. This can be conveniently mounted on a solderless breadboard for experimental work. Remember that semiconductor devices are static sensitive, avoid touching the pins of the IC, use a wrist strap where possible and ensure that your body is not carrying a static charge.

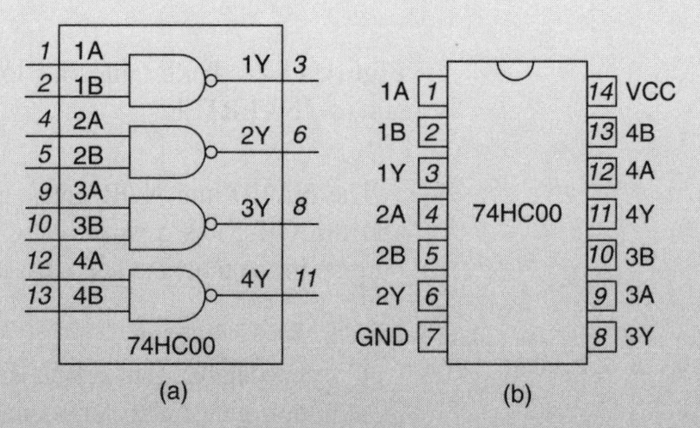


**Figure 13.4** (a) Logic drawing, and (b) package drawing of 74HC00 quad NAND gate IC

With the IC suitably mounted, connect pin 14 to the positive terminal of a regulated 5 V supply, which should be switched off while assembling or modifying circuit connections, and pin 7 to the negative terminal of the supply. Then connect a voltmeter (using its 10 V range) with its positive lead to output of one of the NAND gates, e.g. pin 3, and its negative to earth. Connect 47 k resistors between the +5 V terminal and the input pins of the gate and connect switches between ground and the input pins (e.g. pins 1 and 2), labelling the open-circuit position '1' and the closed position '0'. The reason for labelling

(Continued)

## Practical 13.1 (Continued)

the switches thus is that closing a switch will short the input pin to the ground, resulting in a very low voltage, logic zero; while opening the switch will allow the input pin to be pulled up to +5 V, by the 47 k resistor, giving a logic 1. Connect the inputs of all the unused gates to ground; this is important because if the inputs are left floating the IC may not operate correctly. Label the switches A and B, and fill in the truth table for the NAND gate.

Figure 13.5 shows how NAND gates can be connected so as to provide the actions of NOT, AND, OR and NOR. NOR gates can also be used in combination to provide the output of any other gate, and this equivalence was worked out in theory in De Morgan's theorem some 150 years ago. The practical implication is that digital circuits can be constructed entirely from one simple gate, either NAND or NOR, making the construction of digital ICs easier.

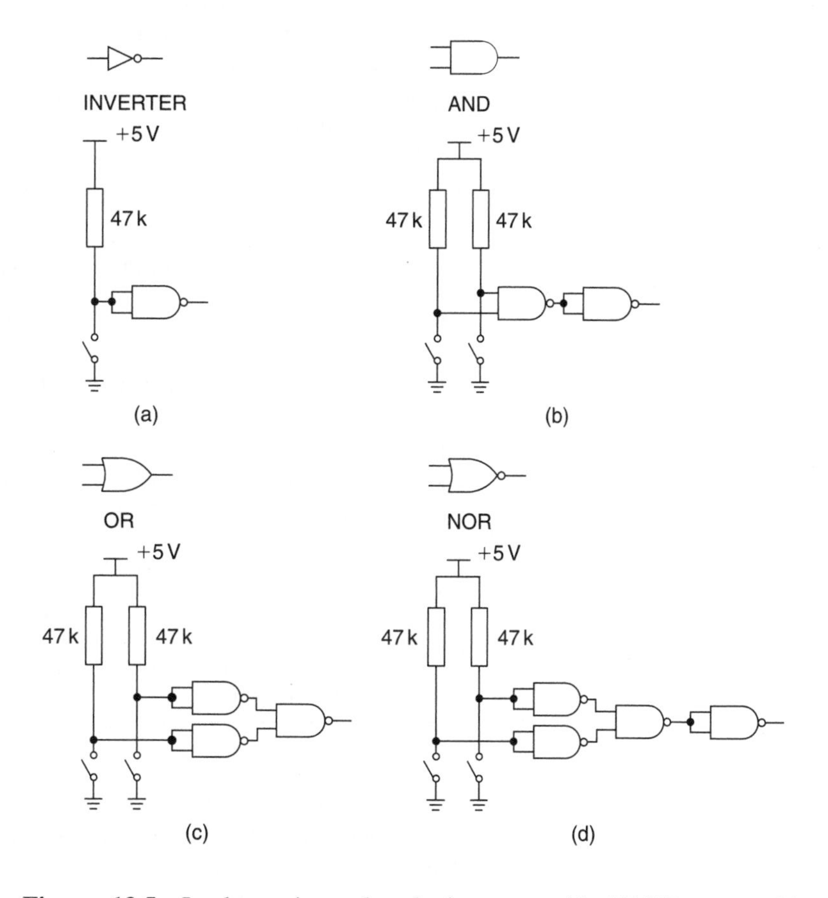


**Figure 13.5** Implementing other logic gates with NAND gates: (a) INVERTER, (b) AND, (c) OR, and (d) NOR gate

## Logic gate families

A set of gates all using the same type of IC construction is called a **family**, and several families of gates are still in use, although some older types are now obsolete. The most enduring form of digital gate IC has been **TTL**, meaning transistor–transistor logic. This type of IC uses bipolar transistors in its internal design, and the input stage is a multiemitter transistor. The original type of TTL system, often called standard TTL, operates with a regulated voltage supply of +5 V (which must never exceed 5.25 V). Any voltage between 0 and +0.8 V will be taken as logic 0 level, and any voltage between +2 and +5.5 V will be taken as logic 1 level. Voltage inputs that lie between +0.8 and +2.0 V must be avoided, because they could be taken as either 0 or 1.

TTL is now manufactured only for replacement purposes, and later versions such as low-power Schottky (**LSTTL**) are now also being phased out. The LSTTL types operate using internal circuits that are the same as TTL, but implemented with Schottky junction transistors to give much lower power dissipation.

The majority of common logic devices are now implemented using metal oxide semiconductor field-effect transistors (MOSFETs). The bipolar transistor TTL family of logic devices has been superseded by several generations of MOS types of IC, some of which can use a wide range of operating voltage, while others are used where low-power dissipation is important. Table 13.2 summarizes important characteristics for some of these IC families.

**Table 13.2** IC families summarized

| | *TTL* | *LSTTL* | *CMOS* | *74HC* | *74AC* |
|---|---|---|---|---|---|
| V+ supply | 5 V | 5 V | 3–15 V | 2–5 V | 2–5 V |
| $I_{max/1}$ | 40 μA | 20 μA | < 0.2 μA | < 1 μA | < 1 μA |
| $I_{max/0}$ | –1.6 mA | –0.4 mA | < 0.2 μA | < 1 μA | < 1 μA |
| $I_{max/out}$ | 16 mA | 8 mA | 1.6 mA* | 8 mA | 15 mA |
| Delay | 11–22 ns | 9–15 ns | 40–250 ns* | 10 ns* | 5 ns |
| Power | 10 mW | 2 mW | 0.6 μW* | < 30 μW | < 30 μW |
| Frequency | 35 MHz | 40 MHz | 5 MHz | 40 MHz | 100 MHz |

* These quantities depend on the supply voltage level.
V+ supply = normal positive supply voltage level; $I_{max/1}$ = maximum input current for logic level 1; $I_{max/0}$ = maximum input current for logic level 0; $I_{max/out}$ = maximum output current; Delay = propagation delay in nanoseconds; a low delay means a fast device; Power = no-signal power dissipation per gate in mW or μW; Frequency = typical operating frequency.

The most important digital gate circuits nowadays are those that combine a very short delay time (fast-acting gates) with very low power, and in these respect the MOS types are much superior to the older bipolar types.

The HC and AC CMOS types have for most purposes replaced the earlier TTL types, and carry the same numbers, so that, for example, 74HC00

(CMOS) carries out the same logic as 7400 (TTL) and 74LS00 (LSTTL). An older CMOS type, designated the 4000 series, is also used where lower speeds can be tolerated and wider supply range (up to 15 V) is required; these are often used in low-cost battery-powered equipment where voltage regulation is not provided. The 4000 series is not usually pin compatible with other families, with the exception of devices that have been added to the 74HC range, such as the 4017 counter, which has an equivalent 74HC4017.

Logic circuits must use a regulated power supply that cannot apply excessive voltage to any unit. Logic circuits also need to be protected from excessive or reverse voltage inputs, and this can be done using diodes. For MOS ICs, these protection diodes are built into the chips.

The d.c. supply voltage level is very important for digital circuits, because a low supply voltage will result in incorrect voltages for level 1, causing 1s to be counted as 0s in the digital ICs. This is therefore the first point to check when a digital circuit ceases to work or starts to work erratically.

You should know how the d.c. supply is regulated, because failure of a regulator chip may be the cause of a low or absent supply voltage. If, however, a regulator has failed in such a way as to cause the supply voltage to rise, it is very likely that this may have destroyed all the ICs, depending on the type used. In this respect, the CMOS ICs are more rugged than some TTL types.

Hybrid digital ICs are so named because they mix more than one technology. Some use a combination of analogue and digital methods, while others use a mixture of MOS and BJT devices. ICs of this type are not grouped into families because they are usually specially designed for a particular purpose rather than for universal application.

## Logic thresholds and noise margin

Digital logic devices require a stable power supply voltage, and in most cases this must be regulated to a set voltage such as 5 V for TTL devices or one of several standardized logic supply voltages such as 2.7 V, 3.3 V or 5 V.

The TTL and compatible devices such as 74HCT series ICs have fixed input thresholds based on a nominal 5 V supply voltage; thus, input voltages between 0 and +0.8 V are treated as logic 0 and those between +2.2 V and the positive supply (5 ± 0.5 V) are treated as logic 1.

The range of voltages between the thresholds, that is +0.8 V and +2.2 V for TTL devices, is undefined in its effects and inputs in this range can produce unpredictable output states. Digital circuits are designed to provide outputs that guarantee to be in the correct part of the input range, so a TTL device output is specified as having a maximum low-level output voltage of 0.5 V and a minimum high-level output voltage of 2.7 V. The difference between the input threshold and the output level is called the noise margin; this is how much noise can be added to a signal without changing its logic level. The low-level noise margin is usually smaller than the high-level noise margin owing to the design of the logic ICs. The low-level noise margin for TTL devices is about 0.3 V (0.8–0.5 V), so it is important to make sure that noise on the signal lines is much less than 0.3 V.

The MOS logic devices, which can operate from a wider range of supply voltages, typically 2.0–5.5 V do not have fixed threshold voltages; rather, the thresholds are set as fractions of the supply voltage, the low-level threshold is typically between 20 and 30% of the supply voltage, and the high level is 70% of the supply voltage (Figure 13.6). The 74HCT family of devices is designed to have fixed TTL-compatible input levels to make interfacing between TTL and CMOS devices easier.

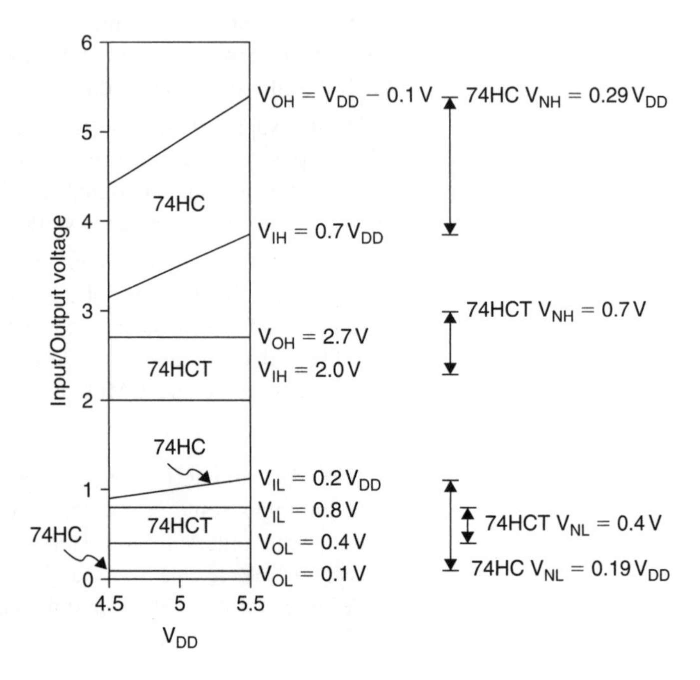


**Figure 13.6** Input switching thresholds and noise margin for 74HC and 74HCT logic gates

Note that because the CMOS logic gates have output voltages that are nearer the supply rail values than those of the TTL devices; that is, maximum low-output voltage is about 0.1 V and minimum high-output voltage is 0.1 V less than the supply voltage, the noise margin when driving a 74HCT device from a 74HC(T) device is better than that for TTL devices.

When using several logic devices together on a circuit board it is important to provide a supply decoupling capacitor for every IC, because every time a gate switches state a pulse of current is drawn from the supply, which may be many times larger than the current taken when no switching is occurring. If a local capacitor is not available to provide the current then the local supply voltage may drop, owing to track resistance, for a short time. In the worst case this may be enough to trigger falsely the input of other gates on the chip or in nearby chips. Logic circuit designs should always ensure good low resistance power and ground connections, and plenty of supply decoupling.

## Practical 13.2

The circuit in Figure 13.7 allows investigation of the threshold levels of a gate of the 74HC00 IC. Set the potentiometer to give the minimum voltage at the gate input, then with the voltmeter connected to the output of the NAND gate as shown, slowly increase the input voltage until the output changes state, and measure the input voltage. Increase the input voltage to maximum and then reduce the input voltage slowly until the output changes; again, measure the voltage at which this occurs.

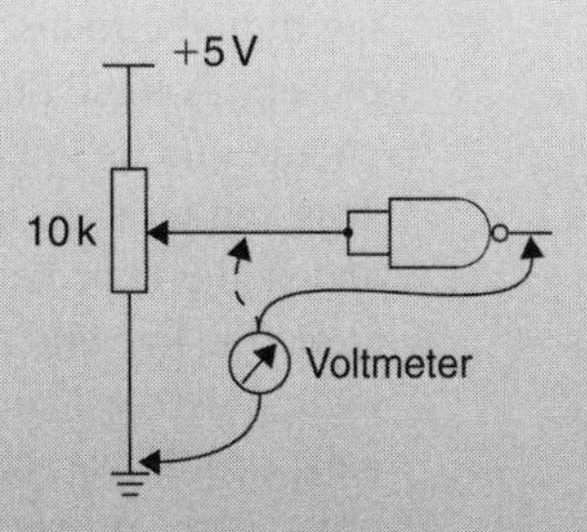


**Figure 13.7** Circuit for investigating logic thresholds

Repeat the exercise with devices from other logic families if they are available, for example 74LS00, 4011 and 74HCT00. Remember to check the pin connections; the 4011 device has different pin connections.

## Power supplies and instruments

All digital logic circuits require a regulated power supply. Some logic families, notably TTL, require a supply voltage that is regulated to +5 V because damage will be caused to the ICs if the voltage exceeds 5.5 V. By contrast, 4000 CMOS circuits can work at higher voltages and are less dependent on one particular voltage level. For all types, however, the supply must be well regulated to prevent any changes in the supply voltage when gates switch on or off. Many circuits incorporate both over-voltage and reverse voltage protection, so that power is shut off in the event of regulation failing, or a battery being connected the wrong way round.

The instruments that are used for measurement and testing in digital circuits include the familiar types that are used also for work on analogue circuits. Measurement of steady voltage levels, such as power supply voltage, can be carried out using multimeters, either analogue (pointer) or digital (number display) types.

Logic circuit action can be carried out using d.c. meters if the logic inputs can be obtained from switches rather than from fast-acting circuits. This makes fault-finding much simpler, because you can observe the inputs to each gate and check that the outputs are as expected. This, however, is very seldom possible with real-life digital circuits.

The oscilloscope is also a useful tool in digital servicing because you can check a clock pulse, and monitor rise and fall times of signals. Note, however, that modern computer servicing needs oscilloscopes of a very high specification because clock frequencies are very high, typically 200 MHz or more. Where more than two logic signals need to be viewed at the same time a logic analyser can be used. Logic analysers are similar to an oscilloscope in that they display logic level vertically versus time on the horizontal axis. Logic analysers are commonly available with 8, 16 and 32 bit wide inputs, because they have byte or wider inputs they can be triggered by matching binary patterns on the inputs to a preselected value.

The most common types of instrument that are used specifically for working digital circuits (and not for analogue circuits) are logic probes and logic pulsers. A logic probe, as the name suggests, is held on to one line of a logic circuit, and its indicator will show if the line voltage is high, low or pulsing. This can be used to check whether the expected signal is present on a line, and if the expected signal is not found you can then check for continuity along the line and for correct operation of the device that controls the line. A logic pulser will inject a signal on to a line, irrespective of any existing signals, so that you can check the effect of 1, 0 or pulsing inputs using a logic probe at the output.

## Multiple-choice revision questions

13.1 What is the binary for the decimal number 37?
(a) 100101
(b) 110100
(c) 001111
(d) 010001.

13.2 How many bits are required to represent the decimal number 7649?
(a) 11
(b) 12
(c) 13
(d) 14.

13.3 What is the decimal number that is equal to 1110 1010 binary?
(a) 1310
(b) 142
(c) 176
(d) 234.

13.4 What is the logic function whose output is 0 only when all its inputs are 1s?
(a) NAND
(b) NOR
(c) XOR
(d) OR.

13.5 What are the logic input thresholds of standard TTL operating from a 5 V supply?
(a) 0.1 and 4.9 V
(b) 0.4 and 3.5 V
(c) 0 and 5 V
(d) 0.8 and 2.2 V.

13.6 What type of transistors are used to make 74HC logic?
(a) Schottky
(b) NMOS
(c) NMOS and PMOS
(d) BJT.

# 14 Digital oscillators, timers and dividers

The most common type of digital oscillator circuit is the **astable multivibrator**, which can be made using transistors of any type, or with integrated circuit (IC) logic gates or even relays. Multivibrator circuits are classed as **astable**, **monostable** or **bistable** according to what they do. If the circuit continuously switches between states without an external signal it has no stable state, so it is an astable; if it changes state when triggered and then switches back after a delay it has one stable state that it always returns to, so it is a monostable; while if it changes state only when triggered it is stable in either state and so is a bistable. The basis of these circuits is usually two units that are cross-coupled, with the output of one fed to the input of the other. We often drop the word 'multivibrator' and refer only to astable, monostable and bistable circuits.

Figure 14.1(a) shows an astable constructed from bipolar junction transistors (BJTs). The circuit is clearly seen to be two RC coupled inverting amplifier stages, with the output of the second stage fed back to the input of the first. Figure 14.1(b) shows the more usual way of drawing the astable with cross-coupling. The frequency of oscillation depends on the values of the time constants, formed here by the capacitors and resistors marked C1, C2, R1 and R3. The output of this circuit is approximately a square wave, and by using more advanced circuitry, the shape of the wave can be almost a perfect square shape, with very low rise and fall times.

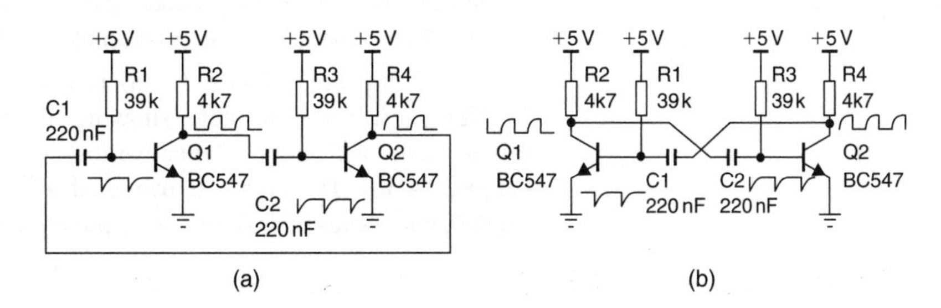


**Figure 14.1** A simple astable circuit using BJTs: (a) drawn as two stages, and (b) shown cross-coupled

A circuit like this can be used to provide timing signals for other digital functions, and it can be crystal controlled so that the timing is very precise. Such a circuit is called a **clock**, since it sets the timing for the other circuits attached to it. Clocked circuits are particularly important for computing.

When a cross-coupled circuit contains one direct coupling and one CR coupling, it is a monostable (Figure 14.2a). The monostable has one stable state, in this example with the second transistor Q2 conducting heavily because its base is connected to the positive supply through resistor R1. When a brief positive pulse turns on transistor Q1, this will also turn off Q2, allowing the collector voltage to rise, and this state will persist until the capacitor C2 charges and allows Q2 to conduct again. The circuit will then rapidly return to its original stable condition. This circuit is used to produce a pulse of a set width (set by the values of C2 and R3) from any brief positive input pulse.

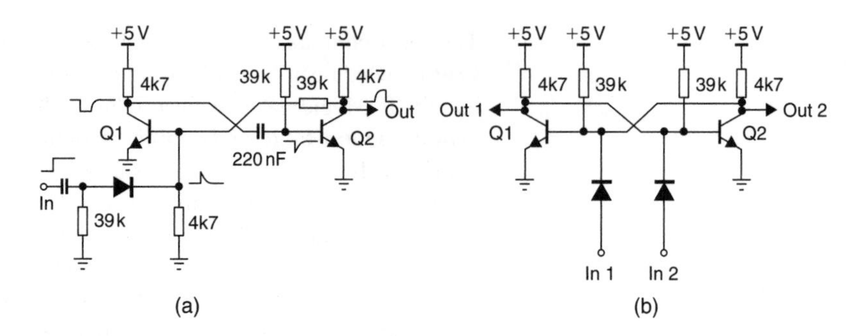


**Figure 14.2** Multivibrators: (a) monostable, and (b) set–reset bistable

The bistable is illustrated, again in transistor form, in Figure 14.2(b). The circuit can be stable with either Q1 or Q2 conducting, but is totally unstable if both are conducting. In this circuit, the transistor that was last triggered, by its input being pulled up via the diode, is the one that is conducting. There is no provision for using a single pulse input, although this can be achieved (not illustrated here) with diodes to steer the pulse to the transistor that is off at the time (**steering diodes**), switching it on and so switching off the other transistor. The next pulse will reverse the process, because of the altered bias on the steering diodes, returning the bistable to its original state. Two pulses at the input will therefore produce one pulse out. Bistables are seldom constructed from separate transistors because the IC form is more convenient.

The RS flip-flop (Figure 14.3a) is a bistable circuit constructed from NAND gates. The state table, showing the output generated for each permitted value of $\overline{R}$ and $\overline{S}$, the active low inputs for this circuit, is shown in (Figure 14.3b). The circuit is triggered when one of its inputs is taken to logic 0, but because the $Q$ and $\overline{Q}$ output are not allowed to be the same state

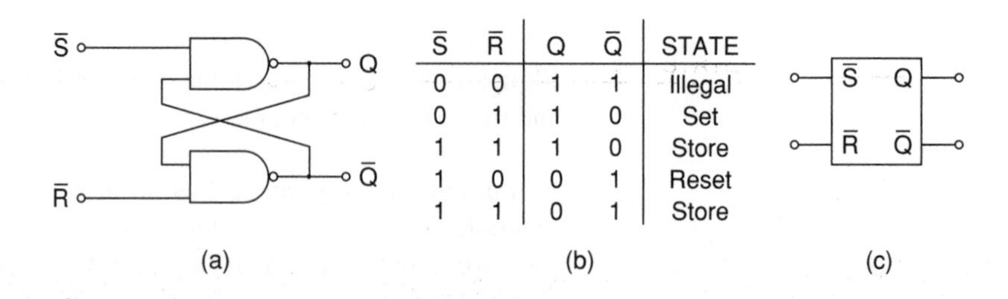


| $\overline{S}$ | $\overline{R}$ | Q | $\overline{Q}$ | STATE |
|---|---|---|---|---|
| 0 | 0 | 1 | 1 | Illegal |
| 0 | 1 | 1 | 0 | Set |
| 1 | 1 | 1 | 0 | Store |
| 1 | 0 | 0 | 1 | Reset |
| 1 | 1 | 0 | 1 | Store |

(b)

**Figure 14.3** RS flip-flop: (a) made from NAND gates, (b) state table, and (c) symbol

the input conditions $\overline{S} = 0, \overline{R} = 0$ must never be used. Note that in this circuit, when the inputs $\overline{R} = 1, \overline{S} = 1$ the circuit is stable in the last state that was set; therefore, it stores an output at the value set. This storing action is also called **latching**. Sets of four, eight or 16 latches can be obtained in IC form and are widely using in microprocessor interface circuits, for example.

Logic gates can be used to construct oscillator circuits, with the advantages that they can provide signals that conform to the standard complementary metal-oxide semiconductor (CMOS) or transistor–transistor logic (TTL) levels. The circuits of Figure 14.4(a) show an oscillator using RC timing, and Figure 14.4(b) one that uses crystal control. The principle is exactly the same as that of analogue oscillator circuits, using a timing circuit as part of the positive feedback loop that causes the oscillation.

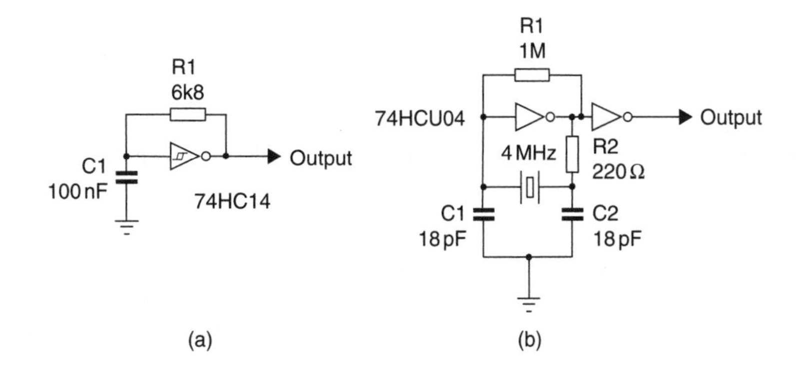


**Figure 14.4** Logic gate oscillator circuits: (a) relaxation oscillator, and (b) crystal oscillator

## The 555 timer

Although it is not strictly a digital circuit, the type 555 IC timer is used extensively as an astable clock generator where precise frequency is not important enough to justify a crystal, but is required to be better than can be achieved with simple discrete or logic gate RC oscillators. The 555 is also useful as a voltage-controlled oscillator, a ramp generator and a monostable timer circuit.

An external CR series circuit is used to set the time constant, and an input to the timer will generate a change of voltage at the output. This voltage will change back to its original values after a time determined by the time constant. Figure 14.5(a) shows a block diagram of the 555 timer circuit, which shows that an RS flip-flop is the core of the circuit. There is also a CMOS version of the circuit, the 7555, which will operate at 2 V and only draws 60 μA from the supply. Recent versions from Micrel and Zetex run from power supplies as low as 1.2 V.

## Binary counter

Frequency division is possible using the type of bistable circuit whose basic action is illustrated in Figure 14.6. As the diagram shows, the output moves from 0 to 1 at only each *second* 1-to-0 change of the voltage at the input. If the input to this circuit is a square wave, the output will be a square wave

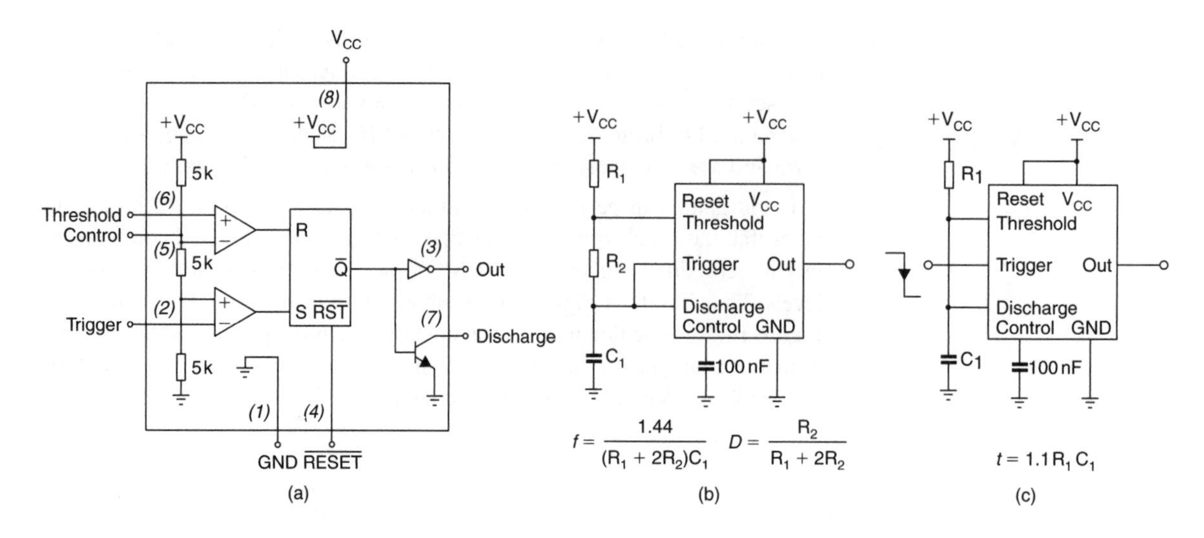


**Figure 14.5** NE555 timer IC: (a) block diagram, (b) circuit for astable oscillator, with frequency and duty cycle formulae, and (c) monostable timer circuit with delay formula

at half the frequency of the input square wave. This is the basis of a **binary counter**, a very important type of circuit. The output of such a divider stage can be taken to the input of another similar stage so as to achieve another halving of frequency. In this way any given frequency can be divided down by any power of 2. The type of bistable that performs this dividing action is a clocked bistable, of which several varieties exist. Each clock pulse into the bistable will cause a change of state (0 to 1 or 1 to 0) at the output.

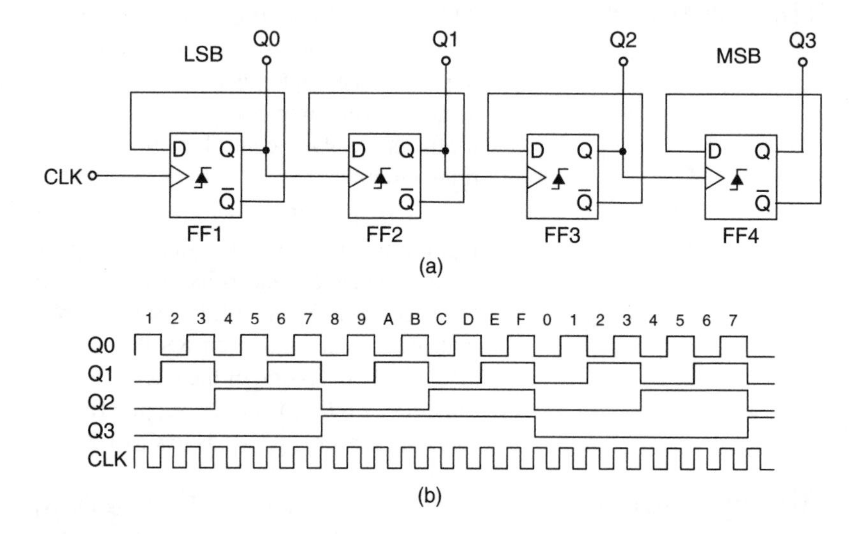


**Figure 14.6** (a) Binary ripple counter, and (b) the clock and output waveforms

If the input pulses to a binary counter, formed from a chain of divider circuits, are controlled by a gate, the outputs of the bistables in the chain will show a count of the number of input pulses passed by the gate. If the gate is enabled for a fixed time, for example 1 s, then the number of pulses counted is by definition the frequency. After a number of pulses the first bistable will have changed state once for every pulse, the next bistable once for every two pulses and so on down the chain. Because each divider divides by 2, each output represents a power of 2, so that taken together the outputs form a binary number equal to the number of input pulses received.

This type of counter is called a **ripple counter**, because the bistables down the chain receive their clock pulse slightly later then the first bistable. Since the outputs of a ripple counter do not all change at the same time, but one after the other, there are some purposes for which a ripple counter cannot be used; for example, if it is necessary to decode the output of the counter and compare it with a preset value, false matches can arise as the count ripples through. The answer to this is another type of counter, the synchronous counter, which uses bistables in a circuit that connects the same clock pulse to all the bistables. The counting is then controlled by other inputs to the bistables.

One of the most useful IC counter circuits is the binary-coded decimal (BCD) counter, illustrated in Figure 14.7 with its outputs driving a seven-segment LED display via a BCD to a seven-segment display driver IC. Input pulses are counted, and the output resets to zero after the ninth count. Therefore, the binary output never exceeds 1001, hence the name binary-coded decimal. On the 10th pulse, when all the outputs reset to zero, the terminal count pin (TC) goes high, which allows stages to be chained together to count numbers higher than 9. If another stage is needed, the TC pin is reset to zero on the first count.

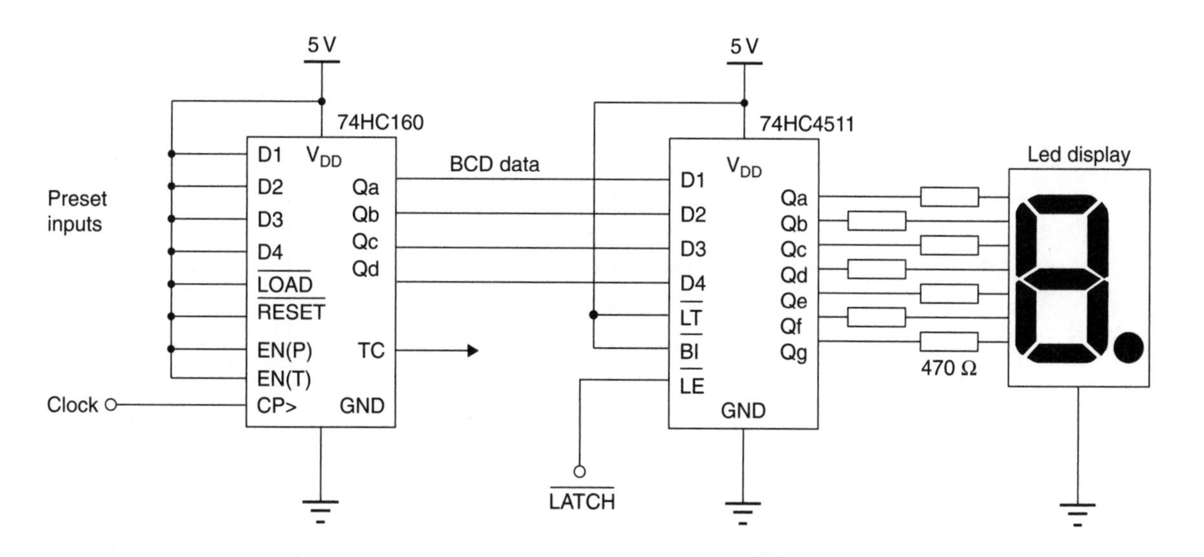


**Figure 14.7** BCD counter with display decoder

The outputs from counters such as the BCD counter can be displayed as decimal figures on seven-segment readouts. A decoder–driver circuit is needed (consisting of logic gates) to ensure that a given binary set of outputs produces the correct decimal number on the readout. The decoder is usually an integrated circuit; and IC chips forming combined counter–decoders, and even counter–decoder–displays, are now available.

## Registers

A register is formed from a set of bistables connected together, and such registers are available in IC form. Registers are of four types, because the inputs can be serial or parallel, and the outputs can also be serial or parallel. Serial means working with one terminal, using pulses in sequence. Parallel means that a set of inputs or outputs will be used, with signals on all of them. Registers are classed as PIPO, SIPO, PISO and SISO, with I meaning input, O meaning output, S meaning serial and P meaning parallel. Registers are manufactured as single ICs and do not have to be made by connecting bistables.

The parallel in, parallel out (PIPO) register can be used as a store for binary digits (bits), typically with the bits at the inputs read into the register on one clock pulse, and fed out from the outputs on the next clock pulse. Figure 14.8(a) shows a PIPO register, the 74HC574 IC, which is 8 bits wide and has an output enable pin as well as a clock input. The 74HC574 has Tri-state outputs; when the output enable pin is set low they act as normal logic outputs, but when it is set high they are set to a high impedance state, and this third state effectively disconnects them from the rest of the circuit. Tri-state outputs allow multiple chips to drive the same line at different times, which is essential in computer buses, etc. The PISO register will accept a set of inputs

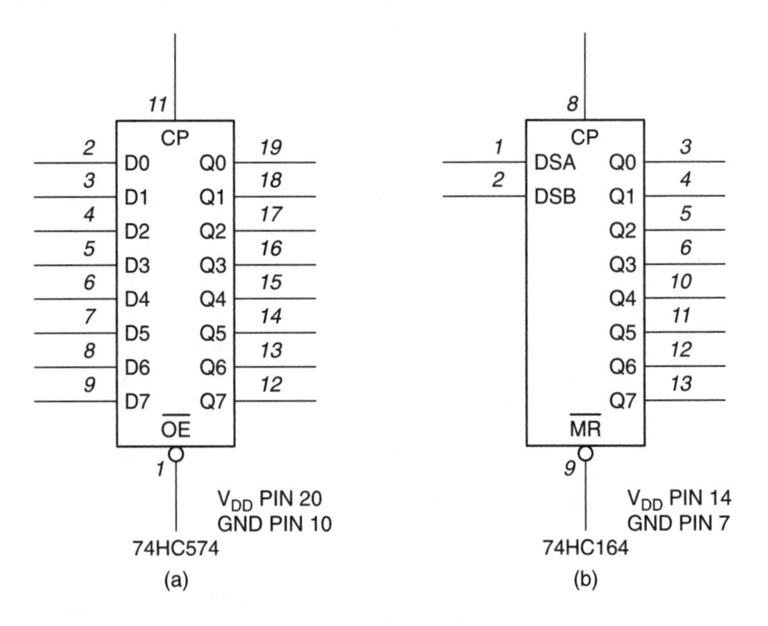


**Figure 14.8** Registers: (a) parallel in, parallel out, and (b) serial in, parallel out

and feed these bits out from the serial output, one at a time, on each clock pulse. The SIPO register will accept an input serial bit at each clock pulse until the full set of bits is available at the outputs. Figure 14.8(b) shows the 74HC164, which is an 8-bit SIPO register with a master reset pin that can be used to clear all the outputs simultaneously. The PISO and SIPO registers are used extensively in serial communications systems. The SISO register will store an input bit and release a stored bit at each clock pulse, and its main application is as a bit delay.

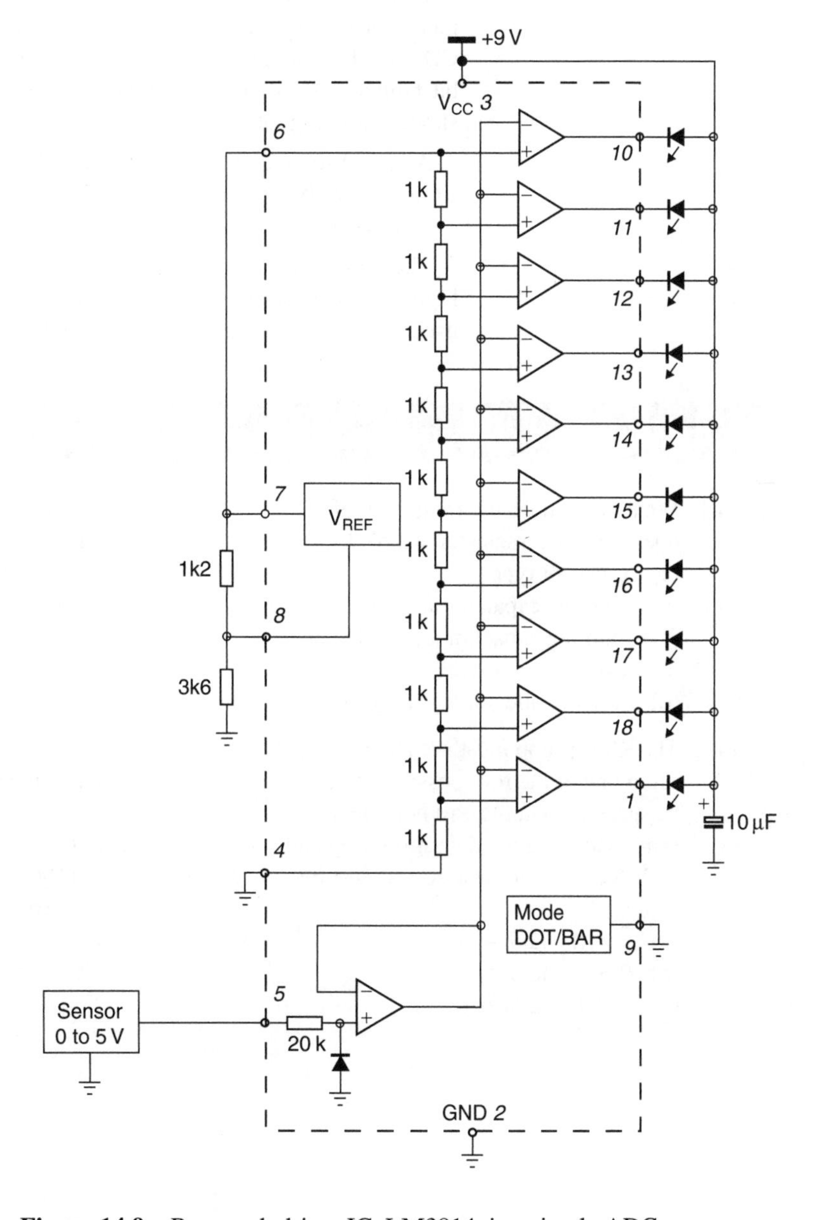


**Figure 14.9** Bar graph driver IC, LM3914, is a simple ADC

## Analogue-to-digital converters

An analogue-to-digital converter (ADC) is a circuit that will convert a voltage level into a binary number. A varying voltage, such as an audio signal, or a slowly changing voltage, such as a temperature measurement voltage, will often need to be converted into a set of digital signals, so that microprocessor-based equipment can use the data. This is done by sampling the voltage of the input signal at intervals, and converting each value of voltage level into a binary number. This action is an essential part of a digital voltmeter and is also used in recording sound on a compact disc (CD).

The clock rate that is needed is set by the number of samples that need to be taken each second, and also by the number of bits that are needed to represent the size. The most difficult ADC task is in converting sound for CD use. The sampling rate is around 44 000 samples per second, and 16-bit numbers are used. Instruments like the digital voltmeter can use much slower sampling rates and a smaller number of bits.

A digital-to-analogue converter (DAC) performs the opposite conversion, from a set of binary numbers into a voltage level. A fast DAC is an essential part of a CD player.

The LM3914 bar graph driver IC is a simple ADC and display driver. Figure 14.9 shows the LM3914 block diagram. Similar ICs are frequently used in audio level indicators and signal strength indicators.

## Multiple-choice revision questions

14.1 What determines the frequency of operation of a transistor astable?
(a) transistor type
(b) coupling capacitor and base resistor
(c) coupling capacitor and collector resistor
(d) ratio of collector to base resistor.

14.2 The RS flip-flop must not have its inputs set to one of the four possible states because the outputs will both be the same, violating the definition of $Q$ and $\bar{Q}$. Which one of these states must be avoided?
(a) $\bar{R} = 0$ and $\bar{S} = 0$
(b) $\bar{R} = 1$ and $\bar{S} = 0$
(c) $\bar{R} = 0$ and $\bar{S} = 1$
(d) $\bar{R} = 1$ and $\bar{S} = 1$.

14.3 A six-stage ripple counter starts in the reset condition. How many clock pulses are required for the last stage output to go high?
(a) 1
(b) 16
(c) 32
(d) 64.

14.4 What is the highest count of a BCD counter?
(a) 1111
(b) 1100
(c) 1010
(d) 1001.

# 15 Digital inputs and outputs

In order for digital systems such as microcontrollers to interact with users and their environment they must be able to receive inputs, for instance, when a switch is closed, and drive output devices such as displays, sounders and motors. How the inputs and outputs are arranged determines the way we see their function, but to the system a seven-segment display is merely a set of arbitrarily ordered lights and the keyboard of a computer is an array of push-buttons to make momentary action switches. When system inputs need to be measurements of physical parameters, analogue-to-digital converters are used to provide the input information to the system.

The problem of obtaining digital signals from switches is that a switch is a mechanical component. When a switch is closed, the contacts should ideally come together and stay together, but in fact they usually bounce. Each bounce will generate another pulse, so that the output can consist of a series of pulses each time the switch is closed (Figure 15.1).

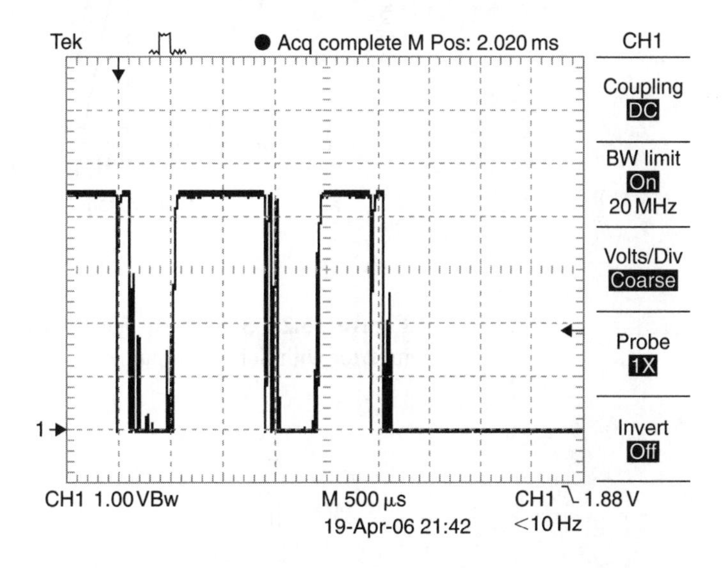


**Figure 15.1** Switch contact bounce (measured with a single-pole lever switch

Switch bounce is caused by mechanical vibration of the contacts on closure, leading to multiple pulses being generated. Some types of switches are more susceptible to this than others, spring lever switches being among the worst and slide switches being better. There is no problem with bounce occurring when a switch is opened, but for a change-over switch, bounce can occur on either direction of switching.

Switch debouncing circuits are used to ensure that only one edge or pulse is generated by a switch being closed. One simple method is to connect each switch input via a low-pass RC input circuit of a Schmitt trigger, so that the input voltage of the gate cannot change quickly. The Schmitt trigger uses positive feedback to provide input hysteresis, meaning that the input thresholds move immediately after switching has occurred, making oscillation less likely (Figure 15.2a). In effect, the threshold for switching moves away from the current input voltage as soon as switching occurs. The use of an RC circuit that does not allow the Schmitt input to change quickly when the switch is opened or closed is illustrated in Figure 15.3(b). Switch debouncing can be carried out very effectively. When the switch closes, the input to the Schmitt trigger follows the discharging capacitor down to zero volts. As soon as the input triggers, the threshold is moved to a higher voltage, so that even if the switch opened again immediately there would be a delay until the

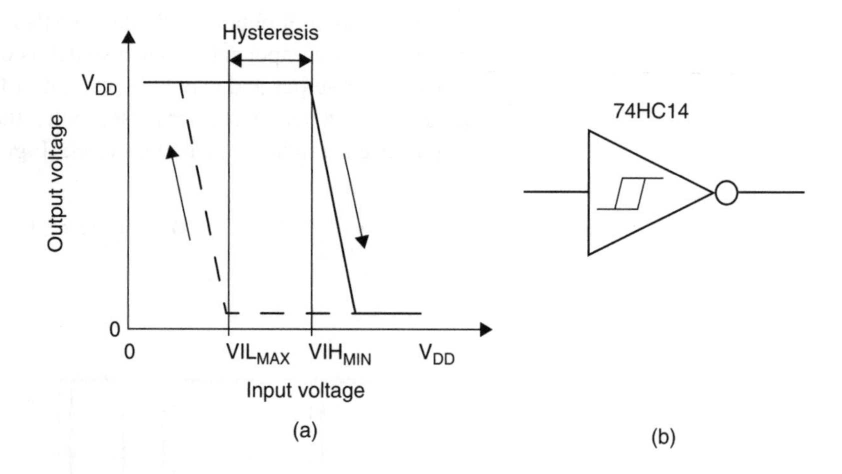


**Figure 15.2** Schmitt trigger: (a) transfer function, and (b) symbol for inverter with Schmitt input

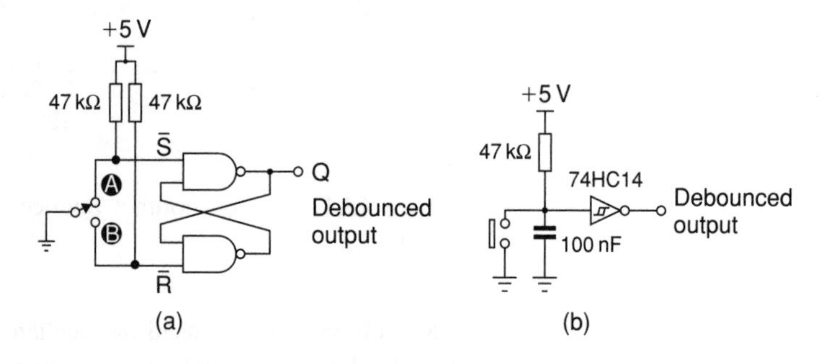


**Figure 15.3** Switch debouncing circuit based on (a) RS flip-flop, and (b) RC network with Schmitt trigger

capacitor had charged via the resistor up to the threshold before the trigger could change state. If the switch contacts bounce, the change in voltage at the input is not rapid, because of the time constant, so the output voltage is unaffected. Buffers and logic gates are obtainable with Schmitt switching characteristics, and microcontrollers often have Schmitt inputs on some pins. Although this can be effective, it is not perfect, and more elaborate methods are often used.

Another method of debouncing switch contacts is to connect the switch to an RS flip-flop. Figure 15.3(a) shows a typical switch debounce circuit. To understand what happens, we need to recall how the RS flip-flop works: the output changes only when one input is at logic 0 and the other is at logic 1. When the inputs are both at logic 1, the output remains unchanged in whatever state it was last set.

When the switch is at position A, the input to the flip-flop is $\overline{R} = 1$, $\overline{S} = 0$, and this gives $Q = 1$ as the output. Now while the switch is being changed over, both inputs will momentarily be at logic 1, but this does not change the output. When the switch contacts are in position B, the inputs to the flip-flop are $\overline{R} = 0$, $\overline{S} = 1$, giving $Q = 0$. If the switch contacts bounce, making $\overline{R} = 1$, $\overline{S} = 1$, the output remains at 0, so that the switch bounce has no effect. This method of debouncing is usually preferred, and is completely effective when properly implemented.

Debouncing for key pads like those used in telephones, calculators and computer keyboards is usually carried out by application-specific integrated circuit (ASIC) keyboard encoders or software run on a microcontroller. The switches will typically be arranged as a matrix, with the controller driving the rows and reading the columns. When a switch is closed, the controller detects that it has been closed and a time delay starts. Changes in the switch setting have no effect until the end of the time interval. This interval can be several tens of milliseconds, since you cannot type quickly enough to operate a key more than once in that time.

### Practical 15.1

Connect a switch and resistor so that when the switch is closed the resistor is connected between a 5 V supply and ground. Use a counter or oscilloscope to measure the pulses caused by bouncing. Assemble a debouncing circuit (either Schmitt or RS flip-flop) and verify that the output of the debouncing circuit changes only once for each switching at its input.

## Digital outputs

The outputs from digital circuits may be used to control devices such as motors, to switch on loads via a relay or to drive display devices (see Figure 15.4). Displays are the most common output devices that allow us to interact with electronic systems.

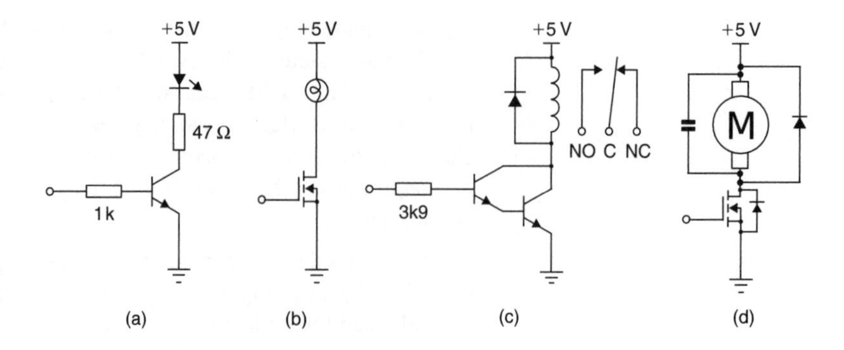


**Figure 15.4** Driving loads from digital outputs: (a) transistor driver for high-current LED, (b) FET drive for an indicator bulb, (c) driving a relay with a Darlington pair (note protection diode), and (d) switching a small motor with a MOSFET (note protection diode and electromagnetic compatability (EMC) suppression capacitor)

**Light-emitting diodes (LEDs)** indicators are diodes that emit light when current passes in the forward direction. Using different semiconductor materials, LEDs can be made to emit light of different colours. Single LEDs can be used to indicate binary numbers by using one LED for each power of two, on for one and off for zero.

Number displays based on segmented displays made with LEDs are commonly used for indicating denary numbers, using integrated circuits (ICs) that convert binary-coded decimal (BCD) to segment drive for a seven-segment type of display (Figure 15.5). As for all LEDs, the amount of current being passed needs to be limited, and this is usually done with a resistor in series with the LED. The brightness of the output is roughly proportional to the current.

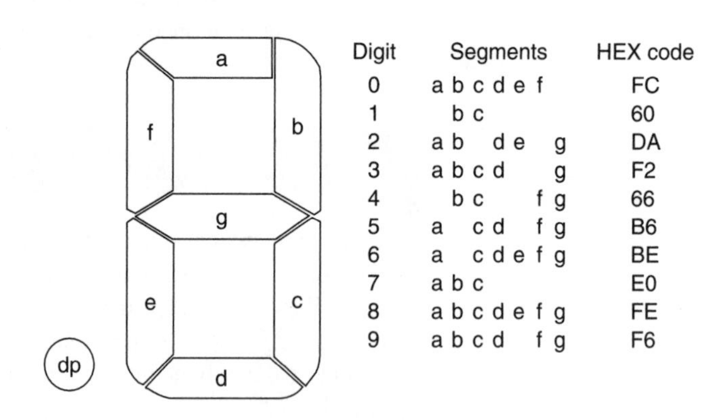


| Digit | Segments | HEX code |
|---|---|---|
| 0 | a b c d e f | FC |
| 1 | b c | 60 |
| 2 | a b d e g | DA |
| 3 | a b c d g | F2 |
| 4 | b c f g | 66 |
| 5 | a c d f g | B6 |
| 6 | a c d e f g | BE |
| 7 | a b c | E0 |
| 8 | a b c d e f g | FE |
| 9 | a b c d f g | F6 |

**Figure 15.5** Seven-segment display

The format of LED indicators can be **common anode** or **common cathode**. For a seven-segment display, for example, each bar can be the cathode of a diode, all of whose anodes are connected, or each bar can be an anode of a diode, with all the cathodes connected. On a common-cathode display, the cathode will be earthed and each anode connection driven to a positive voltage so as to illuminate a bar. On a common-anode display, the anode connection will be made to a positive voltage, and the cathode connections earthed by the driver circuits as required.

**Liquid crystal displays (LCDs)** make use of a quite different principle. They depend on a liquid crystal material that polarizes light when an electrical potential is applied. The display is made by sandwiching this liquid crystal material between transparent metal electrodes deposited on glass substrates (Figure 15.6a), one or both of which are transparent and polarizing. Light passing through it will be polarized by the glass polarizing filter and so the light can be switched on or off by altering the voltage across the liquid crystal, which changes its polarization. The amount of power that is needed is very small, and the brightness of the display is just the brightness of the light that is used. The contrast of the display depends on how well the liquid crystal polarizes; at high and low temperature extremes this may not be as good as at room temperature. This is the principle now widely used for all types of displays, including computer monitors and television displays.

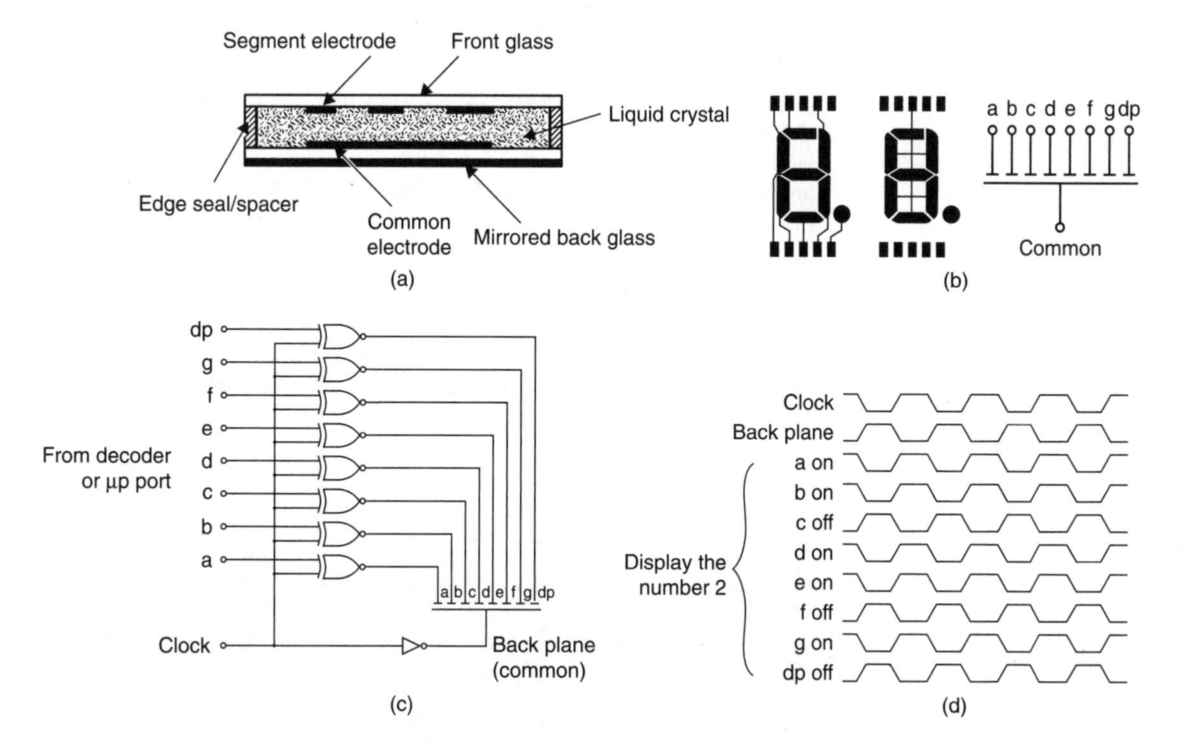


**Figure 15.6** Liquid crystal displays: (a) LCD construction, (b) seven-segment display pattern and connections, (c) drive circuit, and (d) drive waveforms

The drive voltage must be a.c. at a high audio frequency, with no trace of d.c., otherwise the liquid crystal may become permanently polarized and the display will lose contrast. LCD displays like those in watches and calculators are reflective, so that they are most visible in bright light, but backlighting can be used if the displays have to be read in the dark. One noticeable disadvantage is that the display is clearly visible only when you look straight at it. Visibility is much reduced when viewing at an angle.

Transmissive displays (TMCs) are used with a white backlight, and the liquid crystal dots will transmit whatever colour is selected from the whole spectrum of white light. These displays are ideal for use in low-lighting conditions.

Reflective displays (RFCs) are used illuminated by white light, and will reflect whatever colour is selected from each liquid crystal dot. No backlight is needed, so that power requirements are very low, but they can be used only in bright lighting conditions, and the perceived colours will alter if the illumination is not white light.

Transflective displays (TFCs) combine both TMC and RFC principles, and use both a backlight and the ambient light to provide a display that can be read equally easily in the dark or in sunlight. The user can save power by having the backlight off in high ambient light.

Both LED and LCD displays can be made in a variety of patterns, of which the seven-segment display is the most common. Another useful type of display is the dot-matrix, using a set of dots for each character. This is used with LCD and with LED displays. Figure 15.7 shows a typical 5 × 7 matrix display and the schematic arrangement of LEDs. Any character of the alphabet or number digit (alpha-numeric character) can be displayed, and using larger matrices such as 8 × 15 allows for international characters to be displayed. Other formats for displays are starburst patterns, a variant of segmented display pattern that allows for alpha-numeric characters, and bar graphs. Bar-graph displays are very commonly used as audio level meters in hi-fi equipment.

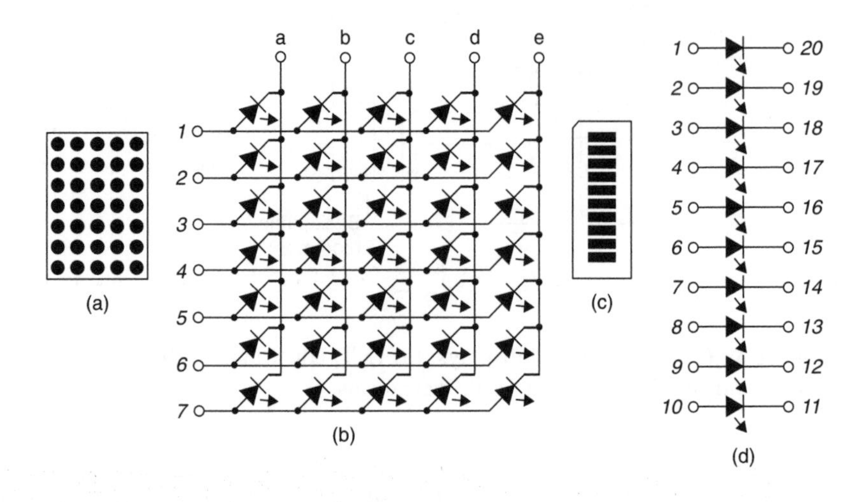


**Figure 15.7** Typical displays: (a) matrix display, (b) typical LED connections, (c) bar graph, and (d) LED connections

All forms of display require driver circuits that will accept normal digital logic level inputs and provide outputs that are suitable for operating the display segments. For number or alphabetical displays, combined decoder–drivers are used to implement the decoding logic required; illustrated in (Figure 15.5). The decoder–driver and display is often combined so that no wiring is needed other than power supplies and inputs.

## Multiple-choice revision questions

15.1 Which switch debouncing method is most appropriate for the type of switch called SPCO (single-pole change-over)?
(a) RC input Schmitt
(b) microprocessor software delay
(c) RS flip-flop
(d) NAND gate.

15.2 When a standard seven-segment LED display is driven, each segment draws about 5 mA from the supply. What is the difference between the maximum and minimum current drawn for the numbers from 0 to 9?
(a) 5 mA
(b) 25 mA
(c) 20 mA
(d) 50 mA.

15.3 What is the effect of a low-value d.c. drive on an LCD?
(a) will damage the display
(b) all the segments are displayed simultaneously
(c) light is emitted
(d) will cause the display to explode.

15.4 What is the main advantage of a dot-matrix display?
(a) low current requirement
(b) can display any pattern
(c) requires fewer resistors
(d) costs less than a seven-segment type.

# Unit 6
## Radio and television systems technology

### Outcomes

1. Demonstrate an understanding of home entertainment systems and apply this knowledge safely in a practical situation
2. Demonstrate an understanding of TV receivers and apply this knowledge safely in a practical situation.

# 16 Home entertainment systems

Home entertainment systems range from relatively small integrated hi-fi units with radio tuner, compact disc (CD) player and tape recorder to systems composed of many separate devices, linked externally by analogue and digital or even fibreoptic interconnects, as well as home cinema systems with separate surround sound processor and amplifier units along with digital versatile disc (DVD) and CD players, digital tuner, tape, CD recorder and possibly a personal computer (PC) or hard disc recorder.

## Audio amplifier system

Figure 16.1 shows the block diagram for a typical audio amplifier system designed for high-quality performance. It is capable of handling a wide range of different input devices ranging from a high-grade CD player to the audio input from a television receiver to form a complete home cinema installation. Because the many different input transducers have different amplitude and frequency characteristics, the initial input stage will have individual sockets with frequency compensation circuits (equalizer stages) to match the parameters to those of the amplifier input preamplifier.

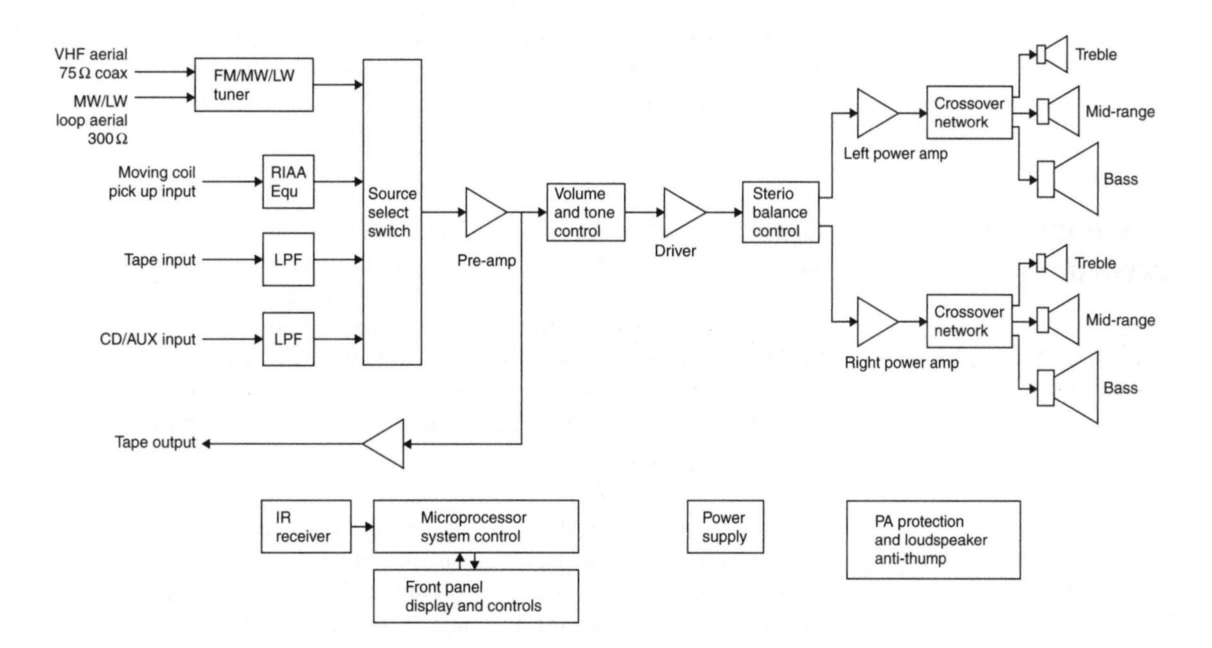


**Figure 16.1** Block diagram for a home audio system

Input signal frequencies will cover between 20 Hz and 20 kHz at signal levels ranging from about 5 mV to perhaps 100 mV. The power output levels will range from about 100 mW for the drive to a pair of headphones for personal listening up to perhaps 100 W per channel for family entertainment. The demands on the power supply section will therefore be quite complex. Because of this, the output load current and voltage will be continually monitored by an electronic circuit to provide protection under overload conditions.

Because the amplifier can be used in a wide range of different applications, the input/output devices are usually selected via a switching stage which is often controlled via an infrared remote system. This system status is often displayed on a liquid crystal display (LCD) or light-emitting diode (LED) display, and by backlight controls.

Typical source inputs for such an audio system are likely to include a radio tuner, either frequency modulation (FM) stereo or digital audio broadcasting (DAB) fed from an outside aerial unless the signal strength locally is very high. For recorded sound, there are usually analogue inputs for devices such as a CD player, and some units provide both digital inputs and output so that a CD recorder can be used in conjunction with the other devices. A mini-disc recorder may also be part of the system and there are sometimes connections such as universal serial bus (USB) so that music can be downloaded to an MP3 player. For a user who has accumulated recordings on cassette, a tape or line input will be provided. Cassette units are now less common following the introduction of recordable CD and MP3 players.

For hi-fi systems it is important that the listening experience should be maximized. This may involve consideration of the room furnishings because hard walls can create unpleasant reflections that exaggerate the reverberation, and curtains and similar soft materials can produce unwanted damping effects. Stereo loudspeakers should be carefully positioned to ensure the best quality of signal distribution throughout the listening space, and this will usually be a compromise. The interconnecting cables should be as short as possible and routed so as not to create aerial effects that give rise to interference.

## Analogue tape recorder

Tape recording depends on permanently changing the magnetization of the magnetic tape in such a way that the changes can be detected later when the tape is played to reproduce the original signal. The tapeheads for record and play are designed to make and detect these changes in magnetism (see Chapter 10), using a material such as Permalloy with a very small gap cut into a ring (Figure 16.2). The ring has a coil wound round it, well away from the gap, and the tape is made to travel a path that feeds it past the gap.

When a signal current (from an amplifier) flows through the coil, a large alternating magnetic flux is created inside the soft magnetic material. Some of this flux emerges at the gap and the magnetic material that is used as a coating on the tape is magnetized by this flux as it passes across the tapehead gap. Because this coating is a hard magnetic material, it retains this magnetism and a permanently magnetized section of tape is produced. Because the tape is moving, each piece of tape has been magnetized by a different part of the incoming signal.

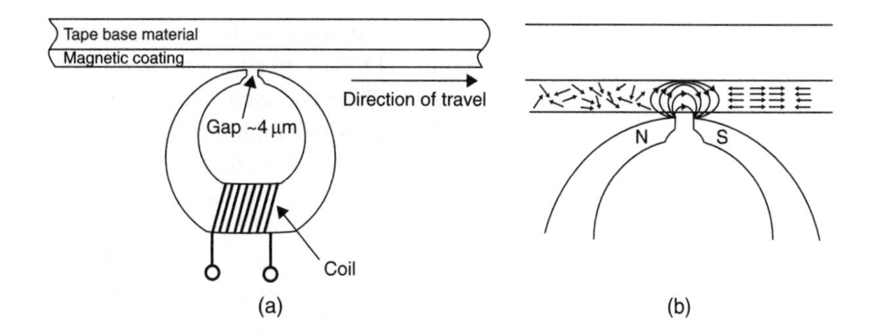


**Figure 16.2** How tape is magnetized in the pattern of a signal wave; (a) the tape is moved past the gap in the head, in close contact, and (b) the magnetic coating of the tape is magnetized by the gap flux, from the record head, which is concentrated in the magnetic coating of the tape in preference to the air

A graph of magnetism plotted against signal strength is not linear because of the hysteresis effect, so a bias signal at about 80 kHz or more must be added to the audio. The bias allows the more slowly changing audio signal to make use of a portion of the graph that is more linear (Figure 16.3). The bias signal is obtained from the erase oscillator, using an attenuator. A filter is also connected between the recording amplifier and the recording head to prevent the erase signal reaching the amplifier circuits.

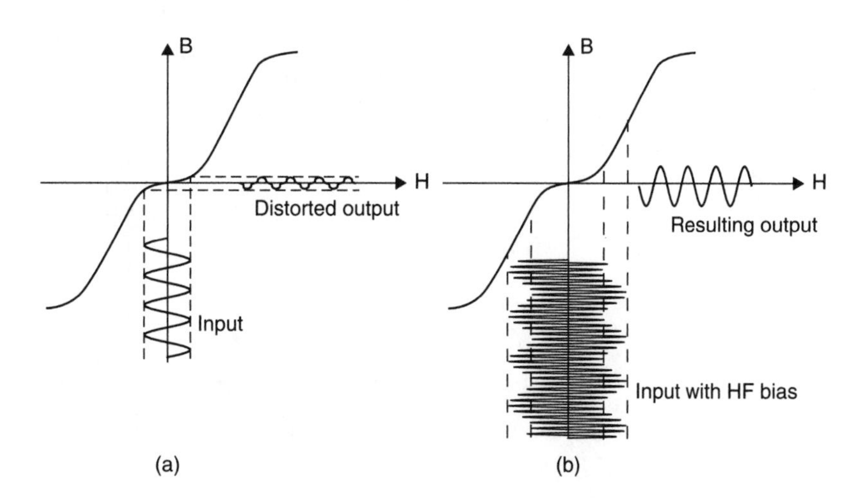


**Figure 16.3** Effect of magnetizing non-linearity on a recorded signal; (a) without high-frequency bias, and (b) with bias

To replay a recorded tape, the tape is moved, at the same speed as when it was recording, past the same head. The changes in magnetic flux from the tape as it moves past the gap induce a corresponding magnetic flux in the head, which in turn induces voltage signals in the coil wound round the head. These signals, which have an amplitude of around 1 mV, are then amplified in the usual way.

In practice, the signals are deliberately predistorted (equalized), both on record and replay, to compensate for imperfections in the tape material and the heads. Tape is erased using a separate erase tapehead, with a larger gap, which is fed with a high-frequency signal of large amplitude from the oscillator circuit that also provides the bias signal. As the tape moves past the relatively large gap on the erase head, it is magnetized in each direction alternately by succeeding cycles of the erase signal. The amplitude of this signal decreases cycle by cycle as the tape moves away from the centre of the gap, and the action eventually leaves the tape completely demagnetized.

Low-cost tape recorders (including the miniature type used for taking notes) erase by means of a flow of d.c. to the erase head, or even by the use of a permanent magnet. Although this method provides erasure, it leaves behind a rather large noise signal on the tape.

In some cheaper tape recorder designs, the record and playback amplifiers make use of the same amplifying stages and in cassette recorders the same head is usually used for record and playback, with only the frequency correction networks being interchanged by switching. Several lower priced designs also make the loudspeaker power output stage serve as the bias oscillator during recording. Figure 16.4 shows a block diagram for a typical tape-recording/replay system.

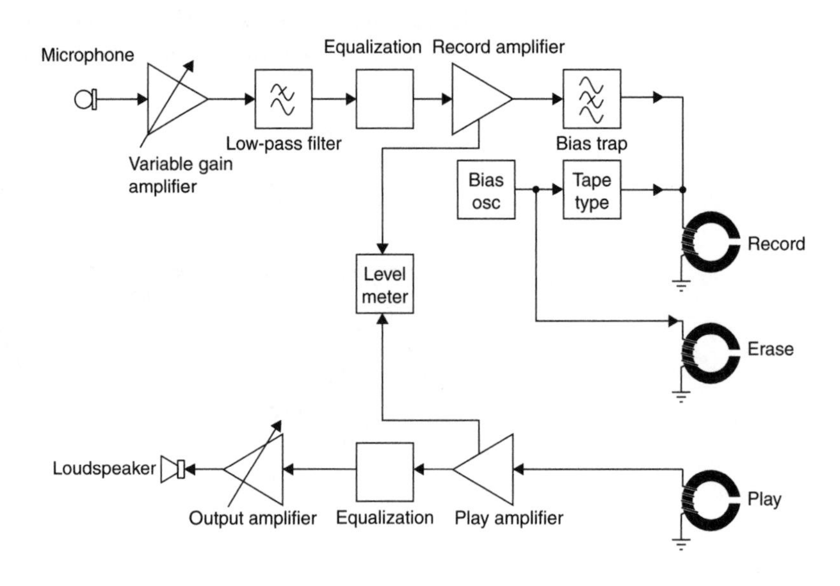


**Figure 16.4** Block diagram for a tape-recorder system. Separate record and play heads and amplifiers are shown for clarity

## Amplitude modulation radio transmission and reception

One of the earliest applications of electronics was in the transmission and reception of radio waves. Any a.c. signal passing along a wire is also radiated into the air from the wire. Waves of very low frequency, such as the 50 Hz used for power supplies, will radiate effectively only if the wire carrying them is extremely long (in the region of several million metres). However, overhead power lines can readily collect and then reradiate higher frequencies as interference. Wave energy of higher radio frequencies (from about 16 kHz upwards) can radiate more easily. This figure of frequency should not be confused with the upper audio frequencies, because the waves are not alike. Audio waves propagate by the longitudinal compression and rarefaction movement of air molecules, and it is these travelling changes in pressure that strike the eardrums. The propagation of sound thus requires air or other material to carry the wave. Radio propagation, however, does not require any material, and passes through a vacuum just as easily as through air. This is because radio signals are transverse electromagnetic waves and they propagate as changing electric and magnetic field potentials that are undetectable by human senses. This electromagnetic energy has two components (electric and magnetic), which act mutually at right angles to each other and to the direction of propagation, which is they are called transverse waves.

Early in the history of radio, it was found that the radiating wire, called an **aerial** or **antenna** (Figure 16.5), would radiate very efficiently if it were cut to a length exactly one-quarter of the wavelength of the signal. This is referred to as a **tuned aerial**. Such a quarter wavelength aerial is said to provide a good match to the impedance of free space, meaning that signals travelling to the aerial from the feeder cable are launched efficiently, with very little loss or reflection.

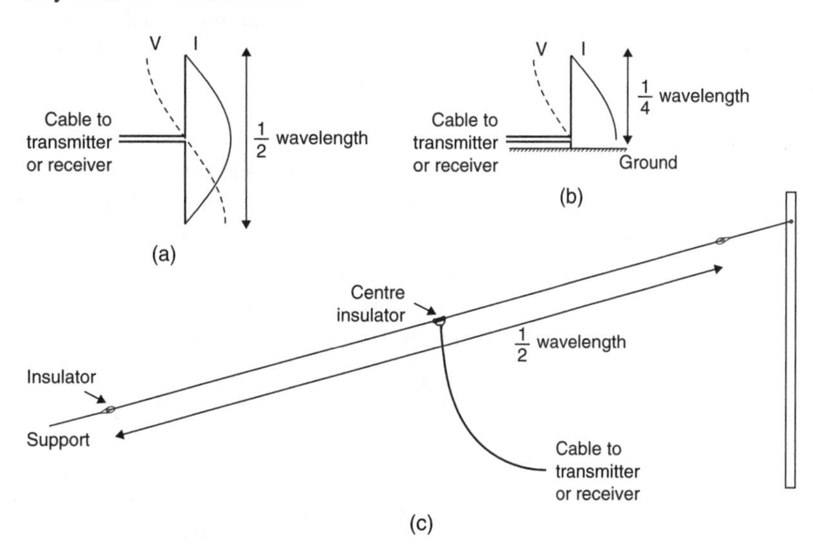


**Figure 16.5** Aerials; (a) half-wave dipole, (b) quarter-wave vertical, and (c) half-wave horizontal dipole wire

The way in which the waves are transmitted depends on how the wire is arranged. If the wire is vertical, the waves are vertically polarized, meaning that the electric part of the wave is vertical. For a horizontal wire, the waves are horizontally polarized, with the electric part of the wave horizontal. The importance of this is that a receiving aerial should be similarly polarized if it is to work at maximum efficiency.

Receiving aerials also benefit from correct tuning. Signal waves passing the aerial are then effectively collected by it, rather than being reflected off it, and the received signals produce voltages and currents in a wire leading from the aerial. The exact length of the receiving aerial determines the efficiency, but in any case only a tiny fraction of the power radiated by a transmitter is ever picked up by the aerial of any one receiver. If you think about radio waves propagating away from the transmitter like an inflating balloon centred on the transmitter, then the surface of the balloon is evenly covered by the transmitted power: the amount of power arriving at a receiver aerial is effectively that proportion of the surface of the expanding balloon that cuts the aerial, often a very small part of the total. For the longer wavelengths, it would be impracticable to use a tuned aerial for a receiver, so short wire or rod aerials are used, which rely on high signal strength and sensitive receivers. Tuned aerials are used for FM and DAB radios and for television reception.

Horizontal and vertical wire and rod aerials detect the electric component of a wave, but loop antennae rely on the magnetic component. A loop aerial has an output signal and efficiency related to its area. Ferrite is a soft magnetic material that concentrates the magnetic lines of flux, so using it to make an antenna makes the antenna physically smaller by concentrating the magnetic component of the wave in a smaller area than would occur in air. A coil wound round the ferrite will convert the changing magnetic flux into a varying voltage at the frequency of the signal waves. Ferrite rod aerials are typically used in portable AM radios.

## Modulation

A transmitted signal that consists wholly of a steady radio frequency is useless except as a way of establishing that a transmitter is operating, or for uses such as radar. To convey a signal, some means of varying the wave must be found.

The oldest known method of carrying information by radio waves is to switch the radio wave on and off in a pattern of long pulses (dashes) and short pulses (dots), known as on–off keying. The main code using this method is called the **Morse code**, and it was used by Samuel Morse for telegraph signalling in 1838. A carrier keyed in this way, whether by Morse, or by any other of the many different types of telegraph code, is usually called a continuous wave (CW) transmission (Figure 16.6).

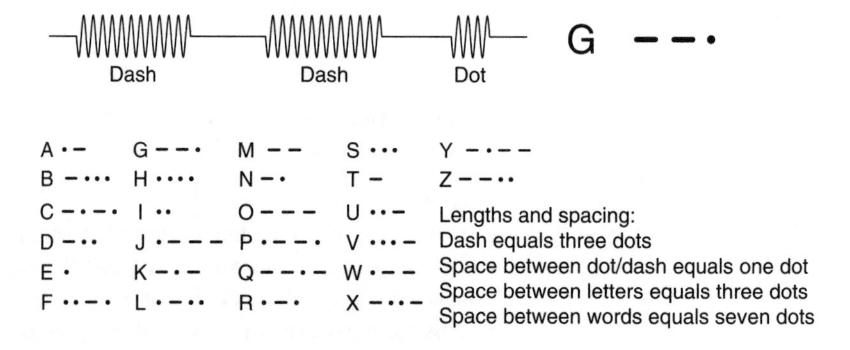


**Figure 16.6** On–off keyed CW transmission of the letter G using Morse code, and the letters of the Morse code

Morse code has been used extensively since the early days of radio for military, marine emergency and radio amateur communications. Although it is a relatively slow method of communication since every letter of the message has to be coded and transmitted separately, it has advantages in that it can provide the greatest propagation range for any given transmitter power.

Since the early 1920s, radio communication has concentrated on varying either the amplitude or the frequency of the carrier wave itself so as to transmit audio and other signals of lower frequency. Such a process is called modulation: the radio frequency carrier wave (RF) is **modulated** by the audio frequency (AF) signal. Figure 16.7 shows a block diagram of a transmitter.

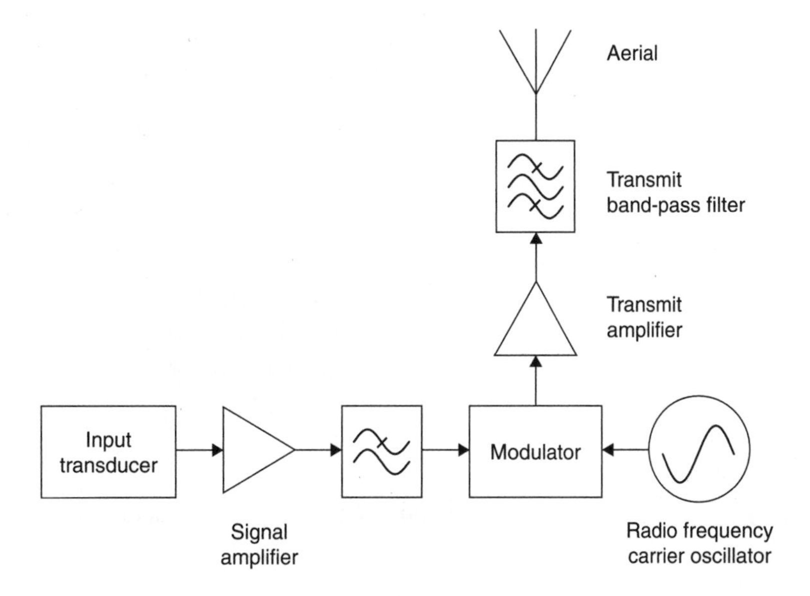


**Figure 16.7** Simplified block diagram of a radio transmitter

High-frequency carriers can carry many different kinds of information, such as audio signals (e.g. speech or music), video signals from a television camera, or other waveforms such as digital signals from computers. The waveforms that provide the information are typically derived from transducer devices, which convert one form of energy into another. The transducers for sound are the microphone and the loudspeaker.

At the receiving end of a radio link, shown in block form in Figure 16.8, the radio system will convert the electrical signals back to the original type of signal, a process called **demodulation**.

One important method of carrying an audio or video signal using radio waves is called amplitude modulation (AM) (Figure 16.9). An amplitude modulator is a circuit into which two signals are fed: the carrier wave at a high (radio) frequency and the modulating signal at a lower (audio or video) frequency. The output of the modulator is a signal at carrier frequency whose amplitude exactly follows the amplitude changes of the modulating

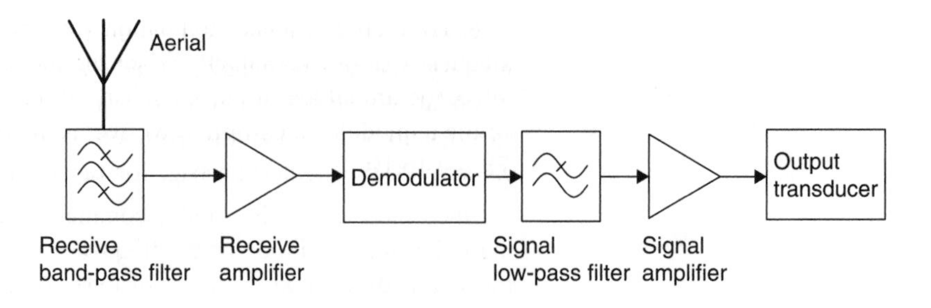


**Figure 16.8** Simplified receiver block diagram

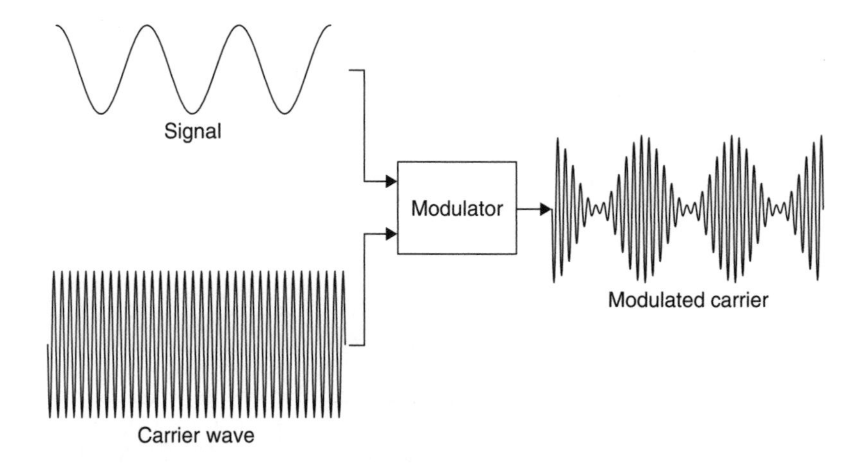


**Figure 16.9** Amplitude-modulated carrier

signal. The greater the amplitude of the modulating signal, the greater the depth of modulation of the carrier. Excessive amplitude of the modulating signal causes over-modulation of the carrier itself, resulting in distortion when the signal is recovered by demodulation.

Another important type of modulation is called frequency modulation (FM). It uses the amplitude of the modulating signal to alter the frequency of the carrier, but the amplitude of the carrier itself remains constant.

## Modulation and sidebands

The simplest possible unmodulated carrier is a sine wave, which has a single value of frequency and carries no useful information. Whenever such a carrier is changed in any way, however, other frequencies can be detected in it. These new frequencies that are caused by modulation are called **sidebands**. Any serviceable receiver must be able to receive these sidebands as well as the carrier wave itself. This is because the sidebands, not the carrier, contain the desired information.

An AM carrier has a simple sideband structure. Imagine a sine-wave carrier at 400 kHz modulated by an audio signal, which is a 2 kHz sine wave. The effect of modulation is to produce two new sideband frequencies, one at 402 kHz (carrier plus modulation frequency) and the other at

398 kHz (carrier minus modulation frequency), in addition to the carrier frequency of 400 kHz itself. When the modulation is at its maximum possible, the amplitude of each sideband should be exactly half the amplitude of the unmodulated carrier. The effect is illustrated (in exaggerated form) in Figure 16.10.

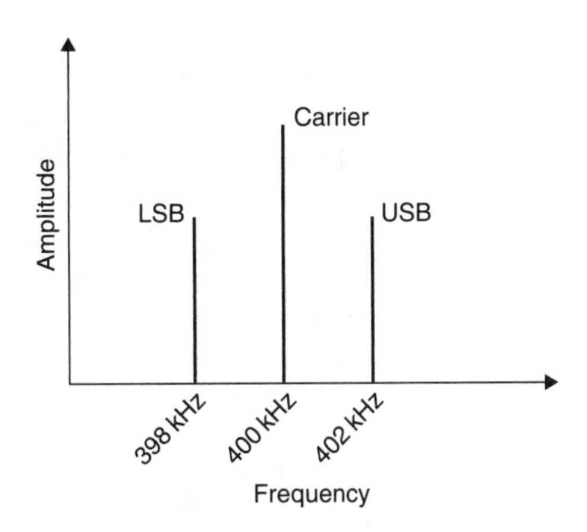


**Figure 16.10** Sidebands of a carrier modulated by a 2 kHz sine wave

**Over-modulation** occurs when the modulating signal has a peak-to-peak amplitude greater than half the peak-to-peak amplitude of the carrier (Figure 16.11). Over-modulation causes severe distortion of the waveform of the modulating signal and is one of the reasons why AM is no longer used to transmit high-quality sound signals. At the transmitter, over-modulation is prevented by means of limiting circuits which reduce the extent of the distortion.

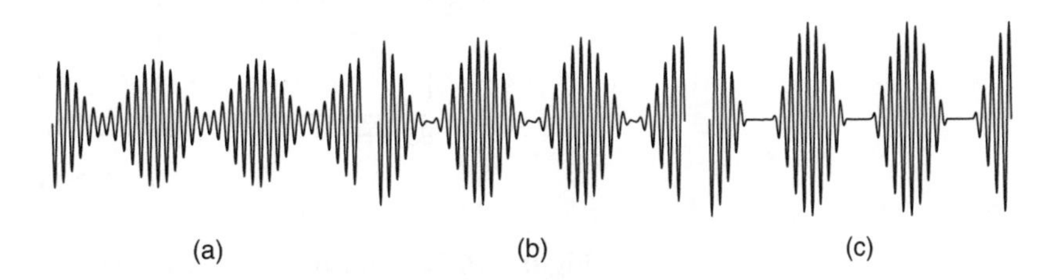


**Figure 16.11** Effect of over-modulation; (a) 70% modulation, (b) 100% modulation, and (c) over-modulated carrier

**Under-modulation** causes no distortion, but it can make the received audio signal very faint and therefore easily drowned out by electrical noise and interference signals. Persistent over- or under-modulation is a clear sign of a faulty or badly adjusted modulator system.

When a carrier is modulated by an audio signal of speech or music, the audio signal that results is not a single sine wave, but a mixture of frequencies. The range of frequencies present in the modulation is called the **bandwidth** of the modulation or the **baseband signal**. Speech can be satisfactorily transmitted over a bandwidth of a mere 3 kHz, but music requires bandwidths up to 20 kHz for high-quality signals to be satisfactorily received and reproduced.

A carrier that has been amplitude modulated by audio or video signals therefore contains sidebands which themselves consist of a mixture of frequencies and extend from $f_c - f_m$ to $f_c + f_m$, where $f_c$ is the carrier frequency and $f_m$ is the highest modulating frequency.

The upper sideband contains all the frequencies between $f_c$ and $f_c + f_m$, while the lower sideband contains all the frequencies from $f_c - f_m$ to $f_c$. Given, for example, a carrier frequency of 855 kHz and modulation frequencies extending to 4 kHz on either side of it, the upper sideband would contain all the frequencies from 855 to 859 kHz, and the lower sideband all the frequencies from 851 to 855 kHz. Figure 16.12(a) shows this sideband structure, which should normally be a mirror image about $f_c$ (855 kHz), as in this example.

A graph that plots the amplitude of signals against a scale of frequency is called a **spectrum**, and an instrument that displays such a graph for any signal is called a **spectrum analyser**. Figure 16.12(b) shows the typical appearance of the trace of a typical AM signal displayed on a spectrum analyser.

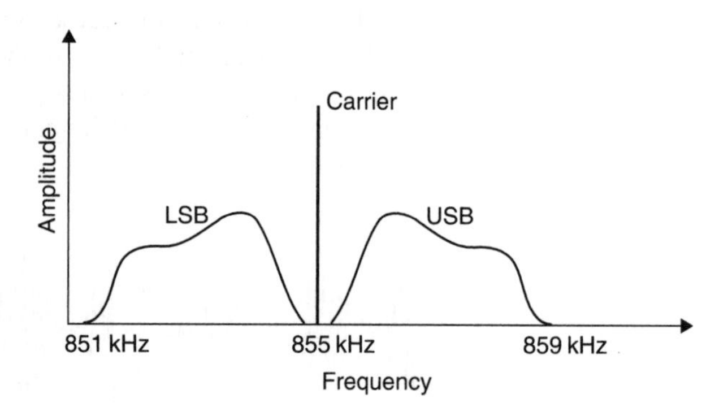


**Figure 16.12** Sidebands of a carrier amplitude-modulated by a typical audio wave

When two AM transmitters in the same geographical area are broadcasting on the same frequency, it will be impossible to receive either transmission clearly because of interference from the other. Even if their broadcasts are on different frequencies, interference will also be caused if the sidebands of the two modulated signals overlap. To avoid interference, the carrier frequencies must be separated by at least twice the maximum frequency of the modulating signals (Figure 16.13a).

Unfortunately, in many parts of the developed world, including Western Europe, there are many transmitters operating on the medium-wave (MW) frequencies. Reasonable reception is made possible, partly by limiting the bandwidth of modulation to 4.5 kHz (which makes the transmission of high-quality sound impossible) and partly by international agreement on the frequency and power output of individual transmitters. Pirate transmitters do not observe these agreements, and are therefore always a cause of interference. Note that in the USA the domestic channel plan uses 10 kHz spacing, but international transmissions such as Voice of America must comply with the 9 kHz international band plan. Figure 16.13(b) shows a typical spectrum of overlapping sidebands caused by transmitter frequencies being too close, which indicates how adjacent channel interference (ACI) arises.

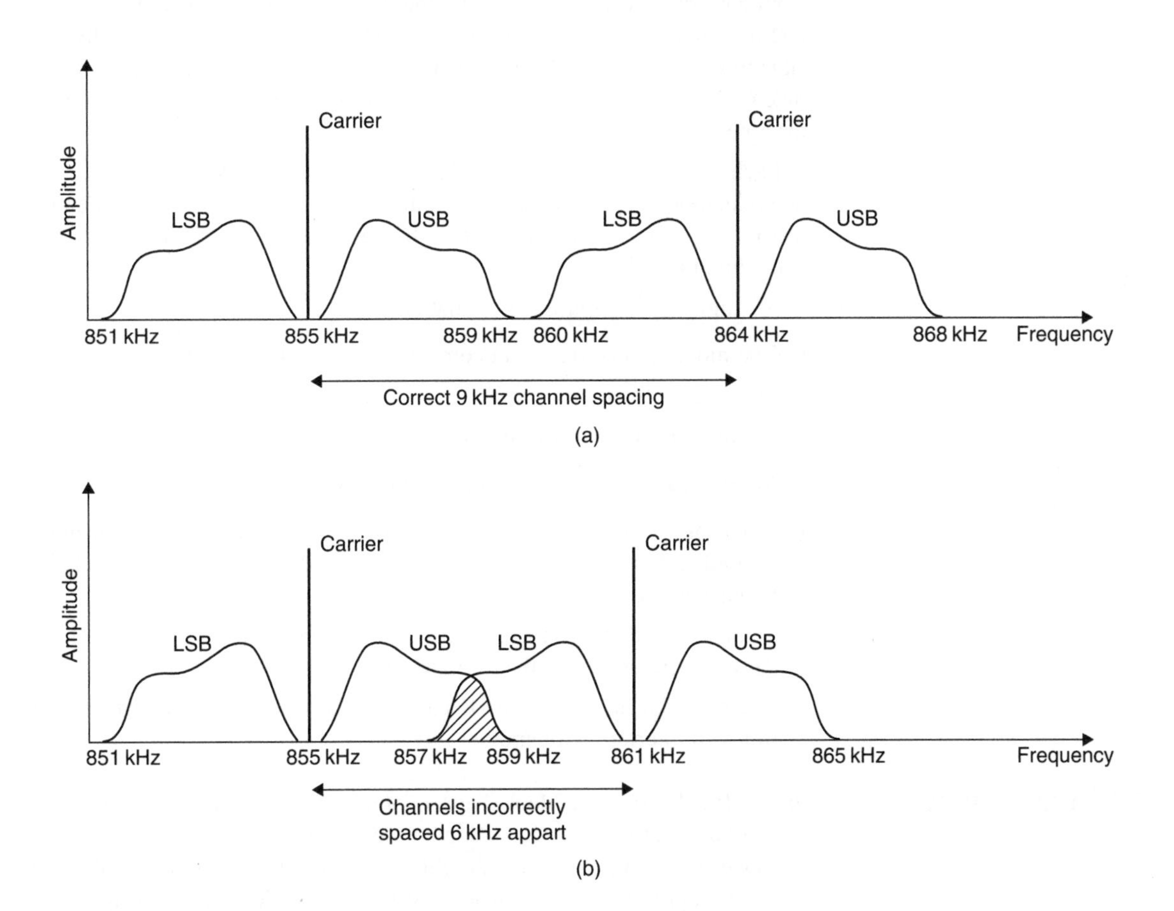


**Figure 16.13** (a) Correct separation of carrier waves, and (b) the sidebands overlap, causing adjacent-channel interference

To try to avoid interference between broadcasts, international agreements exist to allocate frequencies. The allocated bands are shown in Table 16.1. In the short-wave range, some frequencies are allocated for broadcasting, others for military and some for amateur radio and for other purposes.

**Table 16.1** Allocated wavebands for AM broadcasting

| *Name* | *Frequency range* | *Channel spacing* |
|---|---|---|
| Long wave | 153 kHz to 279 kHz | 9 kHz |
| Medium wave | 531 kHz to 1.602 MHz | 9 kHz |
| Short wave | 3.9 MHz to 26.1 MHz | |

## Demodulation

At the receiver, a **demodulator** circuit is used to extract the desired information signal from the modulated carrier. The recovered signal must be free from any trace of the carrier, and a low-pass filter is an essential part of the circuit. Because diodes are used in demodulation circuits, it is usually possible to obtain a steady d.c. voltage that is proportional to the average amplitude of the carrier. This can be used to control the gain of the early stages of the receiver.

Each different type of modulation system requires its own peculiar demodulation circuit to achieve optimum results, and very elaborate demodulators are needed for some types. In this chapter we shall look at AM demodulation only.

Advantages of amplitude modulation (AM):

- The modulation circuits are comparatively simple.
- Demodulation is very simple.
- Calculations of bandwidth are easy.

Disadvantages of amplitude modulation (AM):

- Much of the transmitter power is wasted because only one sideband is needed to carry information, the carrier and the other sideband being unused by the receiver.
- Signals caused by electrical storms or by unsuppressed electrical machines cause interference with the received signals.
- The crowded nature of the allocated bands makes it likely that one station will interfere with another.

## AM transmitter

Figure 16.14 shows the block diagram of an AM transmitter. The audio signal originates at the transducer, a microphone, which converts sound waves into electrical signals. The AF amplifier amplifies these feeble signals (they have an amplitude of 1 mV or less) to the amplitude required to modulate the carrier wave.

At the same time, an oscillator tuned to a high radio frequency generates a carrier wave. The oscillator is usually crystal controlled either directly or via a frequency synthesizer, to ensure that the transmitter frequency is stable and does not drift with changes in temperature, etc. This oscillator stage may be followed by several other stages of frequency multiplication and amplification, but eventually carrier wave and audio signal are combined in the modulator, which produces at its output an amplitude-modulated RF signal.

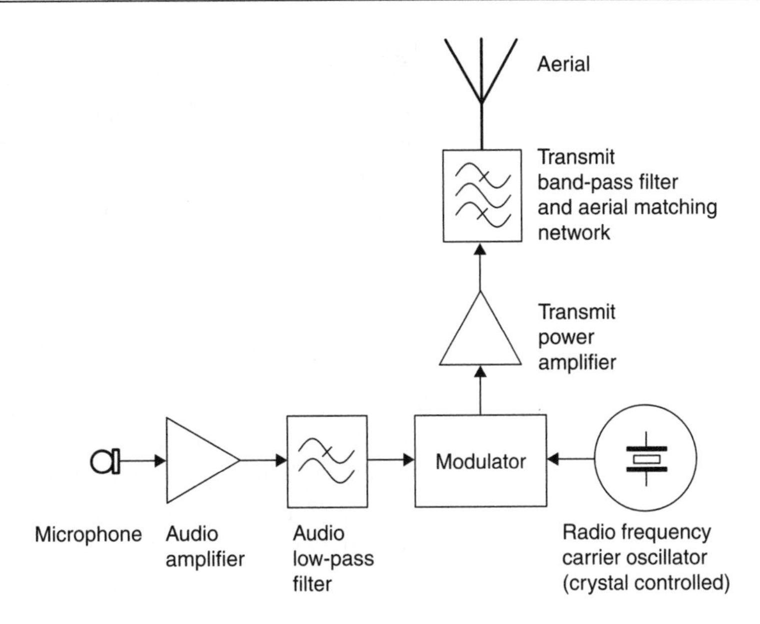


**Figure 16.14** Block diagram of AM transmitter

This modulated RF signal is amplified in the **power amplifier (PA)** stage to produce a signal that can supply the aerial (the output transducer of the transmitter) with an alternating voltage and a large alternating current. The PA stage is required to give the service range required, dissipating it in the form of radio waves of electromagnetic radiation. This power, typically anything from a few hundred watts to several hundred kilowatts, cannot be supplied from an oscillator stage directly, because the loading of the aerial greatly reduces the stability of the oscillator. Many AM transmitters, however, avoid the need to amplify a modulated signal by carrying out modulation at the PA stage.

Apart from the audio amplifier, every amplifying stage of the transmitter is tuned to operate at the frequency of transmission. Filter circuits ensure that the signal fed to the aerial contains only the desired output carrier and its sidebands.

In any transmission channel, noise and interference can be considered killers of information. It is therefore most important that any receiver should not add too much noise to a signal input that may already have been degraded. Interference can be created by other communications channels that occupy the same or adjacent frequencies. Noise is mostly created from other electrical systems, power lines and even the ionosphere. The annoyance factor or quality of the processed signal can be expressed in terms of the system signal-to-noise ratio (S/N). This factor is expressed as: 10log (Signal power/Noise power) dB or 20log (Signal voltage/Noise voltage) dB.

## Superhetrodyne principle

Early AM receivers were of a type called **tuned radio frequency (TRF)** and used several stages of tuned amplification at the selected incoming signal frequency before the demodulator (Figure 16.15a). This often caused problems of insensitivity and instability. A receiver with only one tuned RF stage could not be sensitive enough to pick up faint signals, but a receiver with more than one tuned RF stage ran the risk that any RF radiated from the last RF stage into the first could cause oscillation. This oscillation would not only make reception impossible at that location, but also blot out reception for all the receivers around that were tuned to that frequency. The problem was solved when Edwin Armstrong invented the superheterodyne receiver, often abbreviated to superhet.

In a superhet receiver the selected incoming signal frequency is converted to one fixed value of **intermediate frequency (IF)**, which is usually lower than the RF. This conversion makes amplification easier, because it is simpler to design a high-gain amplifier that is tuned to a set frequency than one that needs to have its tuning altered to receive another transmission. A superhet receiver needs only a few variable tuned circuits, because most of the amplification is carried in the IF amplifier, which is fixed tuned.

Figure 16.15(b) shows the block diagram of an AM receiver typical of a transistor radio. The stage labelled 'Tuneable RF band-pass filter' attenuates

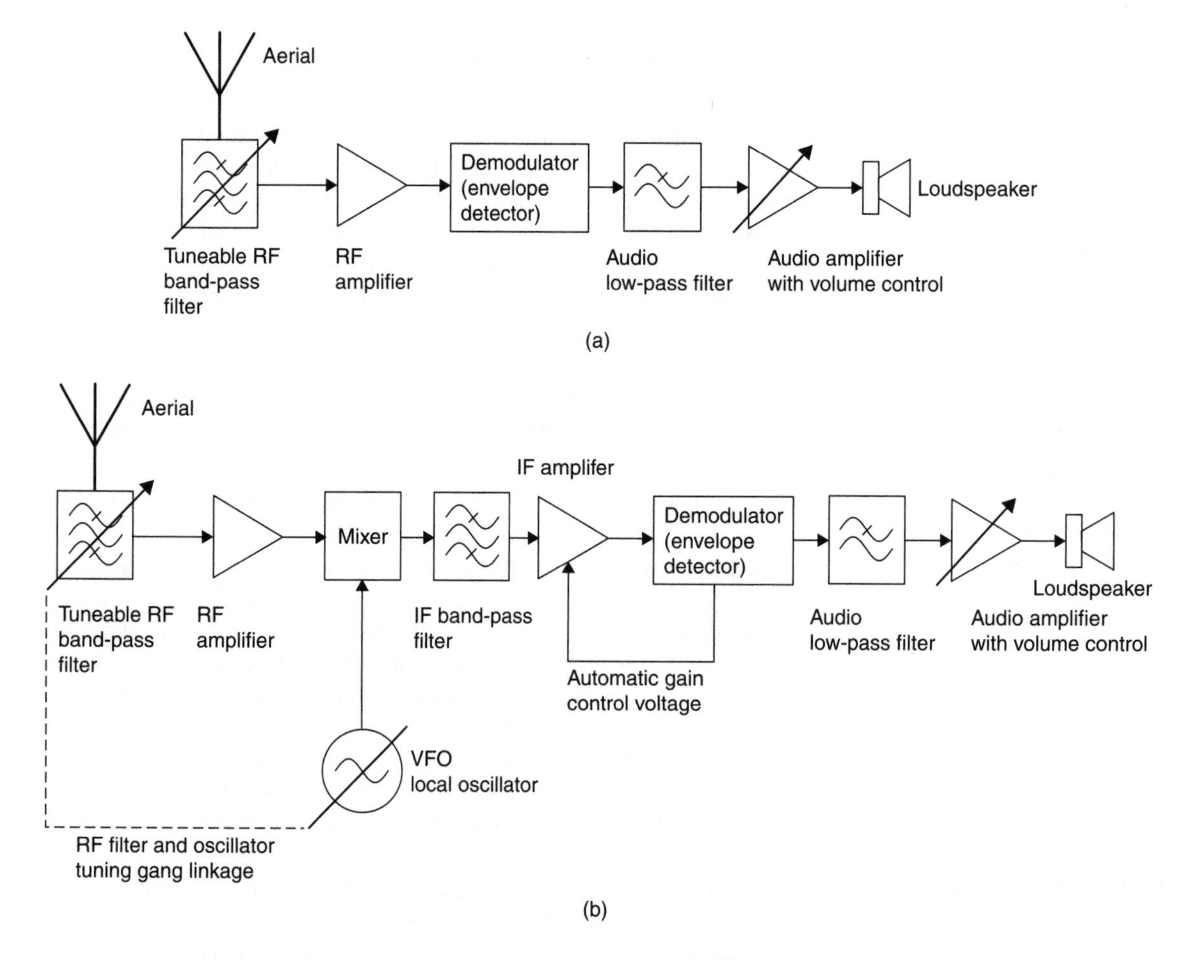


**Figure 16.15** Block diagram of AM receiver; (a) tuned radio frequency, and (b) superhetrodyne

signals either side of the wanted signal. Although most selectivity comes from the IF filter, this preselector is necessary to remove signals that might combine with harmonics of the local oscillator and the image frequency, which is the frequency on the other side of the local oscillator signal, which also produces frequencies in the IF pass band. This filter therefore must have variable tuning. It also acts to reduce the amplitude of any oscillator frequency that might otherwise be reradiated from the aerial.

The variable frequency oscillator (VFO) circuit that forms the **local oscillator** is typically tuned using a variable capacitor that is ganged to the variable capacitor that tunes the RF stage. This ganging, with preset capacitors called **trimmers** and **padders**, ensures that the tuning of the VFO and RF filters tracks as they are varied. The difference between the oscillator frequency and the incoming signal is thus maintained at the intermediate frequency, typically 455 kHz for MW radio. The oscillator is said to be **tracking** the input correctly when the frequency difference, or IF, remains correct over the full range.

The incoming signal and the oscillator sine wave are combined in the **mixer** stage, producing two new frequencies ($f_{LO} + f_{RF}$ and $f_{LO} - f_{RF}$) which still carry modulation. One of these new modulated frequencies, usually the difference frequency in AM radios, is the IF, which is in this example 455 kHz, the difference between the frequencies of the oscillator and of the incoming signal. The second modulated frequency produced in the mixer stage is a signal whose frequency is equal to the sum of the frequencies of the oscillator and of the incoming signal. This is usually known as the **IF image frequency**. If another aerial input signal on the other side of the local oscillator (LO) from the desired frequency can combine with the LO to produce an output signal at the IF frequency, this is known as the **RF image frequency**. For example, if a radio is tuned to receive 648 kHz with a 1.103 MHz LO, the RF filter or preselector has to block various frequencies of the RF image frequency to prevent interference that it would cause (Table 16.2).

**Table 16.2** How image frequency at the aerial can produce an interfering IF signal

| *Signal* | *RF* | *LO* | *IF* | *IF image* |
|---|---|---|---|---|
| Desired RF | 648 kHz | 1103 kHz | 455 kHz | 1751 kHz |
| Image RF | 1558 kHz | 1103 kHz | 455 kHz | 2661 kHz |

The mixer is followed by a tuned amplifier that is tuned to the frequency of the IF, which filters out signals from adjacent channels. In low-cost receivers the mixer and first IF amplifier stage may be combined, and sometimes a self-oscillating mixer circuit uses a single transistor to fulfil all three roles. All the other frequencies of incoming signal, oscillator sine wave and their products are rejected by the IF amplifier, which also greatly increases the amplitude of the IF signal, carrying the same modulation as did the original incoming signal.

The modulated and amplified IF signal is applied to the AM demodulator, which produces two outputs. One is the modulating signal itself, which is at audio frequency and free of any trace of the IF; the other is a d.c. voltage which is proportional to the average amplitude of the IF signal. When a weak or distant transmitter is being received, the incoming signal will produce at the mixer an IF signal of very low amplitude.

Even after amplification in the mixer and IF amplifier stages, this amplitude may still be low. However, a nearby or powerful transmitter may provide a signal that directly breaks through into the IF amplifier stages. This then causes interference or overloading. Both lead to distortion.

The d.c. that is produced in the AM demodulator can be used to minimize the effects of these two extremes, by being fed back to control the gain of the IF amplifier itself. What happens is that a small-amplitude signal at the demodulator gives rise to a small d.c. feedback signal, which permits the IF amplifier to operate at full gain.

A large-amplitude signal at the demodulator gives rise to a large d.c. signal, which is fed as negative bias to the IF amplifier and causes it to operate at much reduced gain. By the use of this **automatic gain control (AGC)** circuit, the signal at the demodulator is kept to an almost constant level even when there are great variations in the amplitudes of incoming signals.

The AF signal from the demodulator passes to an audio frequency voltage amplifier incorporating a volume control that increases or decreases AF gain. The amplifier can also provide tone control, which increases or reduces the gain at low and high frequencies, respectively. The AF signal is then fed to a power amplifier, which boosts it sufficiently to drive the output, a loudspeaker or headphones.

AM radios generally make use of a ferrite rod aerial, which consists of an inductor wound on a high-permeability ferrite core. Some older radios use a coil wound around the inside of the case, known as a frame aerial. These types of aerial are particularly useful for portable radios, because the ferrite rod or frame aerial is directional, and the radio can be turned so as to give the best reception, often by nulling the signal from an interfering station. Car radios typically use a whip aerial, a short length of metal rod with a series inductor to simulate the effect of a longer tuned aerial.

## Multiple-choice revision questions

16.1 Why is a high-frequency bias signal used when recording on magnetic tape?
(a) to avoid hysteresis distortion
(b) to cancel noise
(c) to synchronize the motor
(d) to allow louder audio signals to be recorded.

16.2 If a tape was not erased before recording new information on it, what would be heard on playback?
(a) the old recording
(b) white noise
(c) both recordings at the same time
(d) silence.

16.3 Where is a ferite rod aerial used?
(a) DAB receivers
(b) VHF/FM receivers
(c) CW receivers
(d) medium-wave receivers.

16.4 Which part of a transmitter combines the carrier with the information signal?
(a) Signal amplifier
(b) transmit band-pass filter
(c) demodulator
(d) modulator.

16.5 What channel spacing is used for medium-wave broadcasting in the UK?
(a) 10 kHz
(b) 9 kHz
(c) 855 kHz
(d) 4.5 kHz.

16.6 Which of the following is not an advantage of a superhetrodyne receiver?
(a) fewer variable turned RF circuits
(b) fixed tuned IF circuits
(c) does not need an oscillator
(d) more sensitive for weak signals.

16.7 A superhet radio that uses an IF frequency of 455 kHz must be tuned to receive a signal at 1215 kHz. What oscillator frequency will produce the correct signal at the output of the mixer?
(a) only 760 kHz
(b) 1215 kHz or 455 kHz
(c) 1670 kHz or 760 kHz
(d) only 1670 kHz.

# 17 Frequency modulation

The first public broadcast radio services introduced in the 1920s were in the long-wave (LW) and then medium-wave (MW) band, using amplitude modulation (AM) and limited by the technology of the time. Historically, long- and medium-wave AM broadcasting has used restricted bandwidth, approximately 4 kHz, which allows intelligible speech but is not very good for music broadcasting, since music signals are typically 14–20 kHz in bandwidth. Amplitude modulation also suffers from impulsive interference, such as sparks caused by switches opening and closing, for example thermostats and motor car ignition systems. This is because the information content of the signal is carried by the envelope of the carrier wave, but impulses like sparks are very wide band signals so there is energy at practically all frequencies simultaneously, effectively adding to the envelope of the desired signal.

The problems of using AM on the medium waveband led to a different form of modulation being adopted for high-quality broadcasting. The limited bandwidth available in the medium-wave band means that it would be impractical to increase the bandwidth of the signals there. Even if one accepted the poor interference performance, a channel spacing of 30 kHz or more would be required, giving around 30 channels in the medium-wave band, compared with the 114 available at 9 kHz spacing. This would be a difficult solution because medium-wave propagation characteristics mean that stations can cover very large geographical areas with relatively low transmitter powers, thus increasing the risk of co-channel and adjacent channel interference.

Edwin Armstrong, who invented the superhet, also came up with the idea of frequency modulation, although its advantages were not fully appreciated in his lifetime. The need for high-quality broadcasting, particularly in stereo, and the number of stations using the AM system on medium-wave frequencies, led to the adoption of FM from the 1960s onwards. The frequency band chosen for local broadcasting, approximately 88–108 MHz, allowed 100 channels spaced at 200 kHz intervals. An important advantage is that these very-high frequency (VHF) signals are effectively line-of-sight transmissions, so the coverage area is well defined and channels can be reused quite close by without the risk of co-channel or adjacent channel interference.

In a frequency modulation (FM) system, the amplitude of the carrier remains constant, and the effect of modulation is to change the frequency. The principle is shown in Figure 17.1. The alteration of frequency caused by a modulating signal is called the **frequency deviation**. In wideband FM systems deviations of 75 kHz or more are produced. Frequency modulation is also used for short-range speech communications, such as walkie-talkie handsets used by security guards and citizens' band (CB) radio. These narrowband frequency modulation (NBFM) systems use deviations that are limited to a few kilohertz only, with channel spacing typically around 12.5 kHz.

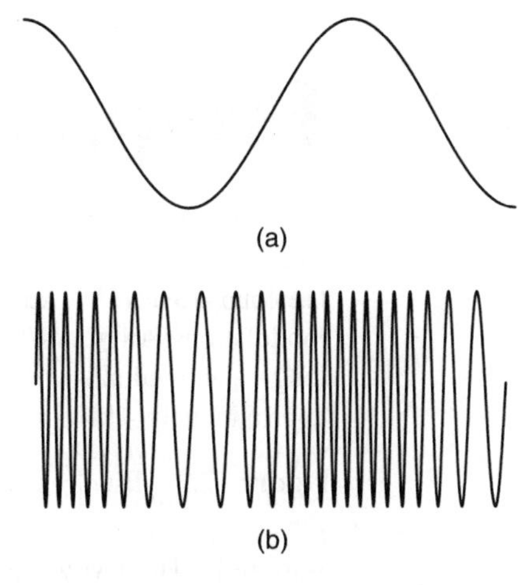


**Figure 17.1** The principle of frequency modulation: (a) signal, and (b) modulated carrier

Frequency modulated signals have a much wider spectrum than AM signals. In the UK, FM broadcasts have an audio bandwidth of 15 kHz, with the modulated bandwidth for each sideband being as wide as 200 kHz. The bandwidth of an FM signal cannot be calculated as easily as that of an AM signal. A very approximate estimate can be made that the bandwidth required is between two and three times the peak deviation. The European VHF radio service uses a peak deviation that has been standardized at $\pm$75 kHz, with a channel spacing of 200 kHz.

Advantages of frequency modulation:

- simple modulation circuits
- carrier amplitude constant, so that transmitter range can be greater than with AM
- freedom from interference because interference signals do not alter the frequency of the transmitted signal.

Disadvantages of frequency modulation:

- more complex demodulator circuits
- wide bandwidth requirement.

In contrast to an AM transmission, no sudden change in waveshape can be caused by over-modulation of the FM carrier. Using AM, a single modulating frequency generates only a single pair of side frequencies in the complex wave, whereas for FM, each single modulating frequency generates many sideband pairs. The FM bandwidth is constant, irrespective of the modulation frequency or amplitude, so that a low modulating frequency gives rise to a larger number of sideband pairs than does a higher modulating frequency.

The block diagram for an FM transmitter (Figure 17.2) shows that the audio signal is still used to modulate the carrier, as in the AM transmitter, but now in such a way as to cause changes in its frequency rather than in its

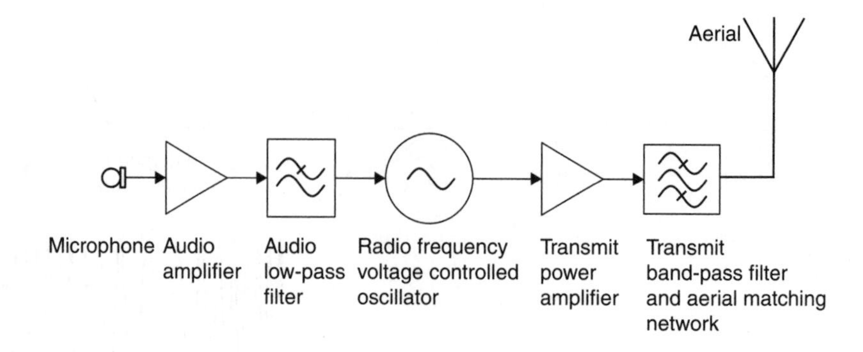


**Figure 17.2** Block diagram of an FM transmitter

amplitude. The FM wave is then amplified in the power stages before being radiated by the aerial system. Table 17.1 summarizes the detailed effects of both types of modulation. The methods used for digital radio transmissions will be looked at later.

**Table 17.1** Summary of effects of modulation

| | *AM* | *FM* |
|---|---|---|
| Amplitude of modulating signal | Varies carrier amplitude | Varies carrier frequency |
| Frequency of modulating signal | Controls rate of change of amplitude | Controls rate of change of frequency |
| System constant | Carrier frequency | Carrier amplitude |

Figure 17.3 shows a block diagram for an FM receiver whose working principles are largely similar to those of the AM receiver. The superheterodyne principle is again used, but because of the much higher wider frequency range (88–108 MHz) and the greater bandwidth (about 200 kHz), an intermediate frequency (IF) of 10.7 MHz is required. The higher IF is necessary to reduce the risk of image channel interference as well as to provide sufficient IF pass band bandwidth for the 200 kHz wide signal.

The mixer, oscillator and IF stages act in much the same way as do their equivalents in a medium-wave AM receiver. The demodulator, however, is a special FM type, usually either a **discriminator** or a **ratio detector**. To reduce the effect of amplitude changes on the received signal, the detector is preceded by a high gain limiting amplifier that ensures that the signal amplitude is constant. This limiting amplifier is responsible for the characteristic rushing noise made by older FM receivers that are not tuned to any channel; modern receivers include mute circuits to disconnect the output when no signal is received.

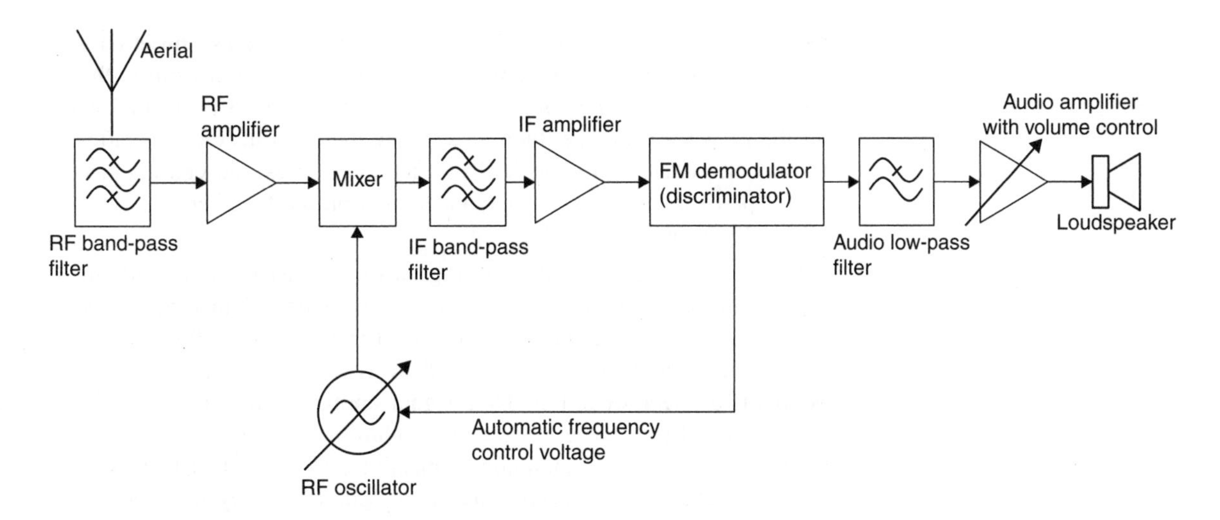


**Figure 17.3** Block diagram of an FM receiver

Since the FM demodulator detects changes in frequency rather than amplitude, the d.c. output from the FM demodulator is of a type different from the d.c. output of the AM demodulator. When the FM demodulator is correctly adjusted, its d.c. output is zero, with the received signal correctly tuned so that it is centred in the IF pass-band. If the receiver is tuned off frequency either side of the centre a d.c. output is obtained from the FM detector. This d.c. signal is used to provide automatic frequency control (AFC), which corrects the tuning of the local oscillator, keeping the receiver correctly tuned.

Automatic frequency control is needed because the frequency of a variable frequency oscillator working in the range 77.3–97.3 MHz (or 98.7–118.7 MHz) is very easily affected by small changes either in temperature or supply voltage. The AFC voltage acts to correct these small changes, so keeping the receiver tuned to the correct transmission frequency.

An outstanding feature of all FM transmission systems is their **capture effect**. An FM receiver will lock on to the modulation of the slightly stronger of two signals of the same frequency and reproduce only the modulation of the stronger signal. This efficient selectivity, combined with the freedom that the FM system gives from interference caused by noise from electrical storms or other electrical equipment, has led to FM being widely used in communications equipment, especially for mobile radio, as well as for high-quality broadcasting.

## Digital audio broadcasting

Digital audio broadcasting (DAB) is a European standard intended to provide many more high-quality radio channels and additional data services. It uses a complex digital modulation method called **orthogonal frequency division multiplexing (OFDM)**, which effectively uses many adjacent narrow-band carriers to transmit data words in parallel. The receivers use custom decoder chips, or fast microcontrollers running software-based decoders, based on the same principles as the Fourier transform to recover the digital data stream. The advantage of the Fourier transform is that it allows the many parallel channels of data to be decoded in parallel without the need for analogue filters to separate them. Its detailed operation is beyond the scope of this book. The UK system uses 256 carriers spaced 1 kHz apart for each of six programmes multiplexed together. This gives a total of 1536 carriers in each group.

The main advantage of digital broadcasting is that it can provide more services in the same bandwidth. Despite advertising claims, that there is no detectable difference between the sound from an FM radio and a DAB radio under identical conditions, and there is still much user debate about the merits of the system. Low DAB signal strength results in a most objectionable sound (best described as like boiling mud) from the loudspeaker. A common feature of digital communication systems is that they provide very good quality until a threshold is crossed, when they fall apart completely. In contrast, AM and FM radio systems tend to degrade gracefully, still providing intelligible signals long after the noise and interference signals have become apparent.

## Reception

The majority of radio listeners in the UK make use of a portable receiver which uses either a ferrite rod aerial (AM) or a telescopic aerial (FM or DAB). The telescopic aerial for FM is shorter than the ideal half-wave size (150 cm for 100 MHz), but DAB portables usually have an aerial that is the optimum length for the band centred around 220 MHz (band III); about 68 cm. Portable broadcast receivers seldom have any provision for connecting an external aerial, so the best signal may have to be selected by moving the receiver until an acceptable sound quality is obtained.

For tuners that can make use of an external aerial, the best aerial for the particular application should be selected, and then orientated in the direction to receive the maximum wanted signal level, consistent with the minimum of interference from other signals in the same waveband. In some districts the circular type of aerial for FM can be used, located in the loft of a house or externally. People in areas of poor FM reception may need to install a high-gain directional (Yagi) aerial (Figure 17.4) to achieve a good signal from a tuner. (In such areas a portable receiver will probably be unable to receive FM in stereo, and drop back to mono reception.) Yagi aerials are also recommended for DAB when using a tuner with an aerial input.

The **radio data system (RDS)** is an additional service carried by a number of VHF FM transmitters to provide programme identification, news and traffic information for motorists. In addition, because national networks such as BBC Radio 1, 2, 3 and 4 have many transmitters simultaneously radiating the same programme across the country, an RDS equipped receiver can automatically retune to the same programme as the vehicle moves from the service area of one transmitter to another. A similar service known as ARI is in use in mainland Europe, but with a different modulation technique. It is expected that DAB will eventually replace the RDS and ARI services in Europe and the UK.

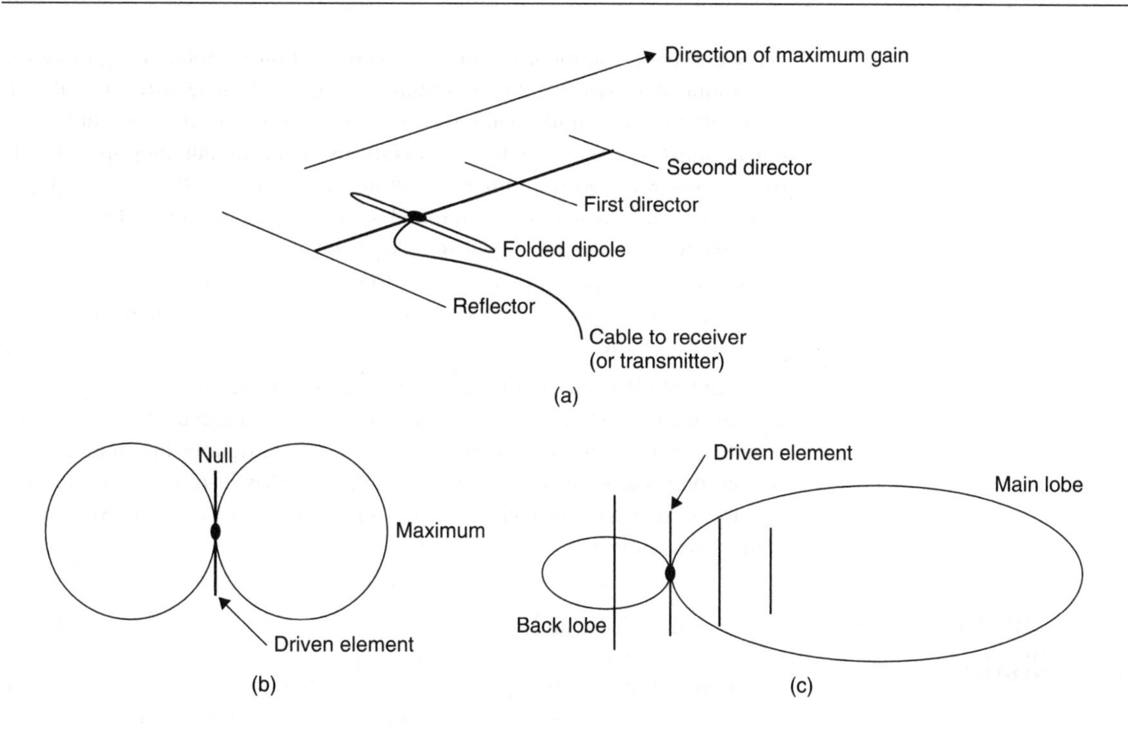


**Figure 17.4** (a) Yagi aerial, (b) simple dipole radiation pattern, and (c) four-element Yagi radiation pattern. For VHF FM this is mounted horizontally.

## Installing a receiver

Installation applies only to more expensive units, usually consisting of a tuner, often combined with an amplifier, with separate loudspeaker. Ideally, the loudspeakers should be placed about 2 m apart, facing down the longer length of a rectangular room. They should be placed above floor level, ideally at the same level as the ears of a seated listener. These ideals often have to be considerably modified to meet the needs of room arrangement.

A suitable aerial should be installed with an outlet point conveniently near to the audio unit. Most tuners now have a coaxial input for FM or DAB input, and where there is an existing television roof aerial and a socket in the room, the FM or DAB signal can use the same coaxial down-lead by using combiner filers at each end. Note, however, that digital television (FreeView in the UK) and DAB are very sensitive to the quality of coaxial cable used for the downlead and the flying lead from the wall plate to the receiver. Where possible, use only the type with foil screen and fully filled outer braid, rather than the older style 75 Ω television coax with a loose, open-braided coaxial screen. Open-braid cable suffers more loss, and also suffers more from noise pickup, so it not usually suitable for digital television except in high signal strength areas. It is also a good idea to minimize the number of joints in the downlead, since each joint introduces a loss. A typical scheme has the aerial downleads taken into the loft of a house to the inputs of a combiner. The combiner output is connected to the cable that feeds the lounge, and at the outlet another combiner separates the television and radio signals. Combiners are readily available for TV/FM, some incorporated into flush outlet sockets. Combiners for DAB are available, but not so easy to find. In some houses, a distribution amplifier is used to feed several rooms, and in such an arrangement, the distribution amplifier should be fed with the combined signal, separating radio from television where needed at the outlet sockets. A distribution amplifier can also be used to boost signal strength and is particularly useful for digital television or radio. Distribution systems are also available for combining terrestrial television, satellite and radio signals.

Once installed, the performance of the tuner should be checked, particularly to ensure that all the required stations are arranged in order for the remote control. If there is an option to ignore faint signals this should be set so that only the strongest are available. On a DAB installation, check that all the available signals can be tuned; in some districts, only the BBC set of signals is available.

## Measurements and fault-finding

Repair and fault-finding of radio equipment can usually be attempted with the aid of a signal generator, oscilloscope and multimeter. Most AM and FM radios follow relatively conventional designs, with many using the same chips from companies such as Philips or Sony. While service manuals are not as easily available as they once were, the data sheets for the chips used in a receiver can make the difference between success and failure.

The key to fault-finding is generally logical thought and understanding of the system, and a good knowledge of amplifiers, filters and oscillators is required. The first place to start is the power supply, ensuring that supplies have the correct voltages and there are no blown fuses. When working on mains equipment it is essential to take suitable precautions, such as using a mains isolating transformer to protect from a mains shock to ground, which can be fatal. The rule with mains equipment is: if in doubt leave well alone.

In portable radio receivers the common faults are often mechanical, including noisy or open circuit volume controls, headphone sockets that have a bent spring contact permanently disconnecting the loudspeaker, and external d.c. input sockets that, having been plugged in repeatedly, have lifted off the circuit board and disconnected the battery as well as external power.

When all else fails, tracing a known signal through the set with an oscilloscope, using a signal generator to provide an input, will usually get results. For AM receivers a carrier modulated with 1 kHz is usually available from signal generators, whereas FM modulators are less common in cheap signal generators. Remember that the signal strength at the input is often as low as a few microvolts, but the audio output will be hundreds of millivolts or more, so you will frequently have to adjust the signal generator and oscilloscope amplitude controls.

## Multiple-choice revision questions

17.1 What is the advantage of using FM for broadcasting?
(a) uses less bandwidth
(b) immune to impulsive interference
(c) works in the VHF band
(d) uses a simpler receiver.

17.2 Which is not an advantage of using VHF for FM broadcasting?
(a) more bandwidth available
(b) can build simple transmitters
(c) lower power required
(d) line of sight propagation allows frequency reuse.

17.3 What sort of external aerial is usually used with DAB receivers?
(a) ferrite rod
(b) long wire
(c) Yagi
(d) loop.

17.4 What type of fault is most common in portable radios?
(a) blown transistors
(b) mechanical faults
(c) leaky capacitors
(d) blown fuses.

# 18 Television systems

Public television broadcasting began experimentally in the 1930s, but it was not until the 1950s that significant numbers of people had access to a television set. Television broadcasting in the 1950s and 1960s was mostly in black and white, although colour systems had been tried experimentally before the Second World War. Standards for both black-and-white and colour systems developed separately in the USA, the UK and France. Different mains frequencies – 50 Hz in the UK and Europe and 60 Hz in the USA – led to different picture frame rates. The number of lines making up the picture was also different, and different colour systems were introduced. Differences are still being introduced as Europe and the USA develop separate, incompatible digital and high-definition broadcast standards.

The analogue phase alternation line (PAL)-I colour television system, as used in the UK, uses amplitude modulation for the vision carrier and frequency modulation for the monophonic sound channel. The sound carrier is positioned 6 MHz above the vision carrier. In addition to these, the carrier for the near-instantaneous companded audio multiplex (NICAM) stereo version of the same audio programme is located 6.552 MHz above the vision carrier. This stereo information in digital format is superimposed on the subcarrier using a modified form of phase modulation. To avoid adjacent channel interference, this wide band signal is allocated a transmission channel that is 8 MHz wide (an 8 MHz channel spacing). The PAL system is used extensively in Western Europe. The National Television System Committee (NTSC) colour system is used in the USA and Japan, and the French Sequential Couleur À Memoire (SECAM) system is used in France, the former USSR and former French colonies.

The current terrestrial digital television (DTV) service in the UK, FreeView, operates with the same channel spacing in the same segments of the frequency spectrum as analogue television to allow the use of the same aerial. These channels are allocated so as not to produce mutual interference. The DTV system uses **COFDM** (coded orthogonal frequency division multiplex), which is a similar technology to that used for DAB. For DTV the signal is carried by 1705 parallel carriers spaced 4 kHz apart. Each carrier is modulated with 16-QAM (quadrature amplitude modulation) or 64-QAM depending on the service provided and the signal level. The group called a multiplex occupies an 8 MHz bandwidth.

The ultra-high frequency (UHF) part of the frequency spectrum used for terrestrial television transmission is subdivided into two bands: band IV from 470 to 582 MHz and band V from 614 to 854 MHz, while in the UK Channel 5 occupies the range from 583 to 599 MHz. Conventionally, these services are covered using Yagi arrays as shown in Table 18.1. The frequency ranges are allocated to different parts of the country to minimize interference. In addition, where there is a possibility of interference, one area may use horizontal polarization and the adjacent area vertical polarization.

**Table 18.1** UHF television aerial groupings

| *Channels* | *Group/band* | *Sub-band* | *Colour code* |
|---|---|---|---|
| 21–37 | A | Band IV | Red |
| 35–53 | B | | Yellow |
| 48–68 | C/D | | Green |
| 39–68 | E | Band V | Brown |
| 21–48 | K | | Grey |
| 21–68 | W | | Black |

Satellite television reception, which occupies microwave frequencies, is almost entirely via reflector antennae (dish aerials) and operates within Europe mainly via **Ku Band**. This band uses the range 10.7–12.75 GHz for the Direct Broadcast Satellite (DBS) service, with a channel spacing of 19.18 MHz.

## Aerials

The basic half-wave dipole has a circular radiation/reception pattern broadside on to its orientation. By adding directors in front and reflectors behind, the pattern becomes directional, with more gain in one direction than any other. The usual way of displaying information about the radiation pattern aerials is on a plot called a polar diagram. Figure 18.1(b) shows how adding a director and a reflector element makes the aerial directional. This type of structure is referred to as a Yagi array. It is highly directive and also has a limited bandwidth, so improving the selectivity of the receiver.

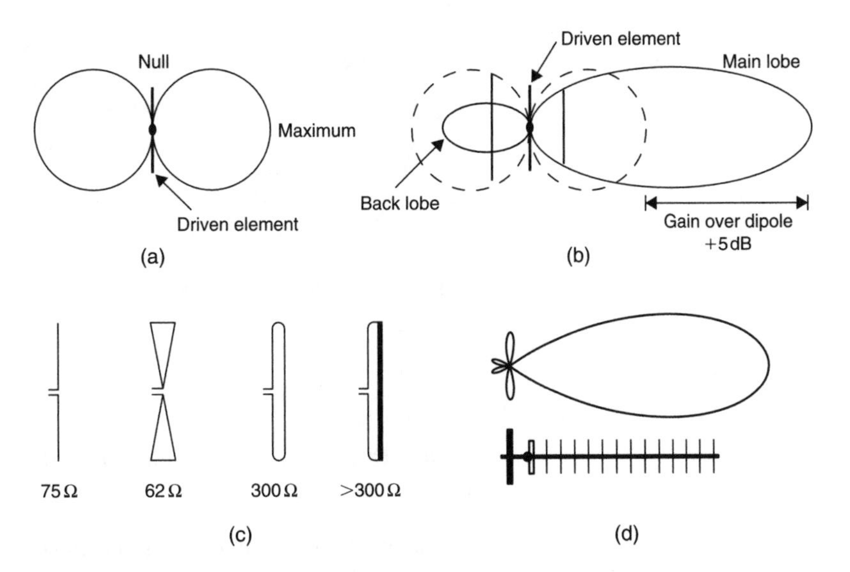


**Figure 18.1** Antenna gain and drive impedance: (a) half-wave dipole, (b) three-element beam (Yagi), (c) dipole feed impedances, and (d) typical television antenna, a 15-element Yagi with its polar gain plot

Figure 18.1(a) shows how these parasitic elements modify the gain of the array. The dipole is usually used as a reference gain against which other aerials can be measured. Thus, a three-element Yagi will be quoted as having a gain of 5 dBd (5 dB with reference to a dipole). By comparison, an 18-element array can have a gain as high 18 dBd. The addition of these parasitic elements has the effect of lowering the dipole impedance (typically 75 Ω), but this can be countered by modifying the dimensions of the dipole, as indicated by Figure 18.1(b). These arrays are commonly used for frequencies up to about 1 GHz. At installation, the aerial array is rotated so that it picks up the maximum wanted signal level, consistent with the minimum of noise. Yagi arrays are usually specified by their forward gain, beam width, front-to-back ratio and bandwidth. The beam width defines how accurately the aerial must be pointed at the transmitter for gain within 3 dB of the maximum; this is also called the half-power beam width. The front-to-back ratio defines how much gain advantage comes from pointing the antenna at the transmitter as opposed to the opposite direction; this is often in the order of 20 dB.

Dish or reflector aerials are more commonly used for microwave frequencies, such as satellite transmission and reception. The common type of antenna is shown in Figure 18.2, and consists of a dish of parabolic shape. The important feature of this particular shape is that all incoming signals arriving at the dish will be reflected to concentrate at the focal point '**a**'. It is this feature that is responsible for the very high gain, which is proportional to both the dish area and the operating frequency.

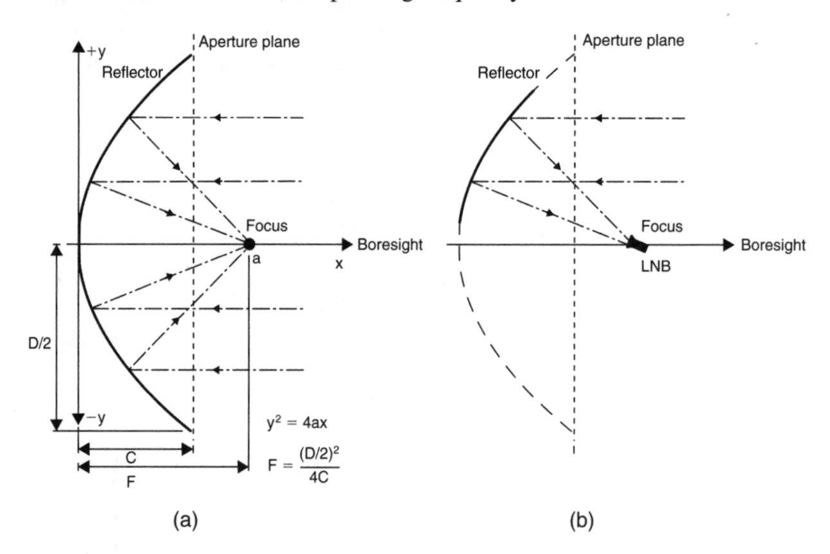


**Figure 18.2** (a) Parabolic reflector, and (b) arrangement for typical offset feed dish antenna

For small dishes the low noise block (LNB) and support block the signal over a significant proportion of the area of the reflector. Therefore, the offset arrangement shown in Figure 18.2(b) is used, since it avoids blocking the signal with the receiving unit and its supports. The aerial fitter will have to aim the aerial at an offset of about 28° to allow for this.

## The television picture

Scanning (Figure 18.3) is a method of obtaining a video signal at a fixed repetition frequency from a light pattern or picture. A television camera tube or sensor produces at its output a signal which is proportional to the brightness of the light reaching its front surface. A camera tube contains an electron beam which is focused to a spot. The position of this spot on the front surface of the tube is the point at which the brightness is sampled by scanning to produce an output signal. A solid-state sensor, a **charge-coupled device (CCD)**, is sampled by activating a set of light-sensitive cells in turn. A lens is used to focus an image on the front surface or faceplate of the camera tube or sensor.

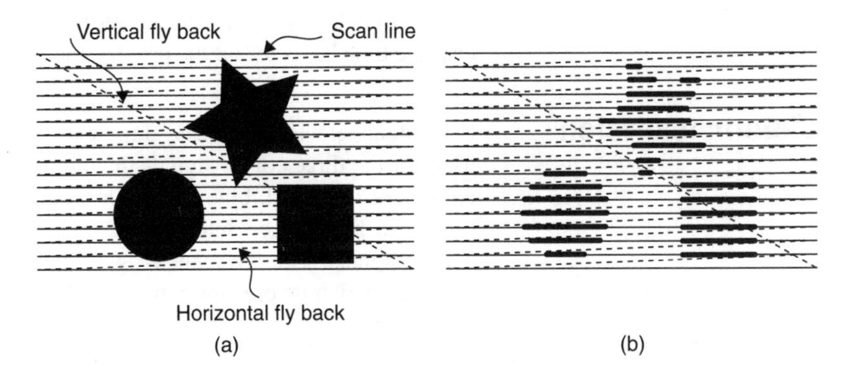


**Figure 18.3** The principle of scanning: (a) the original image is scanned one line at a time, and (b) the display is built up in the same sequence

The scanning action samples all of the surface of the camera tube or sensor. This is done by moving the sample point horizontally (**line scan**), at very high speed, across the sensitive surface and at the same time deflecting it down the surface (**field scan**) from top to bottom at a much lower speed. If a sufficient number of lines is used for each picture, the viewer will not detect the lines unless the viewing distance is too short. If the field repetition rate is high enough, the viewer will not detect any flicker. Television, like the cinema, depends on the features of the human eye that make small detail invisible and rapid repetition appear as a continuous picture.

The signal output from the camera tube or solid-state sensor has a varying amplitude that represents the brightness of every single tiny area surface as it is scanned. The signal contains two main frequencies: the line scanning frequency and the field scanning frequency. Sets of pulses, the **synchronization (sync.) pulses**, are added to this video signal so that the scanning circuits at a receiver will scan at the correct speed and in step with the signal. Television scanning for analogue television is interlaced: the odd numbered lines of a frame (a complete picture) are scanned, followed by the even-numbered lines, so that two scans are needed to cover the screen area. This was implemented in 1936 when the first television system was being designed as a way of saving bandwidth. Digital high-definition pictures do not use interlacing.

The **aspect ratio** of a picture is defined as the ratio of width to height. This was fixed by international agreement at 4:3, largely for the convenience of the cathode-ray tube (CRT) manufacturers. Because of the introduction of better transmission systems and alternative display systems such as plasma and liquid crystal display (LCD), this has now changed. The concept of widescreen television using larger, flat-faced tubes with an aspect ratio of 16:9 produces a new and improved viewing experience approaching that of a cinema presentation. For the older receivers left with the 4:3 aspect ratio, the new format is displayed with very noticeable black bands at the top and bottom of the picture. This so-called letterbox effect can be irritating to the viewer, and a compromise is in operation using an aspect ratio of 14:9, which reduces the height of the banding effect by about 50%. In the true widescreen receiver, a widescreen switching (WSS) flag byte is carried within the Teletext signal. This automatically signals to the receiver to set the amplitudes of the field and line time bases to suit the programme being broadcast.

## Colour television

For colour television signals, three camera tubes or IC light detectors are used for the three colour signals that are needed. Because the three primary colours of light (red, blue and green) add up to white, the signals from the tubes can be changed into a black-and-white (luminance) signal and two colour signals. The luminance signal can be displayed on a monochrome receiver (which ignores the separate colour signals).

> The sensation that we call colour is the effect on the eye of the different frequencies of light. Pure white light is a mixture of all the visible frequencies of light.

The different light frequencies can be separated out from white light by the refracting action of a wedge-shaped piece of glass called a prism (Figure 18.4). A prism produces a spectrum of colours, in the following (ascending) order of light frequency: red, orange, yellow, green, blue, violet (indigo also is sometimes identified as existing between blue and violet). Light frequencies lower than red are called infrared, those higher than violet are called ultraviolet (UV), and both are invisible to the human eye.

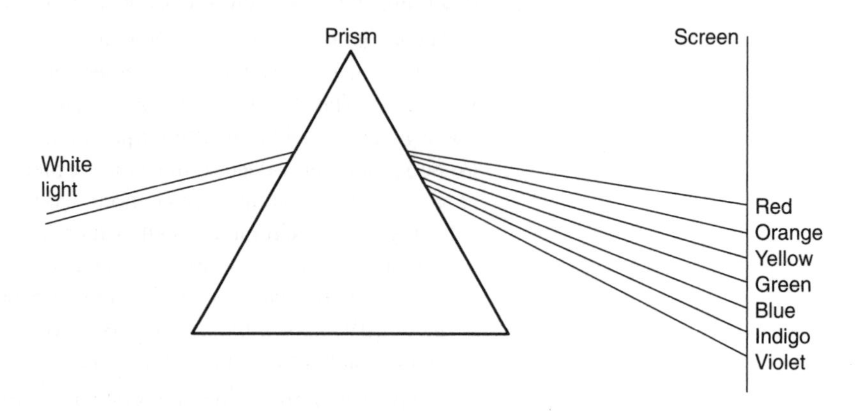


**Figure 18.4** How a prism separates out the component colours of white light

A pure white light can be obtained by mixing together in the correct proportions only three colours of the spectrum, rather than all of the possible colours. These three colours are called **primary colours**. The primary colours used in colour television and some colour photographic processes are red, green and blue (RGB). The mixture of these three in the correct proportions gives a good quality of white light. By appropriate choice of the standard frequencies of the primary colours, a wide range of secondary colours can be obtained by mixing. Yellow and cyan are typical of secondary colours obtained in this way (Figure 18.5).

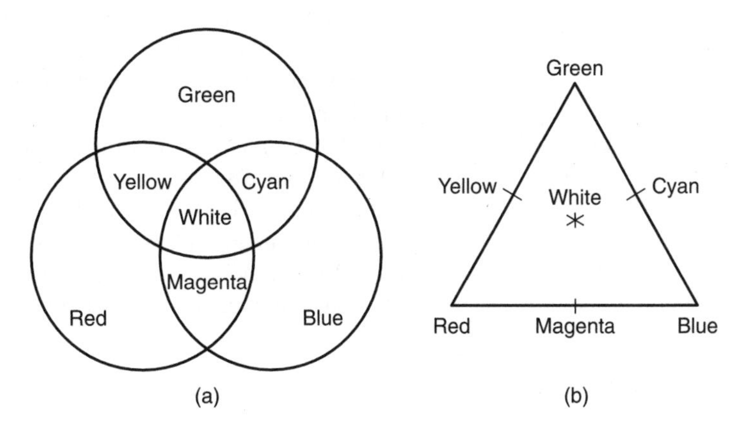


**Figure 18.5** (a) How secondary colours are obtained by mixing, and (b) colour triangle diagram

You can see that yellow can be obtained from an appropriate combination of red and green, cyan from green and blue, magenta from red and blue, and white from a combination of all three primary colours: red, green and blue.

Secondary colours can be obtained either by adding primary colours together (in what are called **additive mixers**) or by subtracting primary colours from white light (in **subtractive mixers**). Additive mixing is the process used in colour television; subtractive mixing is used in most colour photographic systems. The colour triangle diagram of Figure 18.5(b) is often used to show colour addition effects.

The two important quantities required to describe a colour exactly are its **hue** and the degree of its **saturation**. Hue is specified by the proportion of the primary colours that are present in the colour, and describes the colour itself. Saturation is a measure of the amplitude of that colour. Desaturation of a colour with white light produces pastel shades. For example, a blue colour with 50% saturation would be pure blue mixed with the same amplitude of pure white. Saturated colours are never seen naturally, but they can be used in cartoons and in signals derived from computers.

For colour television the hue and the degree of colour saturation of every part of the picture is defined, as well as its brightness (**luminance**) and its exact position in the scene being televised. In the colour television system, the colour information signal is referred to as **chrominance**. The composite video waveform from a television camera contains both the luminance (black-and-white) signal and the chrominance (colour) signals.

The luminance, luma or Y signal component is formed by mixing the outputs, in the correct proportions, from the three colour sensors in the camera. Colour difference signals are then produced to obtain (R-Y) and (B-Y) signals that form the chrominance or **chroma** component of the signal. These two components are then separately amplitude modulated on to two quadrature (90° phase difference) versions of the same carrier frequency and these are added to produce an analogue QAM signal.

The chroma carrier frequency is modulated on to the video signal, forming a subcarrier. The frequency is very precisely chosen so that the modulated chroma signal can be added to the luma component (Figure 18.6). In the PAL system this subcarrier is set at 4.433 618 75 MHz and this allows the luma and chroma components to be separated at the receiver without mutual interference. Because the green (G) signal component is contained within the Y signal, processing in the receiver decoder recovers the third colour difference signal, G-Y. The original R, G and B signal components are then recovered simply by adding the Y signal to each colour difference. Line synchronizing pulses and a 10-cycle burst of chroma subcarrier are then added to produce the composite video waveform as shown. This signal is used to amplitude modulate the final radiating RF carrier.

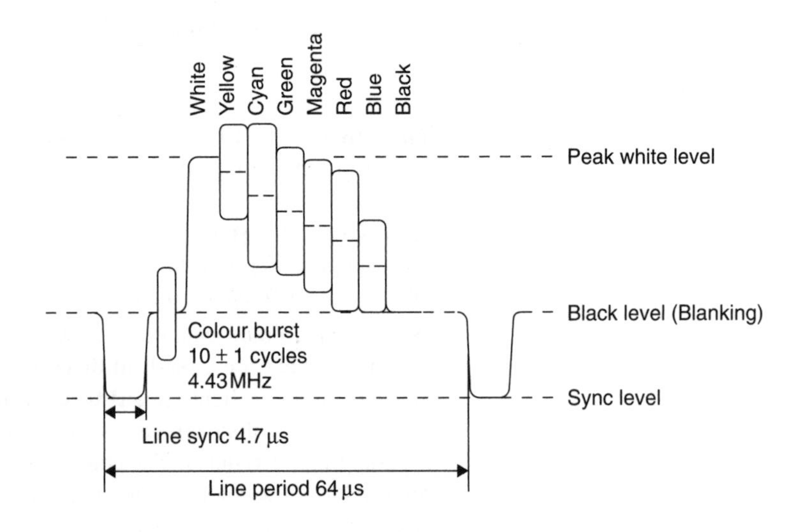


**Figure 18.6** PAL colour bar signal

## Colour CRT

Colour television reception depends on the use of a colour CRT, or an equivalent display device such as a colour LCD or plasma screen. A colour television CRT (Figure 18.7) works on the principle of additive mixing of colours. The three primary colours, red, green and blue, are created at the screen by using three electron guns that are fed with three separate brightness signals. A metal grid, the **shadow mask**, is fixed close to the screen. The pattern of holes in the metal prevents the beam from one gun from striking more than one of the three coloured phosphor dots in each group of three dots on the screen. The beam from one gun lights only the red dots, the beam of the second gun lights only the blue dots and the beam of the third gun lights only the green dots. The sets of dots are so small and close together that when adjacent dots of different colours are lit the eye sees them in the same place and so interprets the colour as a combination of the two, rather than seeing two dots of different colours. The resulting picture as a whole looks continuous, when it is viewed from a reasonable distance from the screen.

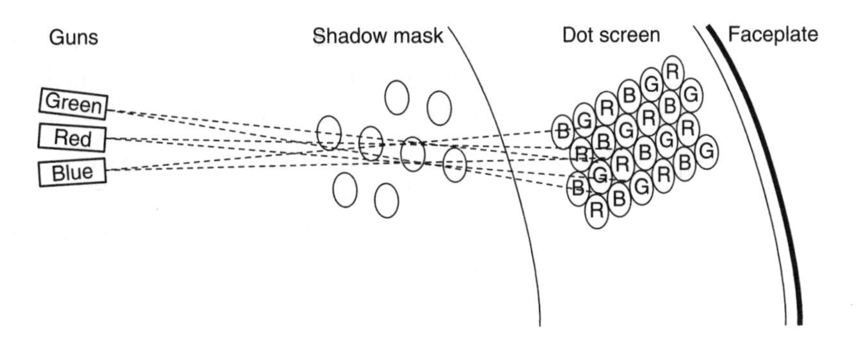


**Figure 18.7** How a shadow mask colour display tube works

Every detail of a colour picture can therefore be reproduced by the beams from the three guns, using voltages at the cathodes of the gun which are in the correct proportions to give the brightness and colour required as the beams are scanned over the face of the tube.

The average beam intensity is varied by altering the voltage to a set of electrodes, the control grids, in the tube. This control is, in practice, a simple potentiometer that alters the d.c. voltage on the grids. The beams are focused by adjusting the voltage on the focusing anodes of the picture tube. Because the three guns are not in the same place but side by side, a set of coils called the convergence coils around the neck of the tube is used to change the deflection during scanning to keep the beams from the colour guns focused on the same part of the screen. The current through these coils produces a magnetic field that deflects the three beams differently to cause them to converge. The scanning is carried out using another set of coils, the **scanning yoke**, which carries signals generated by the timebase circuit of the receiver. The changing current through these coils will deflect the beams at a steady rate, and return the beams rapidly at the end of each scan. There are two sets of deflection coils in the yoke, one for horizontal (line) scanning and the other for vertical (field) scanning. Additional coils or magnets are used for convergence, meaning that they ensure that all three beams meet at a single area on the screen.

Typical CRT faults include the following:

- Purity faults occur when the beams fall on the wrong stripes. A purity fault is obvious when one gun alone is used. On a white screen picture, the single colour will have patches of other colours.
- Convergence faults cause colour fringes to appear. This is particularly noticeable on a black-and-white picture.
- Poor grey scale occurs when the guns are not equally sensitive.

Many faults can be caused in an old CRT when one gun has low electron emission or fails altogether. Receivers provide for switching each gun off so that the effect can be checked.

Convergence and purity adjustments may be limited with the more modern receivers. This is because the scan coil assembly, and focus and convergence components, will have been matched to the tube during manufacture and then fixed firmly into position.

The decoded video signal is usually applied to the cathodes and the level of this signal affects the picture contrast. In the case of the colour tube, the level of the three signals, red, green and blue, need to be accurately matched.

- The overall brightness is dependent on the level of the d.c. setting of the grid voltage.
- The first anode normally operates at a fixed d.c. voltage level.
- The variable focus anode voltage provides a means of obtaining overall image sharpness.
- The final anode voltage is chiefly responsible for the overall image brightness.

Television CRT voltages are compared in Table 18.2.

Warning: While most voltages in a television receiver can be measured with a general-purpose multirange meter, the extra high-tension (EHT) voltage should only be measured using a special high-impedance electrostatic volt meter.

**Table 18.2** Comparison of monochrome and colour television tube voltages

| *CRT element* | *Mono tube* | *Colour tube* |
|---|---|---|
| Heater | 6.3 or 12 V | 6.3 or 12 V |
| Cathode | +80 V | +150 V |
| Grid | +20 V | +40 V |
| First anode | +300 V | +1500 V |
| Focus anode | Up to 350 V | Up to 4 kV |
| Final anode (EHT) | 10–15 kV | 25 kV |

## Replacing a CRT

The procedure for replacing a faulty or damaged tube varies from one tube type to another and the manufacturer's handbook should *always* be consulted.

If you wear a diamond ring or anything that can scratch glass, remove it.

The glassware is most vulnerable at the tube neck and base connector where the glass changes shape most rapidly. The rim band not only provides a point of common electrical earth and a means of mounting the tube, it also provides a restraint against the force of the atmosphere pressing on the faceplate (several tons). If the tube is damaged (even just scratched) it may implode and cause sharp glass fragments to fly at dangerous speeds.

The following general notes are useful.

1. Always wear safety goggles and gloves to avoid injury from flying glass; there is the risk of implosion.
2. Cover the workbench surface with a blanket or similar material. Place the television on a clean bench, with sufficient space available to lay the CRT alongside it once it has been removed.
3. Disconnect the receiver from the power supply, and discharge all capacitors. Any outer coating of carbon on the tube should be treated with respect, as it may have acquired a significant charge which needs to be earthed for safety.
4. Remove carefully the tube base, the high-voltage connectors and the leads to the scanning assembly. Take off the scanning yoke after noting carefully its exact position and orientation so that you can replace it correctly on the new tube.
5. Turn the receiver on to its face and carefully remove the tube clamps. Lift the tube out carefully, holding the ends of the screen and protecting the neck of the tube from any impact. Place the tube on the blanket face down. Watch your back when lifting the tube, and never attempt to lift a tube by its neck.
6. Insert the new tube in the same way as you removed the old one and then complete the reassembly of the items around the new tube.
7. Observe all safety precautions until the receiver is checked out and the old tube is safely packed into its crate for disposal.
8. Finally, test and recalibrate.

**Plasma display panels (PDPs)** are manufactured as large, typically 30–50 inch diagonal, slim replacements for the colour television CRT. Their main features are high brightness, good colour and contrast scale, and large viewing angle. The lifetime tends to be a trade-off between brightness, contrast and power consumption. Displays capable of providing more than 1024 lines at 1920 pixels per line are currently available, ready for possible future broadcasts of high-definition television (HDTV). Digital drive signals to the display panel controller provide 8 bits per RGB pixel to generate about 16.7 million possible different pixel colours.

A **plasma** is the region in an electrical discharge path in which the numbers of positive and negative ions is approximately equal, so that the path is electrically neutral and highly conductive. The plasma is often described as the fourth state of matter (the other three being solid, liquid and gaseous).

The panel itself consists of an array of phosphor-coated cells sandwiched between two glass plates, with a matrix of electrodes to control the lighting of the cells. A computer breaks the video image into cell data and then generates column and row addresses for the correct cell to fire (Figure 18.8). The electrical discharge in each cell creates both visible and UV light. The visible light is typically a dull orange glow like a neon, while the UV causes a much brighter light to be emitted by the phosphor coating of the cell, and this colour dominates the output colour.

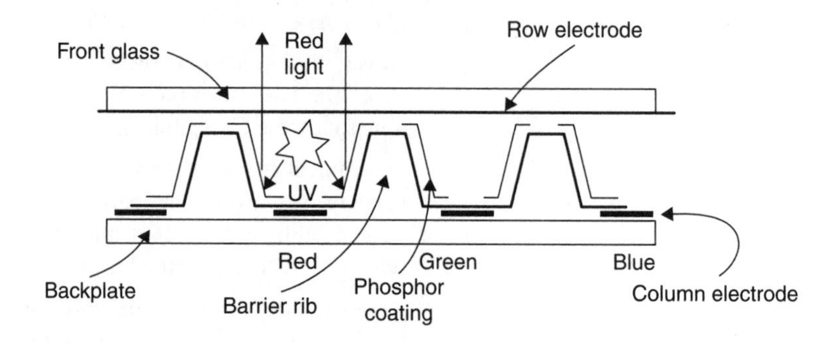


**Figure 18.8** Colour plasma display panel structure

## LCD computer and television displays

Unlike CRT and plasma displays, LCDs require an external light source. This can have advantages for power consumption, particularly if they can use reflected ambient light. One of the reasons that they are so popular in handheld devices and notebook computers is that the display itself does not require high voltages or high currents to operate. Liquid crystal displays for television and computer monitors are driven as a column and row matrix in much the same way as for plasma displays. This requires a microprocessor or fast logic circuit to break the scan lines of the television picture into pixels and generate the column and row drive signals.

Liquid crystal television and computer monitor displays are often described as thin-film transistors (TFTs); they consist of driver circuits made from thin-film silicon transistors fabricated on the glass of the display along with the display pixel electrodes (Figure 18.9). Sometimes described as active matrix displays, the pixel address logic is fabricated on the glass of the LCD substrate, decoding columns and rows at each pixel. Because the LCD does not have persistence like the phosphor of a CRT, a capacitor is included to keep the cell turned on until the next refresh. Until recently, LCD displays were quite slow.

Since the liquid crystal of the LCD would be damaged if it were exposed to a d.c. bias, the display is driven with alternate polarities of drive, either on a pixel-by-pixel basis, or line by line or frame by frame. Pixel by pixel is flicker free, whereas the frame-inversion and line-inversion drives lead to a degree of flicker.

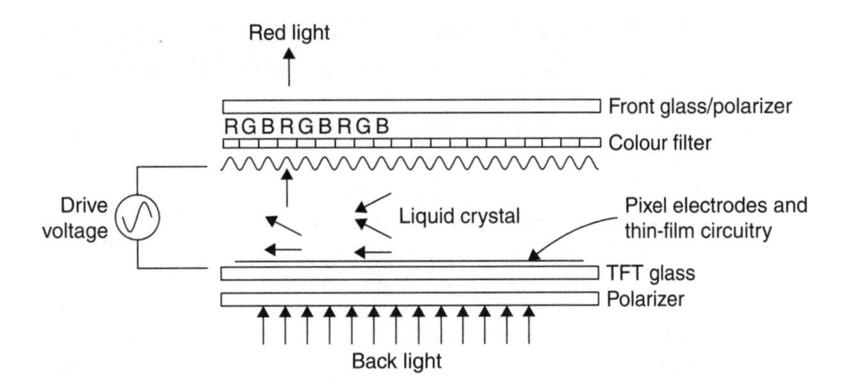


**Figure 18.9** LCD

Colour LCDs are available in three forms. Transmissive displays (TMCs) are used with a white backlight, and the liquid crystal dots will transmit whatever colour is selected from the whole spectrum of white light. These displays are ideal for use in low lighting conditions, and are typically used in laptop computers and LCD television and computer monitors.

Reflective displays (RFCs) are used illuminated by white light, and will reflect whatever colour is selected from each liquid-crystal dot. No back-light is needed, so power requirements are very low, but they can be used only in bright lighting conditions, and the perceived colours will alter if the illumination is not white light.

Transflective displays (TFCs) combine both TMC and RFC principles, and use both a backlight and the ambient light to provide a display that can be read equally easily in the dark or in sunlight. The user can save power by having the backlight off in high ambient light. These displays are used in PDAs, digital cameras and some mobile phones.

The most common backlighting methods use light-emitting diode (LED) arrays or cold-cathode fluorescent (CFL) tubes. Electroluminescent (EL) lamps are also sometimes used. EL backlights are slim and require low power, but have relatively limited life. The LED backlights can provide high brightness, uniform appearance, long life and low cost. The CFL backlights have very uniform brightness and appearance, long life and low power consumption, and are the preferred option for computer monitors and LCD televisions.

## Multiple-choice revision questions

18.1 In the UK analogue television system the mono sound carrier is:
(a) 5.5 MHz above the vision signal
(b) 4.443 MHz below the vision signal
(c) 6 MHz above the vision signal
(d) 6.552 MHz above the vision signal.

18.2 White light can be obtained by adding:
(a) red, green and blue light
(b) red, yellow and green light
(c) red, magenta and blue light
(d) cyan, green and yellow light.

18.3 What is the line period of a UK PAL television signal?
(a) 4.7 μs
(b) 5.5 μs
(c) 64 μs
(d) 20 m.

18.4 In a plasma television set the picture is produced:
(a) by scanning the phosphor coating of the panel with UV light
(b) with a shadow mask
(c) by UV emission from the addressed pixels
(d) by reflecting coloured light.

18.5 What type of backlight do LCD television displays usually use?
(a) LED lamp
(b) cold-cathode fluorescent lamp
(c) electroluminescent lamp
(d) incandescent filament lamp.

# 19 Television receivers

Before the introduction of home computers and other complex electronic systems, the television set was the most complicated piece of electronic equipment in common domestic use. An analogue colour television set consists of an ultra-high frequency (UHF) superhetrodyne radio receiver, AM and FM demodulators to recover the vision signals and sound, a high-voltage power supply to provide the voltages for the display tube, a phase-locked loop (PLL) and crystal oscillator for recovery of the colour information, and a great number of ancillary circuits and presetable adjusters.

A simplified block diagram of a typical phase alternation line (PAL) analogue colour television receiver is shown in Figure 19.1. A wide band antenna, typically of the Yagi type, is pointed at a local transmitter to receive, usually, five analogue channels. This aerial is linked to the set by a 75 Ω coaxial cable.

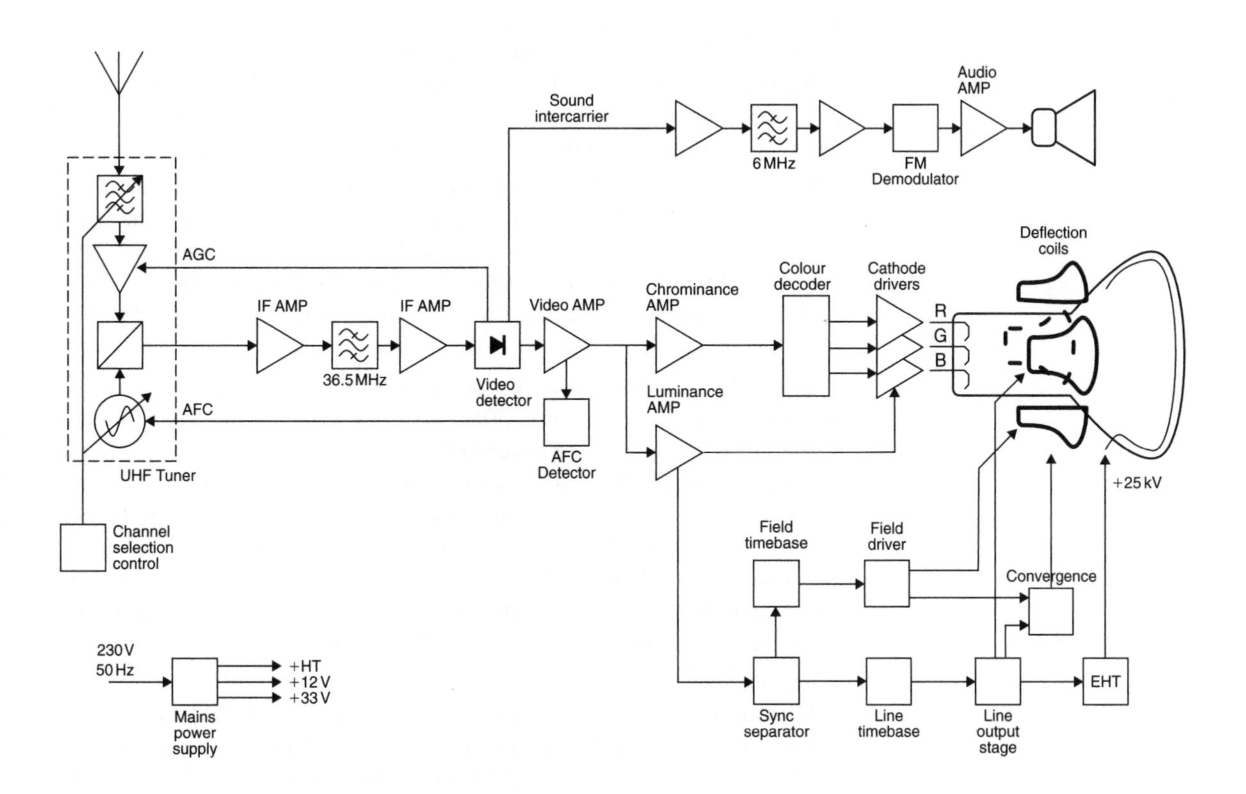


**Figure 19.1** Simplified block diagram for a CRT colour television receiver

## Tuner

The tuner, in the set, selects one of the available channels, and down-converts the UHF signal to an intermediate frequency (IF) of about 39.5 MHz. The IF bandwidth is about 5.5 MHz, sufficient for the video signal and the intercarrier sound signal. In operation, the tuner is typically a single conversion superhet receiver. A tuneable band-pass filter (preselector) and radio frequency (RF) amplifier with automatic gain control (AGC) provide the signal to the mixer, where mixing with a sine wave from the local oscillator produces sum and difference outputs, and the difference output forms the IF signal. The choice of 39.5 MHz for the IF pass-band is determined by the need to provide good image channel rejection and sufficient bandwidth. When the IF is centred on 39.5 MHz the image channel is 79 MHz away from the required one, so the front end filter must attenuate these signals significantly. The local oscillator and RF preselector filter are tuned using variable capacitance (varactor) diodes, the alteration of capacitance being achieved by altering the reverse d.c. bias applied to the diodes.

## Intermediate frequency and sound

The IF signal is then amplified in the IF amplifier. A band-pass filter with very steep skirts, the slope between the pass-band and the stop-band, often using a surface acoustic wave (SAW) filter, removes any out-of-channel signals. The tuner **automatic frequency control (AFC)** circuit obtains its control signal from the demodulated IF signal. The effect of this d.c. signal is to correct the frequency of the oscillator to compensate for frequency drift. An advantage of varactor tuning is that it allows AFC (voltage) to be applied to the oscillator tuning input to maintain station turning. Older mechanically tuned television tuners were usually very heavily constructed to make them as thermally and mechanically stable as possible, often using a die-cast chassis.

The signal from the IF amplifier is supplied to the vision demodulator. The luminance signal, with synchronization (sync.) pulses, which forms the shape and grey tones of the picture, is recovered by a simple amplitude demodulator, usually called the video detector. **Automatic gain control** is used to keep the average amplitude of the received signal as constant as possible. The AGC voltage is derived from the detected video signal.

The sound signal is frequency modulated on to a separate carrier separated by 6 MHz from the vision carrier. Because of the frequency modulation of the sound signal, it cannot be demodulated by the amplitude demodulator; but the presence of both the 39.5 MHz vision IF signal and 33.5 MHz sound IF signal in the video detector circuit causes mixing to take place, which produces a frequency-modulated difference frequency at 6 MHz. This signal is called the **intercarrier sound** signal. It is amplified by the intercarrier amplifier, and then frequency-demodulated to produce the audio frequency (AF) sound signal, which is amplified and supplied to the loudspeaker. The volume control is incorporated into the voltage amplifier part of this audio section.

Most analogue receivers currently for sale incorporate NICAM (near-instantaneous companded audio multiplex). The normal monophonic sound carrier for UK PAL television is set 6 MHz above the vision carrier. The NICAM digital stereo signal, which is transmitted at a bit rate of 728 kbits/s using a form of phase modulation, is located at 6.552 MHz offset from the vision carrier (6.552 MHz is the ninth harmonic of the bit rate). The NICAM signal can carry either a stereo pair or alternative monophonic language channels.

## Digital and satellite receivers

A digital receiver will include the same set of blocks for the UHF tuner part of the receiver, but typically the extraction of digital signals is carried out by one or more large and complex integrated circuits (ICs). In liquid crystal display (LCD) and plasma televisions the decoding of the video signal and drive to the display are handled together by the same chip set. If the receiver is a set-top box (STB), or part of a cathode-ray tube (CRT)-based television set, then the circuitry to generate horizontal and vertical sync. pulses and RGB colour video signals is required.

Reception of direct broadcast satellite programmes relies on very weak microwave signals typically picked up by a dish antenna. The dish focuses these signals on to a microwave **low noise block (LNB)** converter mounted at the focal point of the dish. The LNB is effectively a single conversion superhet receiver. It provides preselection, amplification and down-conversion of the Ku band signal to frequencies in the range 950–2050 MHz. These signals are carried down the coaxial cable from the LNB to be processed through an STB or integrated receiver.

The wide band of frequencies, 1.1 GHz, which is about five times the bandwidth available for terrestrial broadcasting, is effectively doubled by the very effective separation of horizontal and vertical polarized signals at the microwave frequencies used, which accounts for the much greater number of channels available for the Direct Broadcast Satellite (DBS) service.

## Vision system

The output of the vision demodulator, the composite video signal, consists of the luminance signal whose waveform depends on the picture information, and horizontal and vertical sync. pulses, which have a repetition rate of 15.625 kHz, together with a phase-modulated colour subcarrier. The luminance signal and sync. information are sufficient to produce a black-and-white picture. The system was designed this way to allow colour to be introduced without making obsolete all the existing monochrome television sets. Information about colour is phase modulated on to the 4.43 MHz subcarrier. In very weak signal conditions the receiver is designed to detect that the colour information is unreliable and switch to black and white.

## Synchronization

The luminance amplifier increases the amplitude of these waveforms to drive the following stages. Part of the amplified composite signal is supplied to the sync. separator, in which the combined synchronizing pulses (Figure 19.2) are separated from the luminance signal and also from one another. The field sync. pulse at a rate of 50 per second is used to synchronize the field timebase. This drives the field scan coils to deflect the CRT spot vertically from the top to the bottom of the screen face.

The field timebase circuit drives a low-frequency sawtooth waveform (50 Hz) through the low-impedance scan coils which generate a magnetic filed, deflecting the path of the electron beam from the top to the bottom of

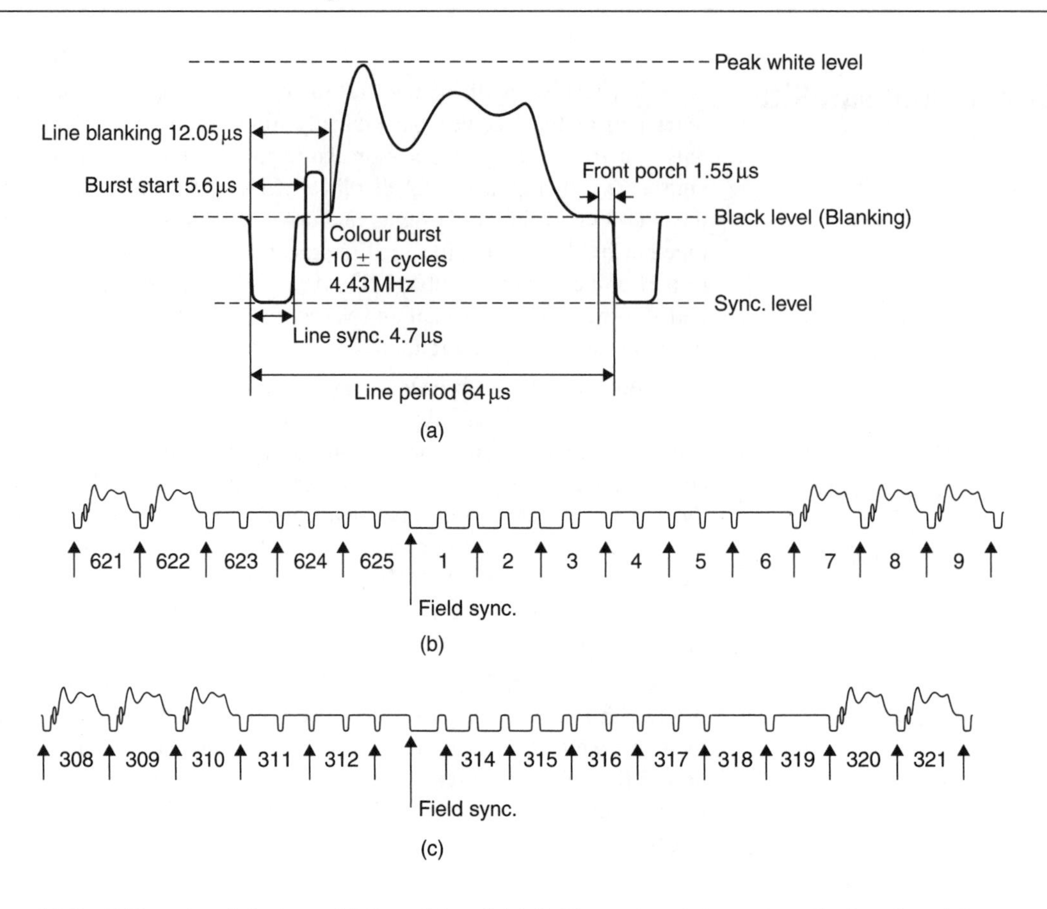


**Figure 19.2** Video signal timings: (a) line, (b) and (c) field synchronizing pulses for interlaced frame

the picture tube. Because of the low frequency these coils almost behave in a resistive manner. Since the peak-to-peak current is in the order of 3 A at a maximum of about 25 V, the circuit behaves very much like an audio amplifier system, which can typically be implemented as a single IC. The vertical field sync. is typically detected by integrating the mixed sync. signal (Figure 19.3). In digital sets this function is counter based rather than using an RC integrator. A protection circuit is provided that will shut down the tube voltages in the event of timebase failure, to prevent burning of a line across the tube phosphors. Linearity of the scan is assured by using a negative feedback loop that encloses the field scan coils. The circuit, usually part of an IC, typically includes two preset controls, one to set the field oscillator frequency and the other to adjust the amplitude or height of the scanning waveform.

The line sync. pulse is used to synchronize the line timebase. It runs at 15.625 kHz, and drives the line scan coils which deflect the spots across the screen to form lines.

The line scan circuit is the more complex of the two timebases (Figure 19.4). Largely because of the higher frequency (15.625 kHz), the line scan

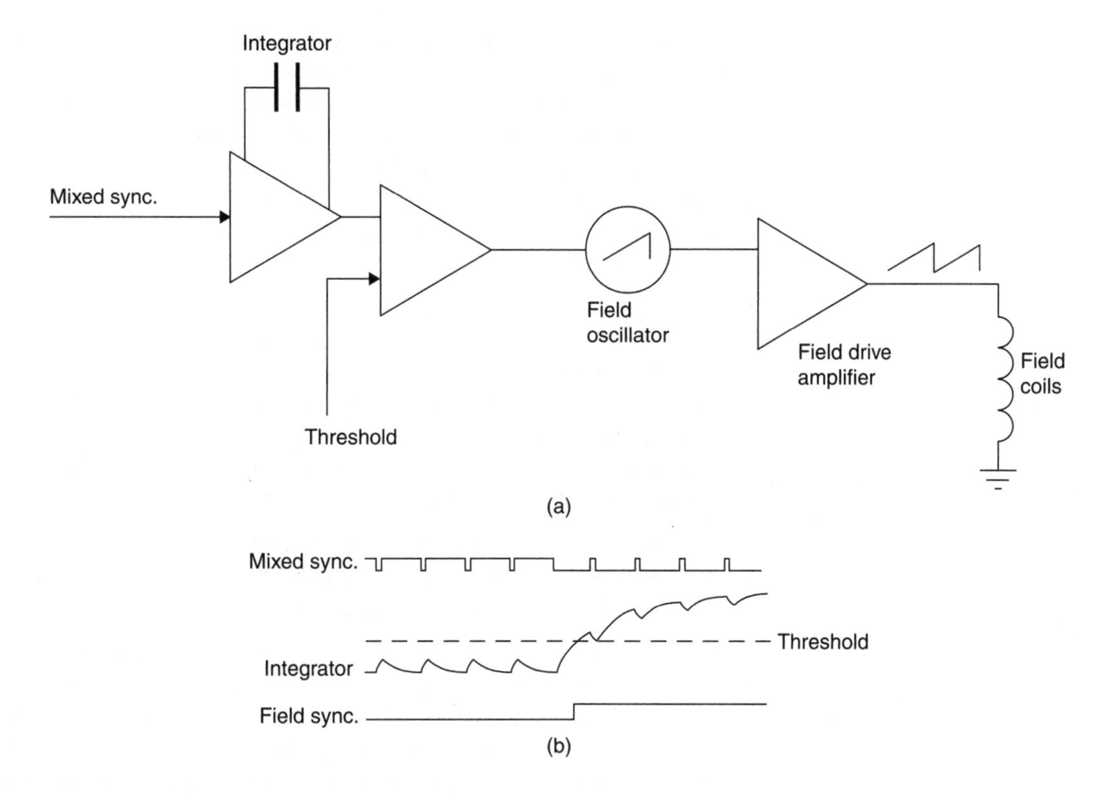


**Figure 19.3** (a) Block diagram of field sync. and vertical drive circuit, and (b) associated waveforms

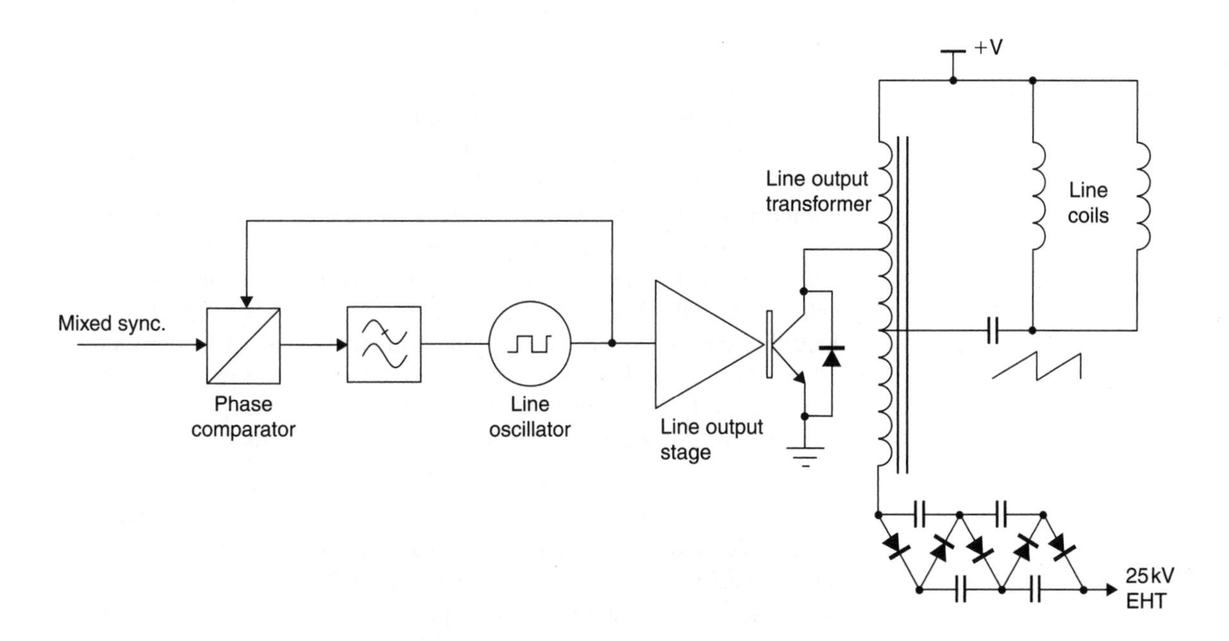


**Figure 19.4** Block diagram of line timebase circuit

coil impedance is almost entirely inductive and the line scan peak power can be quite high, with peak currents of about 6 A at more than 1000 V requiring a transformer drive circuit; these features combine to produce some very high back electromotive forces. This energy is recovered and used to drive the extra high-tension (EHT) circuits of the television receiver. The repetition frequency is locked to the line sync. pulses using a **PLL** circuit with a long time constant. Thus, the circuit will remain in lock even if the sync. pulses are lost for a period of time.

The line output transformer produces a pulse of high voltage, which is stepped up further by a diode capacitor charge pump voltage multiplier circuit, to produce the EHT voltage which is needed to accelerate the electron beam towards the screen of the tube. An EHT voltage of around 24 kV is needed for a colour tube. This is much higher than the EHT of a monochrome receiver, and much of the beam energy is wasted as heat because of the electrons that strike the shadow mask instead of the screen.

The important features of this circuit are the high voltage levels and large magnetic fields that can arise, requiring effective screening and insulation. The high power levels also mean that there is a need for good ventilation.

Typically, the line timebase circuit has preset controls for frequency, amplitude or height and linearity. Frequency control is usually affected through an iron dust cored inductor associated with the oscillator stage of the PLL, and amplitude usually depends on the level of signal provided by the drive amplifier. The scan linearity can usually be preset using a saturable reactor. This consists of a coil and permanent magnet assembly where the relative position of the magnet affects the control action. Because the line output stage runs at a high frequency and with a switching action, this stage is commonly linked with a **switched mode power supply** that powers the rest of the receiver.

## Colour system

The 4.43 MHz subcarrier signals extracted from the composite video signal are amplified by the chrominance amplifier. They are then demodulated by mixing them with a crystal-controlled signal of exactly the subcarrier frequency and in the correct phase. This demodulation (called **synchronous demodulation**) produces two signals which are called **colour-difference** signals. If there is no subcarrier, a **colour killer** circuit biases off all the colour signals so that no colour fringes (caused by noise) appear on monochrome signals. The voltage-controlled crystal oscillator signal is phase locked to the colour burst signal at the beginning of each scan line. A crystal oscillator is used because the $Q$ of the crystal resonator is so high that once locked it will remain in synchronism with the carrier phase for at least the whole line; this ensures that the colour signal is correctly decoded.

The colour-difference signals are then mixed with the luminance signal in part of the colour decoder stage, in which signals representing two of the three basic colours – red, green and blue – are subtracted from the luminance signal to produce the third colour signal. These three separate colour signals are now applied to the appropriate cathodes of the colour tube. In this way, every part of the picture has the correct luminance (brightness) and colour balance.

The highly simplified account of colour television reception given above applies to all three of the colour television systems used in the world: NTSC in the USA, SECAM in France and Russia, and PAL in Britain, Germany and the rest of Western Europe. The differences between the three systems concern only the details of chrominance amplifier and the colour decoder blocks.

## Integrated digital receivers

Integrated digital receivers now available for receiving free-to-air and pay-per-view terrestrial digital television services can use the same aerial as analogue television where signal strengths are sufficient. The UHF digital television (DTV) service provides up to six channels multiplexed on each main UHF carrier frequency. Set-top boxes can be used to convert earlier analogue television and video-recorder receivers. There is a growing number of integrated DTV, DVD and hard-drive personal video recorders (PVR) for recording digital signals. These offer the possibility of pausing live television, for example, because they can record and play simultaneously, allowing the programme to be 'buffered' in a file on the hard disk.

## Servicing problems

Until recently, the average television installation consisted of little more than a television and video-cassette recorder (VCR), with a occasional hi-fi amplifier for NICAM sound, but in the last few years a whole range of new features has been offered. These may require the system to include digital versatile disc (DVD) players and recorders, home cinema surround sound system, games console with an asymmetric digital subscriber line (ADSL) modem and possibly a WiFi link to a personal computer. Setting up such systems can be a time-consuming job.

When something breaks, the problems for service departments can be very complex indeed, matching equipment and in many cases operating software from multiple vendors.

In the past, when television receivers were constructed almost entirely with discrete components, faults often occurred repetitively, giving rise to the term stock faults. Many service departments established a good reputation for fast and efficient repair work based on such work. Today, much of the complexity of the television receiver is hidden within a few large ICs which now provide a much higher degree of reliability. Stock faults still occur, but they are less frequent, so that people may fail to recognize them as quickly as in the past.

*Television*, a journal published by Reed Business Publications, lists sources of service data for consumer electronic devices. Alternatively, a search of the Internet will usually produce useful suppliers of information. Free file-sharing websites such as http://www.eserviceinfo.com list many service manuals and schematics that may be downloaded.

Many robust, portable battery- and mains-operated items of test gear are now available for use by the service technician. These range from television

**Figure 19.5** Components marked with this symbol in the service schematic are safety critical and should be replaced with identical types, as directed by the service manual

pattern generators, through digital oscilloscopes with the facilities for the display of both time- and frequency-based signals, to signal strength meters suitable for selecting the required transmissions and accurately aligning aerial systems.

Most consumer devices are mains powered, and when work is carried out on such equipment components the demanding safety requirements of mains connection must be met.

Particular care must be taken over **critical components**, usually marked out on a service sheet (Figure 19.5). These must be replaced with factory-approved components because a replacement may otherwise present a safety hazard. The repairer would be held liable for damage caused as the result (of a fire, for example) caused because a critical component had been replaced by an unapproved substitute.

## Video recording

The recording of analogue video signals on magnetic tape follows the same basic principles as analogue audio recording. However, the very wide bandwidth of the video signal requires that the tape pass the record head at a much greater speed than is possible with a conventional analogue tape machine. To achieve the required speed of tape past the head, two or more heads are mounted on a spinning drum which is mounted to scan the head across the tape at a shallow angle. This allows the tape to move relatively slowly past the spinning drum. The helical scan system has been under development since the late 1950s, with the VHS system introduced in the early 1980s and largely unchanged today. Additional features include multiple heads to provide better still-frame playback and long play/record performance, and there have been massive reductions in unit cost owing to improved manufacturing and electronic integration. The general arrangement of helical scan **VCR** tape path is shown in Figure 19.6.

When the cassette is inserted in the machine, the first operation is threading; this is feeding the tape through the tape path around the drum so that it can be played or recorded. To achieve this, the front cover of the cassette is lifted as the cassette drops into position. The unthreaded position of the loading arms means that the slant poles, guide rollers, tension pole and capstan spindle are behind the tape; these are all mounted on movable loading arms. The threading operation uses the loading arms to move the slant poles, guide rollers, etc., to their operating position (Figure 19.5), carrying the tape around the circumference of the head drum and past the audio, control and erase heads. The slant poles now perform their main function, which is to guide the tape on to the drum at the correct angle for it to be scanned by the passing of the heads.

The analogue audio signal is recorded on a narrow strip along one edge of the tape, while field sync. pulses are recorded on the opposite edge. The heads and tape travel in the same direction to minimize frictional losses and to extend the mechanical lifetime.

The luminance signal is frequency modulated on to a carrier of about 4.3 MHz, which swings through the range of about 3.8–4.8 MHz between

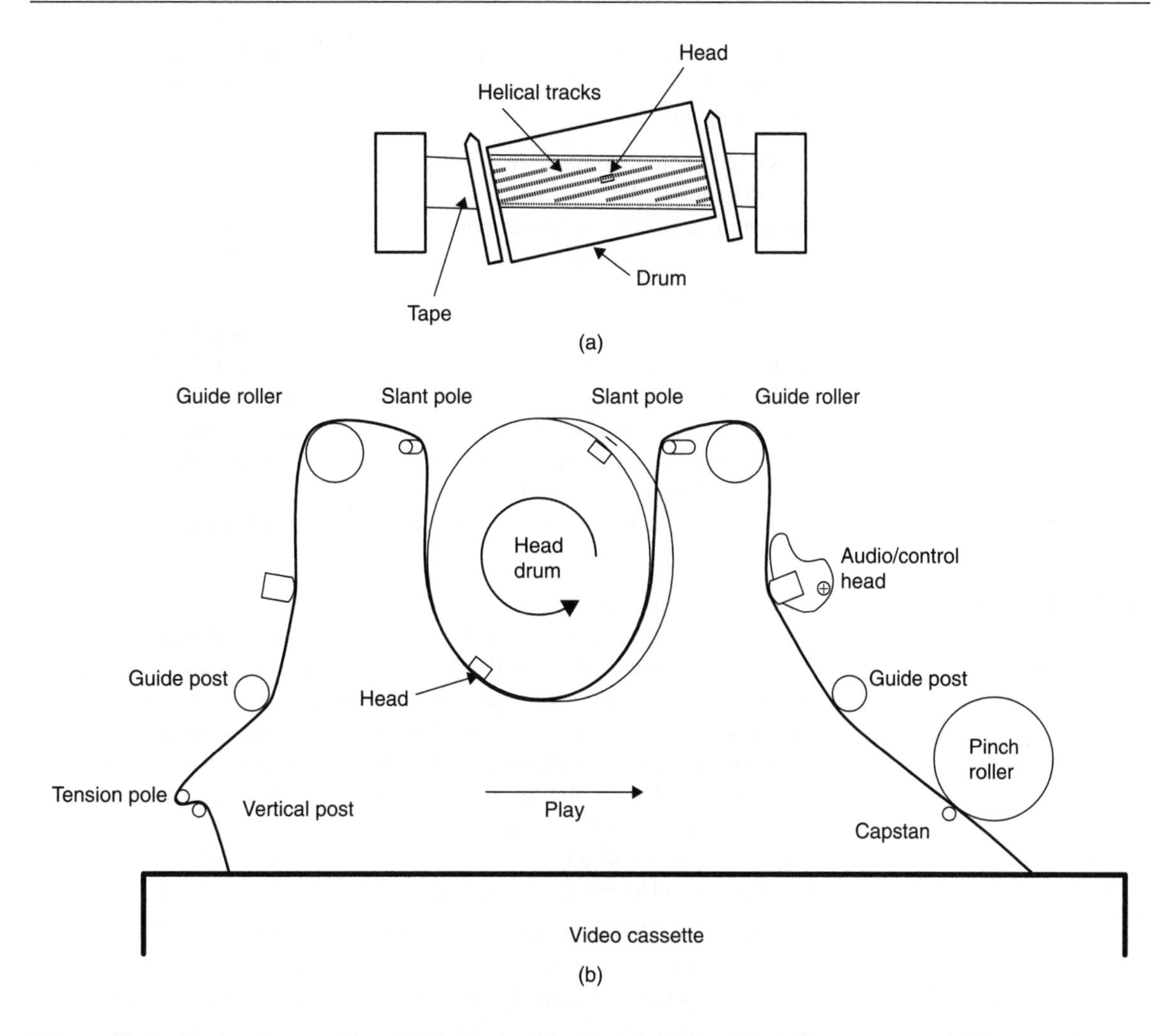


**Figure 19.6** Tape transport of typical helical scan video recorder: (a) tracks on tape, and (b) tape path

sync. pulse tip and peak white. Because this FM signal is of constant amplitude, it can be recorded on tape without needing a higher frequency bias signal as used in audio recording, since it is not affected by the amplitude distortion introduced by the magnetic hysteresis of the tape. The FM luminance signal is used as the recording bias signal for the chrominance component, which is down-converted to a **colour under frequency** of about 625 kHz and added to the luminance signal for recording. The replay follows the inverse process; the tape speed is locked to the required vertical sync. rate, the luminance and chroma signals are recovered along with the sound signal. These are typically available via the SCART connector and on some new machines and most older ones remodulated onto a UHF carrier for input to the television receiver aerial socket.

To ensure synchronism, the tape capstan drive motor is locked to the vertical sync. pulses, while the head drum rotation is locked to the horizontal sync. pulses. Despite all the compromises that have to be made to the signal bandwidth, the output from a VCR can be almost as satisfactory to the eye as the original.

NICAM sound-compatible machines convert the NICAM subcarrier and its sidebands so that they can be recorded along with the video signal by the rotating heads. The standard mono track is also laid down, and on replay the output will be taken from the mono track if the NICAM subcarrier is not detected.

The very low cost of consumer electronics, particularly VCRs and DVD players, has made them effectively cheaper to replace than repair; however, at the time of writing, VHS video recorders are rapidly being replaced by digital recorders, either DVD or hard disk types, and prerecorded VHS tapes, for example video rental, are becoming rare. This means that the requirement for VHS video recorder servicing may increase, owing to the cost of replacement machines rising as fewer manufacturers offer them.

## Installation

Most television receivers are sold without installation, the market being dominated by the supermarkets and out-of-town electrical chains. This has been made possible both by the low cost and because they can be set up automatically. Most models offer auto setup, in which the receiver scans the available signals and uses Teletext ID data to identify them and sort them into order so that, for example, pressing the button 1 on the remote will select BBC 1. This applies to analogue terrestrial signals. In the future (after 2012), when the analogue service has been turned off, all receivers will be digital and they will all be able to set up automatically. This will rely on a correctly installed aerial to provide the required UHF signal strength.

Where installation is needed, the service engineer should be able to advise on the best site for the receiver with regard to the available television outlet point and room lighting. If the receiver is large and heavy, help should be available for lifting it from the van to the room. The manufacturer's recommended setup procedure should be followed, and the remote control functions explained to the user.

## Multiple-choice revision questions

19.1 What method is used to modulate the mono sound subcarrier in the UK television system?
(a) AM
(b) FM
(c) COFDM
(d) PM.

19.2 If the received signal strength is low what vision effect may be noticed on a correctly adjusted set?
(a) no picture
(b) black-and-white picture
(c) colour fringes on lines and edges
(d) letterbox picture.

19.3 Some components are marked on the service schematic and PCB with an exclamation mark in a triangle. What does this indicate?
(a) do not touch
(b) component runs hot
(c) replace when servicing
(d) safety critical component.

19.4 What is the main function of the slant poles in a VCR tape transport?
(a) carry the tape during threading
(b) slow the tape down
(c) guide the tape on to the drum
(d) keep the guide rollers away from the head drum.

19.5 In the VHS system the luminance signal is used to provide the bias for recording the colour signal. This works because:
(a) the luminance signal is constant amplitude
(b) the colour under signal is FM modulated
(c) the speed of the tape is controlled
(d) the colour signal is at 4.443 MHz.

19.6 What is the frequency of the field sync. in the UK television system?
(a) 100 Hz
(b) 60 Hz
(c) 50 Hz
(d) 25 Hz.

19.7 How many frames are transmitted every second in a UK television signal?
(a) 25
(b) 50
(c) 75
(d) 100.

# Unit 7
## PC technology

### Outcomes

1. Demonstrate an understanding of basic PC systems and apply this knowledge in a practical situation
2. Demonstrate an understanding of basic input/output devices and apply this knowledge in a practical situation
3. Demonstrate an understanding of data storage modules and apply this knowledge in a practical situation
4. Demonstrate an understanding of current printers and apply this knowledge in a practical situation.

# 20 The personal computer

## The computer

A computer is a logic system that is programmable, meaning that the actions it carries out can be altered by feeding in program instructions in the form of binary codes. Domestic computers are classed as **microcomputers** and the heart of such a computer is the **microprocessor**, which is a very complex programmable logic chip.

The type of machine that we now describe as a **PC** means one that is modelled on the IBM PC (Personal Computer) type of machine that first appeared in 1981. The reason that this type of machine has become dominant is the simple one of continuity: programs that will work on the original IBM PC machine will work on later versions and some (but certainly not all) will still work even on today's much more developed machines. By maintaining compatibility, the designers have ensured that when you change computer, keeping to a PC type of machine, you do not immediately need to change software (programs). Since the value of your software is much greater than the value of the hardware (the computer itself), this has ensured that the PC type of machine has become dominant in business and other serious applications. Other types of computer are not compatible with the PC or with each other, and have less choice of software and more expensive components.

In short, a real PC machine is currently identified by the following points:

- It uses a microprocessor which is compatible with the Intel or AMD designs. Current examples are the Intel Pentium-4 or Celeron, and the AMD Athlon and Semperon.
- It uses a program called MS-DOS as a master controlling system (an **operating system**), to enable it to load and run all other programs. Other operating systems can be used, but are less common. The MS-DOS action is made easy to use by a **front-end** program called *Windows*. The effect of a front-end program is to make commands simpler, using the now-familiar mouse and icon system rather than by typing in word commands.
- It is modular, meaning that it can be **expanded** (its facilities can be increased), by plugging in additional circuit boards into slots provided on the main motherboard.
- It maintains compatibility with the previous models of PC.

The IBM PC was announced in November 1981, about 16 months after work had started on the design of the machine. The Intel 8088 microprocessor was used along with a set of support chips from Intel, and even by the standards of 1981 the specification of the machine was not particularly impressive. IBM did not initially see the PC as a business machine; even the name Personal Computer indicated that the intended market was the home user, and the price ensured that only the home user in the USA could afford the machine.

These first PCs were not of a particularly advanced design even for 1981; they had a small memory size and no magnetic disks were used; storage depended on connecting a cassette recorder. They were, however, build on a **modular** design, so making it easy to replace units and also for any other manufacturer to produce the machines using the same modules. By contrast, the Apple machines, the only significant rival, have remained one-manufacturer products. In 1984 the IBM machine was redesigned and relaunched as the PC/AT (Figure 20.1), with much more memory, one or two magnetic disk drives and a hard disk option, and optional graphics display boards that could make use of either a colour monitor or a high-resolution monochrome monitor.

**Figure 20.1** The original form of the IBM PC/AT computer

This was the form of the PC machine that was to become the standard for business use from the 1980s onward, and which was so extensively copied that the word **clone** entered the computer user's vocabulary. From the time of the launch of the XT, disks made by other computers were dubbed as either compatible or non-compatible. The subsequent history of desktop computing has been of the gradual submergence of the non-compatible except where specialized markets could make an incompatible machine viable. The PC may be manufactured by anyone (including the home user), but incompatible machines are each the product of a single supplier.

- The format of the AT machine is now known as ISA, meaning industry standard architecture. This is by now obsolete, and modern fast machines use a more developed form known as EISA, E meaning *extended*.

- Computers are generally replaced at frequent intervals by more modern designs, so that you are less likely to be required to service old models. Servicing an old model could easily cost more than replacing it.

A modern desktop microcomputer system consists of three main units: the system unit or central processing unit (CPU), which can be a desktop box or tower casing, the monitor and the keyboard/mouse. Other units called **peripherals** are normally connected. The peripherals are units that need additional programs, called **drivers**, to operate, and they extend the use of the computer. Typical input peripherals are the mouse and scanner; typical output peripherals are the printer, loudspeakers and external drives. There are also peripherals such as the **modem** which provide for both input and output of data through telephone lines. The type of casing currently in use is distinguished by the letters ATX, a redesign of the original AT type. In addition, the trend has been towards smaller desktop machines (small form factor, or **SFF**) such as the Shuttle. Laptop, notepad and other portable machines should be referred to specialists for servicing. Machines that use the older types of casing are not likely to be worth servicing.

Peripherals are normally connected through **port** connectors on the main system unit. A port is a complex circuit that interfaces the high data speeds of the computer to the lower speeds of other units, and also changes data voltage levels if required. In the past, these port connectors have been specialized, one type used for the printer, one for the modem, and separate connectors for keyboard and mouse. More recently, a form of connection, the universal serial bus (**USB**) has been devised that allows a wide variety of peripherals to be connected to any of the sockets on the system unit. A particular feature of USB is that peripherals can be **hot-plugged**, meaning that they can be plugged or unplugged while the computer is running. Once again, a computer that does not use USB is likely to be too old to make servicing viable.

The computer is totally useless without **software**, and in the early days this meant a program that would service the elementary needs of input (keyboard) and output (monitor and printer) and disk drives. Such a resident program became known as **DOS**, meaning disk operating system. Early DOS programs needed some skill to use, because each instruction had to be typed in the correct format, but such programs have been refined into a visual icon (symbol) and mouse type of system such as Windows.

## PC block diagram

Figure 20.2 shows a simplified block diagram of the PC type of digital computer. The system unit (or CPU) is the main processing unit in a desktop or tower case, and is connected to the monitor and the keyboard. The keyboard is the main input unit, and the monitor is the main output unit. The programs are read from disk storage (the **hard drive**, see Chapter 23, that is contained in the main unit) into the memory of the system unit so as to control the processing of inputs into outputs. As a simple example, the action of pressing the ‘A’ key on the keyboard will generate binary codes that the system unit converts into signals that can produce an image of the letter A on the monitor screen, and a code for the letter A can also be stored in the memory (as part of a **file**).

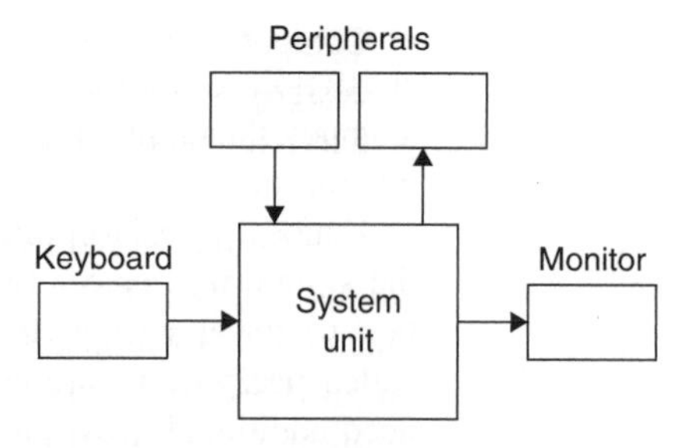


**Figure 20.2** Block diagram for microcomputer system

Computer memory is cleared each time the machine is switched off, so that data that is needed for later use must be stored on some permanent form of memory.

On very early microcomputers this was a tape cassette, but later machines used removable disks (floppy disks) and, later, built-in hard drives. The name 'hard drive' derives from the hard materials (such as glass or metal) used for the disks (platters) that are coated with a magnetic material.

The data unit is the **byte**, a set of 8 bits. This is also the unit that is sufficient to code one letter of the alphabet. The larger units listed below are also used; note that the ratio of sizes is 1024 rather than 1000, because 1024 is $2^{10}$. Note that B is used as the abbreviation for byte, keeping 'b' for bit.

1 kilobyte (**KB**) = 1024 bytes
1 megabyte (**MB**) = 1024 kilobytes = 1 048 576 bytes
1 gigabyte (**GB**) = 1024 MB = 1 048 576 KB = 1 073 741 824 bytes.

As a guide, typical memory size is around 512 MB, and typical hard drive size is about 120 GB.

The terms hardware, software and firmware are used frequently in computing. **Hardware** means the physical devices such as disk-drives, printers, computers and even the integrated circuits from which a computer is constructed. **Software** consists of the physically intangible items like the programs without which the system cannot operate. If the software programs are stored in a hardware (non-volatile) device such as a read-only memory (ROM) chip or a compact disc (CD)-ROM, this composite item is referred to as **firmware**. Both software and firmware can be updated without opening the computer, but when the internal hardware needs replacement or updating the cover must be removed and modules replaced.

## Software

A computer system is totally useless without the programs that we collectively call **software**. Software consists of a collection of bytes that form instructions and data for the microprocessor in the system unit. To run a program, the bytes of software must be placed into the memory of the computer, and the location of the start of that memory block notified to the microprocessor.

One form of system software is described as **BIOS**, meaning basic input/output system. This is a comparatively short set of programs that allow the microprocessor unit to read from the keyboard and the disk drive(s), and to provide outputs to the monitor. This software is permanently held in the form of a chip, the BIOS chip, so that it is always available when the computer is switched on. On modern machines this chip can be rewritten so that the BIOS can be updated.

The other important piece of system software is the operating system (OS). The older PC machines used **MS-DOS**, the Microsoft disk operating system, which is compact enough to need very little of the memory, but which can carry out only one action at a time. In addition, using DOS requires the user to type in a command word, followed by pressing the Enter/Return key, for each command, often with supplementary commands following. Later machines have used the **Windows** operating system. This makes use of MS-DOS, but in a way that is easier to learn and use. The commands are selected by using the mouse to move a pointer on the screen, and clicking a button on the mouse when the pointer is over a command that is usually one of a set in a menu. Clicking on the name of a program will run that program, and clicking on the name of a document will run the program that created the document (if available) with the document loaded for editing.

DOS is still available on modern PC machines to allow the use of actions that cannot be handled by Windows. An alternative is the Linux operating system, which uses typed commands, but which can be controlled by front-end programs resembling Windows.

Each program can create **documents**, items of data. Typical documents are text (from a word processor or editor), worksheets (from a spreadsheet program), drawings (from a graphics program), pictures (still or video) from a digital camera, or digital sound. Each document is stored in binary form as a file on the hard drive.

## Hardware

The **motherboard** is the main printed circuit board of the PC, usually consisting of several layers of circuitry. The motherboard (Figure 20.3) contains a socket for the microprocessor, along with sockets for memory, power supplies, input/output connectors and expansion cards.

As PC technology changes rapidly, any photograph of a motherboard will be out of date in a year or so, but the general layout is fairly constant. In the illustration, the older type of microprocessor socket is shown, but the more modern types differ only in the number of pinholes. The microprocessor socket is always of a **ZIF** variety, meaning zero insertion force. A lever is used to release clamps around the pinholes so that a microprocessor can be

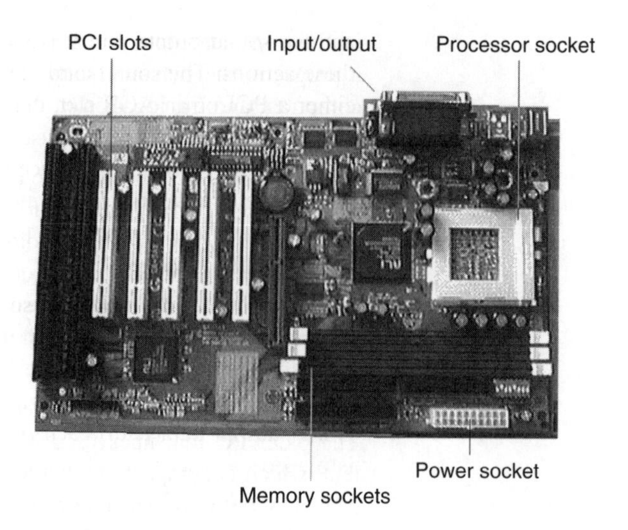


**Figure 20.3** A typical motherboard from a machine of a few years ago with some important parts labelled

dropped into place. The lever is then moved to its lock position so that the chip is securely held. Modern sockets use 423–930 pins, depending on the processor type, but for some time it was more fashionable to use a slot-fitting, with the microprocessor soldered to a small board along with memory chips (the cache, see later) that fitted into a slot. The use of ZIF sockets makes it easy to perform an upgrade by changing to a more powerful compatible processor. Note that Intel and AMD processors are **not** compatible with each other.

A battery, once a nickel–hydride type but now almost certain to be a rechargeable lithium cell, is clamped to the motherboard. This provides power backup for a small portion of memory, called **CMOS-RAM**, that holds information on the setup of the computer (such as the hard drive values). Without the battery backup, the information would be lost when the computer was switched off. If a user complains that the machine has not started up correctly, a faulty battery is one possible hardware cause. If this memory is cleared by disconnecting the battery, the computer cannot be used until the values are either typed in or found by starting an automatic scan of equipment.

The **expansion slots** are shown as PCI (PC interface), replacing the older ISA. Modern motherboards also include a slot that can cope with bus signals at higher speeds (typically 66 MHz or more). This AGP (advanced graphics processing) slot is used for the video graphics board that interfaces the computer to the monitor. The memory slots illustrated are for the DIMM type of memory board.

At one time, connections for such items as input/output and interfaces for the floppy disk and the hard disk were made using plug-in cards, but nowadays these features, and the CD interface, are included as part of the motherboard. Some motherboards also include the interfaces for the monitor and for sound output, but it is still more common to use separate cards for these actions. The sound card fits into a PCI slot, and the graphics card into either a PCI or an AGP slot, depending on the card design. At the time of writing, most graphics cards used the AGP fitting, although a new standard, PCI Express, is being introduced. The serial and parallel interfaces, along with sockets for keyboard and mouse, are all built into the motherboard. The trend at the time of writing is to use a single type of connection, the USB (see later) for all connections.

Data, in the form of bytes, can be transferred on either serial or parallel paths. In a serial path, each bit of data is placed in a single wire in turn, at high speed. This requires some form of synchronization to ensure that the bytes can be distinguished from each other at the other end of the wire. On a parallel data path, each bit of a set (now usually a set of 32 or 64) is placed on a separate conductor, so that a set of bytes can be transferred in one action. Parallel paths are used on the motherboard, but connections to hard drives can be serial (called SATA) or parallel (called PATA). The PATA type is still more common.

Computer technology is fast-moving, so that descriptions of motherboards and other components soon become out of date. In general, both speed and capacity (memory and drives) increase continually. Keep yourself up to date by reading the computing magazines.

## Power supply

The power supply unit (PSU) for a PC is contained in a sealed box inside the main system case, with connections to the main switch on the casing. The power supply has in the past been a 200 W unit, but the fastest modern computers (1 GHz and above) require a 300–600 W unit. This is a switch-mode form of power supply, which makes servicing difficult, but the cost of replacement PSUs is so low that repair of a faulty unit is seldom economical. The PSU contains a fan, often thermostatically controlled using software commands.

Typical supply voltage and current levels are:

| | |
|---|---|
| +5 V | 20–25 A |
| −5 V | 0.5 A |
| +12 V | 4–7 A |
| −12 V | 7 A |

Note that the current output of the +5 V supply is very high, so that care must be taken against short-circuiting this supply. Most PSU designs incorporate fold-back protection, meaning that the output voltage will be reduced to almost zero in the event of a short-circuit.

## The microprocessor

This one-chip block contains three important sections:

- a **control section** to generate all the necessary timing functions
- an **arithmetic and logic unit** (ALU) to carry out the required calculations in either binary arithmetic or binary logic form
- an **accumulator** (register) to hold the results of any operation before it becomes convenient to pass them to the memory, either for storage or for output.

The action of the microprocessor is to read in one or more bytes of instructions (called **opcodes**) from the memory, and to carry out these instructions. Typically this will require some data to be read from the memory, and the microprocessor will then carry out logic actions on the data and store any results back in the memory. At one time many microprocessor actions required no more than reading data from one part of the memory and storing the same data in another part of the memory. Actions of this type are now carried out by another chip, the **DMA** (direct memory access) chip, rather than by the microprocessor.

All the actions of a modern computer are the result of logic actions carried out one at a time by the microprocessor under the control of the program. The program must contain the command bytes for each instruction together with bytes that provide a location in memory for reading or writing data.

The capabilities of a modern computer are largely due to its high speed. The microprocessor itself is governed by clock pulses, generated from a pulse generator circuit that is part of the microprocessor chip. Whereas early microprocessors used clock rates of around 4 MHz, modern types can run at much higher clock rates, typically in the 1–3 GHz region. The microprocessor clock rate determines how quickly the actions of the processor can run, but the overall speed of a computer depends also on other factors, such as:

- how quickly data can be read from or written to memory, typically around 133 MHz
- how quickly signals can be passed to the other cards on the motherboard, typically 66 MHz
- how quickly signals can be read from or written to a magnetic disc, typically 15–66 MHz
- how quickly the data can be transformed into a display on the monitor, typically 66 MHz.

Note: These speeds have been quoted in MHz rather than the more usual MB/s (megabytes per second) to show the typical frequencies that can be encountered.

The speed of a particular set of instructions depends on what percentage of the actions are of the slower type, so that computers are often compared by using **benchmarks**, meaning programs that contain a large number of actions of one particular type, testing each of these different speeds.

Early computers used the byte of 8 bits as the unit of data, and microprocessors at that type could work with one byte at a time. Since then, the capabilities of microprocessors and computers have steadily expanded, and units of 4 (32-bit) or 8 (64-bit) bytes are more common. A data unit of more than one byte is called a **word**, and the usual word size for a modern PC is a 4-byte word (32 bits). The use of 8 bytes (64 bits) is rapidly becoming the established word size, although at the time of writing the new version of Windows, called Vista, was not in general use. It is possible that a word size of 16 bytes (128 bits) may be introduced in the lifetime of this book. The advantage of using a larger word is that more processing can be carried out in each microprocessor clock time if the operating system can cope with the word size.

To use the other parts of the computer system, the microprocessor is connected with three sets of parallel lines, the data bus, the address bus and the control bus. The data bus transfers data bytes between the microprocessor and other units, in either direction. The address bus locates positions in memory or in other units, and operates in one direction only, so that an address is placed on the bus lines by the microprocessor alone. The control bus ensures that the address bus and the data bus are used correctly; for example, when the address bus and data bus are being used, the control bus will specify which way data is to be copied. Each bus (set of lines) contains as many lines as are needed for the data size, so that the data bus will typically use 32 or 64 lines on modern machines.

## Memory

In addition to containing the program that controls the way in which the computer operates, the memory has to hold or store the data items that are waiting to be processed. This requires two types of memory device. One is called a read-only memory or **ROM**. The ROM will retain its information even when the power is switched off and it carries the most basic operating instructions. The second type of memory has to be capable of being written to and read from, and is known as a random access memory or **RAM**.

Because of its retentive memory, the ROM is described as being **non-volatile**, while the RAM, which loses the data when the power is switched off, is said to be **volatile**. RAM is made up of MOS IC chips, and it can be read or written in very short times, typically 9–50 ns. The size of RAM for a modern PC is typically 512 MB or more. While the computer is being used, data is being read from the hard drive into the memory, and if the data is changed during the use of the computer it must be written back to the hard drive in its altered condition.

When a PC machine is switched on, program instructions are read from the BIOS ROM, and these are used to load in the main operating system (such as Windows) from the hard drive. During this time, the instructions in the BIOS are usually copied to the RAM because RAM is much faster than the ROM used for BIOS. The whole of this process is described as **booting**.

Another type of memory that is used in modern PC machines is **cache memory**. Cache memory is volatile but very fast, and its use considerably increases the operating speed of the PC. The principle is to read a block of data from the hard drive, and then allow the microprocessor to read from or write to this cache of data. This works because reading or writing a block of data (typically 128 Mbytes) is much faster than carrying out individual read/write actions on the same amount of data on the hard drive. The microprocessor can then read from or write to the cache at a much higher speed. Sometimes the data that the microprocessor needs is not present in the cache, and another block must be read, but with good program planning, some 90% of the data that is being used at a particular time will be held in the cache. Modern computers use more than one cache, with a very fast cache (the **primary cache**) built in as part of the microprocessor chip, and another (the **secondary cache**) obtained by using part of the main RAM memory.

## Data storage devices

These devices act as an extension of, or backup to, the working of the computer's internal memory. During normal working, the computer may need to perform a variety of tasks, each requiring a different program. These **service programs**, as they are called, are conveniently stored on magnetic disks. Unlike the RAM of the computer, data stored on magnetic disks is non-volatile, so that it is available whenever the computer is switched on. Magnetic recording of data is not like analogue tape recording, and several systems are in use. All, however, rely on digital saturation recording, meaning that the magnetic material is magnetized to the maximum extent possible, with the direction of magnetism used to indicate a 1 or a 0 bit. The systems that are now used are more elaborate, but the principles are the same.

Disk (or tape) units are also valuable for storing masses of data of the type that might be gathered during such applications as process control or quality control. Since both these forms of backup devices handle binary data in a way that is not normally acceptable to the computer as it stands, they also need interfacing in the same way as did the input.

The main type of disk for a modern PC type of computer is the **hard drive**, and details of this and other drives are dealt with in Chapter 23.

Hard drives are mechanical components that are continually running while a PC is active, so that they have a limited life. Inevitably, a hard drive will fail, although this may be after many years of use, and the computer may have been scrapped and all the data transferred before a drive failure occurs. The most serious failure is one that affects the main drive, because this makes it impossible to load the operating system. If a separate data drive fails it cannot be used, but the computer remains useable. Some machines make it possible to use a RAID (random array of independent disks) system with several hard drives and the data distributed so that the failure of one drive does not affect the operation of the machine or the integrity of the data.

Typical hard-drive failure problems concern either electromechanical components or disk corruption. The main drive motor can fail, making it impossible to use the drive, so that the operating system will not load. This type of failure is easy to diagnose because you will not hear the drive motor start when the computer is switched on. Another form of electromechanical failure affects the voice-coil drive for the heads, so that the tracks cannot be correctly located. In this type of failure also the operating system will not load, and the usual clicking noise of the head system will be absent.

Failure of the electromechanical system will require the drive to be replaced, but the data that was recorded on the drive will still be intact. Specialist firms can open the drive in a clean room and copy the data to a backup drive so that the data can be restored. Remember that the value of the data on a computer system may be very much greater than the value of the computer itself.

Drive failure that causes corruption is much more serious, particularly on a disk that contains data. A typical cause is that a motor fault slows the platters down and the 'flying' heads scrape against the platter surfaces, tearing the magnetic material. It is very difficult, or impossible, to recover anything more than fragments of data from this type of head crash.

## Backup

Because the data is precious, you cannot assume that a hard drive will be totally reliable, and backups are needed. A **backup** is a copy of important data on a hard drive, and a good backup system will ensure that data is copied to another medium at frequent intervals and can be easily copied back if a replacement hard drive is needed. Typical backup systems include floppy disks (1.4 Mbytes), CD (up to 700 Mbytes) and DVD (up to 4.7 Gbytes on single-layer types and 8.5 Gbytes on double-layer types) writing drives, and also removable hard drives (which run only during backup and are removed from a holder after backup is complete). The BIOS of modern computers contains the firmware needed for controlling both floppy disks and CD-ROM. Older machines contain only the floppy drivers, and a special floppy (boot disk) has to be used to install CD-ROM drivers and the essential MS-DOS programs preparatory to installing Windows from a CD-ROM.

The floppy-disk drive allows the use of 3.5 inch magnetic disks with a capacity of around 1.4 Mb. These are a hangover from older systems, and on many modern machines the floppy drive is omitted. On older machines with no CD-ROM drive, the floppy drive is essential because it allows a computer to be commissioned with a new unformatted hard drive. A system floppy, usually included with a copy of Windows for original equipment manufacturer (OEM) use, contains the MS-DOS command files along with drivers for CD-ROM drives, allowing you to format a hard drive, copy files to it and make the CD-ROM drive available for installing Windows. On modern machines, Windows is installed entirely from a CD-ROM.

> If a floppy has to be used on a modern machine, it is simpler to use a USB floppy drive, and to connect it as required. The power for the drive motor is supplied through the USB cable.

The floppy uses the same scheme of tracks and sectors as the hard drive, but the recording system is standardized because a floppy recorded on one PC must be readable on another (hard drives are usually fixed and used by one machine only, and the platters cannot be separated). Interfacing for floppy drives is, on modern computers, included as part of the BIOS, replacing several older types of interfacing that are now obsolete and which required a separate interface card. The connector for the floppy drives is on the motherboard, and uses a 34-way flat ribbon cable with the usual strip to identify the Pin 1 side. Cables usually allow for two floppy drives, but only one is normally used if a floppy drive is fitted. The floppy drive also uses the same type of power connector as the hard drive, although a miniature version is sometimes used.

The CD-ROM drive allows the use of CDs that carry computer codes, and these are extensively used to contain programs and other information (such as multimedia text). Older CD-ROM drives are read-only, but a later generation of drives permit both reading and writing. The conventional CD-ROM will hold about 700 MB of data. The data cable is of the same type as for a hard drive with a 40-pin connector, and it is preferable to use separate connections for these, usually the IDE1 connector on the motherboard for the hard drive and the IDE2 connector for the CD-ROM (or, more usually now, the DVD-RW drive). See Chapter 23 for details of optical drives (CD-ROM and DVD) used on modern computers.

The PC needs to provide for connections to and from peripheral devices such as printers, scanners, modems and other units. Until recently, keyboard and mouse connections were made using small sockets of the DIN type, but more recently the USB type of connectors has come into favour.

## Port connections

Until recently, the main input and output (I/O) port connectors for the PC were either parallel (Centronics) or serial (RS232 or RS423). The parallel port transmits eight data bits at a time, and is used for the printer, since printers at one time predominantly used this type of connection. The data cable for a parallel port requires more than eight lines, because others are used for synchronization. Typically, 17 lines are used for signals, and the other lines are used for earth returns.

The serial ports were at one time used for modem connection, but the extensive use of internal modems (as a PCI card) and USB modems has eliminated the need for this. The serial port transmits 1 bit at a time, and can be used as a mouse connection if no dedicated mouse port is supplied on a motherboard. More recently, all of these types of connection have been phased out in favour of the USB, particularly now that the faster USB-2 is available.

The parallel port is often termed a **Centronics port** because its standardized format was due to the printer manufacturers Centronics who devised this form of port in the 1970s. The Centronics port socket on the PC uses a 25-pin D-type female connector at the PC end of the cable, and the 36-pin Amphenol type at the printer end. This can be confusing, because the older serial port (COM1) connector on the PC is the 25-pin D-type male connector.

> Where ribbon cable is used, the line corresponding to Pin 1 is marked in some way, usually with a red stripe.

Serial ports of the older type were used for modems, for serial mice and for linking PCs together into simple networks, but they are seldom used on modern machines. These ports send or receive 1 bit at a time, and at their simplest they need only a single connection (and earth return) between the devices that are connected. Serial ports are seldom quite so simple, however.

> The keyboard and PS/2 mouse use serial ports, but these are dedicated types, meaning that you cannot use them for other devices. The serial ports we are dealing with here are the COM types used for connecting other peripherals.

The port type that is fitted on all modern computers is the USB. It has been designed to be **hot-plugged**, meaning that devices can be connected and disconnected with the computer switched on and working. This is possible only if the system is supported by the computer, the peripheral device and the operating system. At the time of writing, the later version, USB-2, is fitted to motherboards, but this is compatible with the earlier version, USB-1.

USB-2 permits communications between devices that are equipped with suitable interfaces at serial data rates of around 480 Mb/s. This is very much faster than the old-style serial port system (and the USB-1 standard). The interconnecting cable can have a maximum length of 5 m and consists of two twisted pair cables, one pair for power and the other for signalling. The distance can be extended to about 30 m by using a hub terminal device as a line repeater.

Terminal devices that use USB, such as keyboard, mouse and printer, are added to the basic PC in a daisy-chain fashion and each is identified by using a 7-bit address code. This allows up to 127 devices, in theory at least, to be connected. In practice, not all devices allow daisy-chain connection (picture a mouse with two tails!), and so the computer needs more than one USB connector. In practice, the computer will be fitted with two to six USB connectors of the flat type, and each peripheral will use one connector of the square type (Figure 20.4).

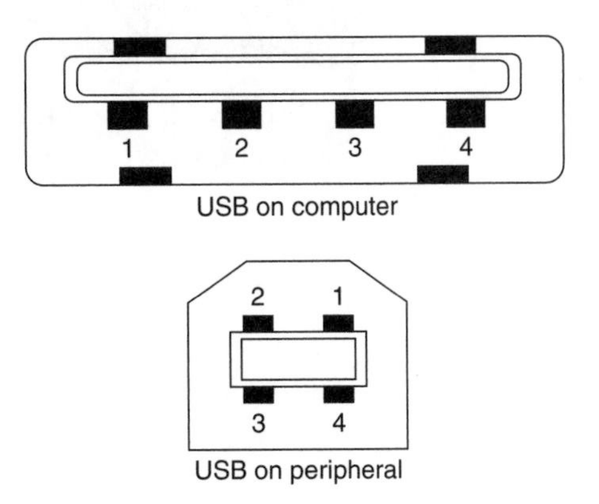


**Figure 20.4** Standard forms of USB connectors. Some devices, notably digital cameras and MP3 players, use a smaller connector at one end of a custom-made cable

Other terminal devices can include scanners, fax machines, telephone and integrated services digital network (ISDN) lines, multimedia display and recording systems, and industrial data acquisition devices. ISDN (broadband) lines, however, normally use Ethernet connectors.

Cables are always a potential source of problems in PC equipment. It is rare for a cable to be faulty when it is unwrapped, but faults can develop, mainly due to excessive flexing of cables when cables are fitted or when items of equipment are moved with the cables left connected. Where a cable connector provides for anchoring by a clip, this should be done, and anchoring is particularly important for parallel printer cables, particularly for the 36-pin connector at the printer end. Since USB cables make no provision for anchoring, care needs to be taken to avoid pulling out these small connectors when equipment is moved.

Cables can be checked by continuity testing or by substitution. The parallel type is most easily tested by substitution, and this is a preferred method because it involves the least amount of handling. If no substitute is available, you can check cables by checking for continuity, but this can be very time consuming, and you risk damaging a cable in the process of checking connections manually. A much more satisfactory method is to use automatic testing equipment which connects to each end of the cable and then carries out a set of continuity checks on each corresponding pair of pins.

Note that data cables for computers are rated for low voltages only, and must not be tested using any type of unit that employs high voltages.

## Practical 20.1

Test parallel and USB cables for continuity and insulation.

## Multiple-choice revision questions

20.1 An operating system such as DOS controls:
(a) all of the computer
(b) how software is installed
(c) the timing of the processor
(d) the use of keyboard, disk drives, printer and monitor.

20.2 The BIOS is:
(a) a program that you have to run when you use the computer
(b) a chip that controls inputs and outputs
(c) a device that controls the monitor
(d) a device that controls the use of memory.

20.3 The controller for a hard disk drive is:
(a) built into the drive itself
(b) built into the motherboard of the computer
(c) added as an extra board to the computer
(d) not needed on modern machines.

20.4 The PC microprocessor:
(a) can carry out several tasks simultaneously
(b) contains the main memory
(c) can carry out program steps in sequence
(d) retains information when the PC is switched off.

20.5 RAM is:
(a) slow and non-volatile
(b) fast and non-volatile
(c) slow and volatile
(d) fast and volatile.

20.6 Cables connecting to peripherals are now mainly of the type:
(a) Centronics parallel
(b) USB
(c) Firewire
(d) 25-pin serial.

# 21 Installing a PC

## Unpacking

A new PC will be delivered from the factory or wholesalers in one or more large cardboard boxes, using polystyrene foam to protect the computer from impact. The box containing the main processor unit should be placed on the floor, correct way up, and opened, taking care with large staples (which can rip fingers). Packing materials should be removed and kept in the (rare) event that the machine has to be returned, and the computer components can then be removed in turn. The mains lead is usually on top, and other items that may be easily accessible are the keyboard and mouse, although these are sometimes packed separately. The monitor is almost always in a separate package if it is a cathode-ray tube (CRT) type of monitor, but the flat liquid crystal display (LCD) types can be packed with the main processor.

With all the units unpacked and on the bench, look for **transit packing** that is used to protect moving parts. A typical example is a tape placed over a CD/DVD drive to prevent the drive opening. Check with the documents that are packed with the unit for other pieces of transit (or shipping) packing that have to be removed. If the monitor is of the LCD type, handle it with great care, because the screen is vulnerable to impact or even light pressure. Keep the transparent covering over the LCD screen until the machine is put into use; some users like to keep this covering as a dust shield so that the screen seldom or never needs to be dusted. A set of CD-ROMs should also be included. Even if the software from these discs has been installed you may need to extract other programs from them, or use them in an emergency. One of the discs should be an authentic copy of Windows XP, while others will carry driver programs for various units.

Some manufacturers now use part of the hard disk (a **partition**) to store a copy of the operating system; this is particularly true of laptops and some desktop machines now also follow this practice. Although this makes it easy to reinstall Windows if needed, it makes it impossible to restore Windows on a replacement drive. If you need to replace a hard drive on such a machine you will need to find a legitimate copy of the correct version of Windows on a CD-ROM.

Check with the manual or other documents that come with the machine that the CMOS-RAM (a small piece of memory kept active by a battery) is active. On some machines, a jumper connection is set so that the CMOS-RAM is not powered, so that this jumper must be reset (connecting the battery) before the machine can be used. If this is not done, then the machine will have to have its settings re-entered each time it is switched on.

Check also whether the hard drive is partitioned and formatted, and with Windows installed. Some machines are supplied with none of this done, and if it is required then a specialist book should be consulted (for example, *Build and upgrade your own PC*, Ian Sinclair, 4th edition, Newnes).

For bench testing, the units can be plugged together with no need to follow some particular arrangement. The important point for testing is that the sections should all be accessible, and the monitor is best placed temporarily on one side of the main casing. When the system has to be fitted into place for a customer, usually on a computer desk, more care should be taken over how the components are arranged.

At one time, the main unit would always have been of the desktop type with a flip-top lid, with the monitor placed on top of it, and this type of arrangement is likely to return as main units become flatter and LCD monitors replace the heavy CRT. For the moment, though, the predominant arrangement is the **tower** form of construction, particularly the mini or midi tower, owing to the small *footprint* it makes on a desk (Figure 21.1). A tower can sit on the edge of a desk, allowing the monitor and keyboard to be arranged more centrally, and an alternative is to place the main tower under the desk; this is essential for a full-size tower which would be too large on a desktop. Another advantage is that using a tower for the PC box allows space for a larger monitor now that such units are reasonably priced. It also leaves more space for other items such as a printer and a scanner.

**Figure 21.1** Typical tower case arrangement

Some mice of anonymous (or anonymouse?) manufacture have very short connecting leads, preventing you from placing the main unit under a desk. Microsoft mice all seem to come with a good length of cable. Extension cables for the mouse can be bought, although you have to search for suppliers.

If you are tempted by cordless mice or keyboards remember that they have to be able to pass signals in other ways, so that your computer will have to be fitted with a card that sends and receives these signals, which can be optical or radio.

You can place the mouse to the left or to the right, and software will allow you to interchange the functions of the mouse switches to allow for left-hand or right-hand use. Both mouse and keyboard should come with leads that are long enough to give you considerable choice about where you place them relative to the main casing.

The use of a small-format computer such as the *Shuttle* allows a full set of equipment to be placed on the working surface of a small computer desk, even allowing for a printer, scanner and broadband router on the desk, along with a 17-inch LCD monitor, without restricting access to the front and rear of the computer casing, and room to remove the lid for internal work.

The connections to the back of the main unit should be made before the main unit is in its final place, because they may be more difficult to reach afterwards. These are, in particular, the mouse and keyboard connections, because they are recessed and care is needed to ensure that the tiny plugs are inserted the correct way round, something that is much easier if you are facing the back of the main unit with good illumination. Other rear-case plugs are the monitor plug (usually 15 pins in three rows, but some new units use a smaller type of plug) and the mains cable (Euroconnector). Items such as printer and scanners will normally use the USB sockets on the rear of the casing; any sockets on the front should be reserved for peripherals that are connected only when required, such as a digital camera (to download image files).

If problems are found at a very early stage in commissioning a computer it can be returned using the packing that was supplied with it. This is seldom applicable until the computer is started for the first time, because it is unlikely that anything short of a severe impact would cause damage that would be visible. When a computer is first switched on, lack of indicators (such as lights or the sound of the hard-drive motor) would be a cause for returning the unit, although the monitor, keyboard and mouse might be retained. Other faults may be found once the hard drive is formatted (see later) and Windows installed. Once a customer's programs and data have been installed the decision to return a unit becomes very difficult, because it is almost always less trouble and expense to carry out repairs on the spot. For example, if the customer has just installed his accounting system on the hard drive with all its data, returning the machine would require all the sensitive information to be totally deleted, and this is not so simple as it sounds. Simply using the delete action of Windows does not delete data, it simply makes it difficult to find by the normal methods. Other methods, however, can reveal such data unless the whole of the hard drive has been cleared using software intended for that purpose.

**Initial checking**

For a machine bought with Windows installed (usually less costly than buying one without an operating system and installing Windows for yourself), everything should be set to default values that allow the machine to be used almost immediately. The act of switching on is called **booting up** (or **booting**) and when mains power is applied, the power supply unit (PSU) of the computer is in standby. The On/Off switch does not handle mains power, only a low-voltage supply from the PSU. To switch on, press the button and release it. Switching off should be done only by using Windows (click the mouse on the **Start** button shape on the screen, and then on the '**Turn off computer**' notice that appears). If you do not follow this procedure there is a risk that some data that you have entered may be lost.

The first booting action is to run a set of tests. These are known as **POST**, meaning power-on self-test, and they carry out a set of checks that ensure that the voltage levels are correct and the microprocessor is functioning correctly. After POST, some MS-DOS programs run, following which Windows XP (or whatever type of operating system is currently used) will start to run.

Once Windows has started you may need to run programs (drivers) that control the mainboard. This may have been done already for you, but if not, you need to insert the CD-ROM of main board utilities that accompanies the machine and follow the instructions that appear on screen. One of these utilities will probably be a **virus checker**.

If a virus checker is supplied, it may be the excellent AVG, which is free to use, or one of the others that requires regular payment. If the virus checker is already installed it is most likely to be one that requires registration and periodic payment. You should make sure that the customer understands this and is aware of good free alternatives such as AVG. The virus checker should be set to scan all the existing software, and to load into memory each time the machine is booted so that it will detect a virus in any data that is loaded in, either from discs or over the Internet (the most usual source of virus infection). Emphasize to a customer that discs from unknown sources must **always** be checked for virus content before any attempt is made to use them. Because viruses are being continually created and distributed, virus software must be continually updated, and an Internet connection is **essential**. The virus checker can be configured to look on the Internet for updates each time the PC is switched on.

## Practical 21.1

Connect a computer main casing (with Windows installed) to a monitor, keyboard and mouse, and start the machine running, checking that there is a monitor display and that the keyboard and mouse are active.

## Practical 21.2

Install Windows on a computer that is known to be working.

### Adding software and hardware

When new hardware is added on a modern PC it will make use of the **plug and play** system. This means that when the peripheral is connected (whether by USB or by any other form) the PC can detect the connection and will find the driver software that is needed for it. In some cases, this can conflict with the provisions that the manufacturer of a peripheral has made for installation, so that you need to read very carefully the instructions for installing an unfamiliar unit. For such units, you are usually informed that you need to reject the offer (onscreen) to install automatically, and use the CD-ROM that came with the peripheral. Another method is to install the driver software before plugging in the peripheral. Some driver software will serve for more than one peripheral; for example, the driver for a memory stick/pen drive (USB fitting) will usually allow transfer of data to or from a MP3 player, but you may need to install further software for some MP3 player actions.

New software is added either by using a CD-ROM or DVD-ROM or by downloading from the Internet, and the normal action of the PC is to detect the insertion of a disc and automatically start the installation of the software. This depends on the default settings, and if these have been changed you may have to start installation manually. This is done by starting Windows Explorer and finding the CD/DVD-ROM drive (often the **D** drive). By clicking on this, you will see the files, and one will usually be names SETUP.EXE. Clicking on this file will start the installation, and the remainder of the process will usually be automatic. Software downloaded from the Internet should be saved on disc, checked for virus content, and then run to install itself.

## Servicing procedures

### Flowchart diagnosis

Diagnosing faults in an unfamiliar system can often be speeded by using a flowchart provided by the manufacturer. This is equivalent to the well-established 'divide and conquer' method of electronics diagnosis. Working from a starting point, you answer a question with a YES or NO, and this leads you to another box with a question that also must be answered in the same way. Eventually, this will provide a limited number of possible causes of a fault and allow you to find which is applicable. A very simple example is illustrated in Figure 21.2, for the fault condition that switching on the computer has no effect, with no lamp indicators lit and no sound from the drives.

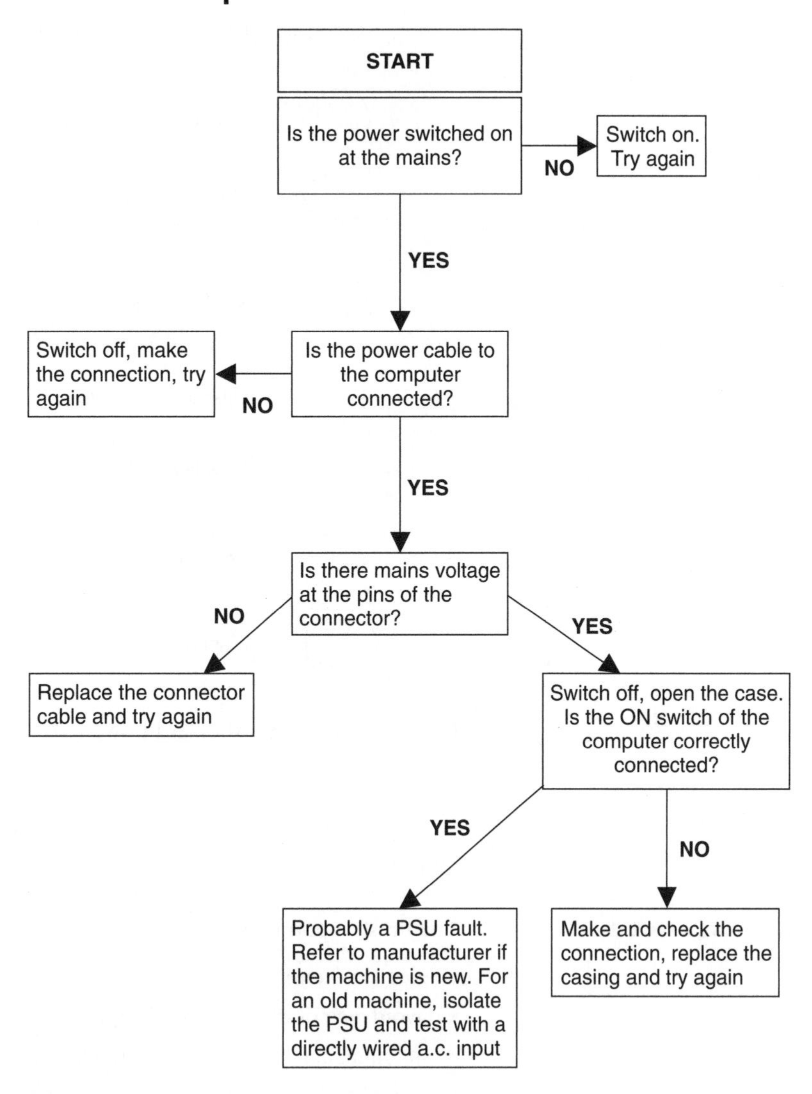


**Figure 21.2** A simple type of flowchart, not using standard symbols

Manufacturers will usually supply flowcharts to recognized service agents, but they are not widely available (on the Internet, for example) to anyone else. The simple example that has been shown here does not use the standard symbols that are illustrated in Figure 21.3, but for fault diagnosis most of

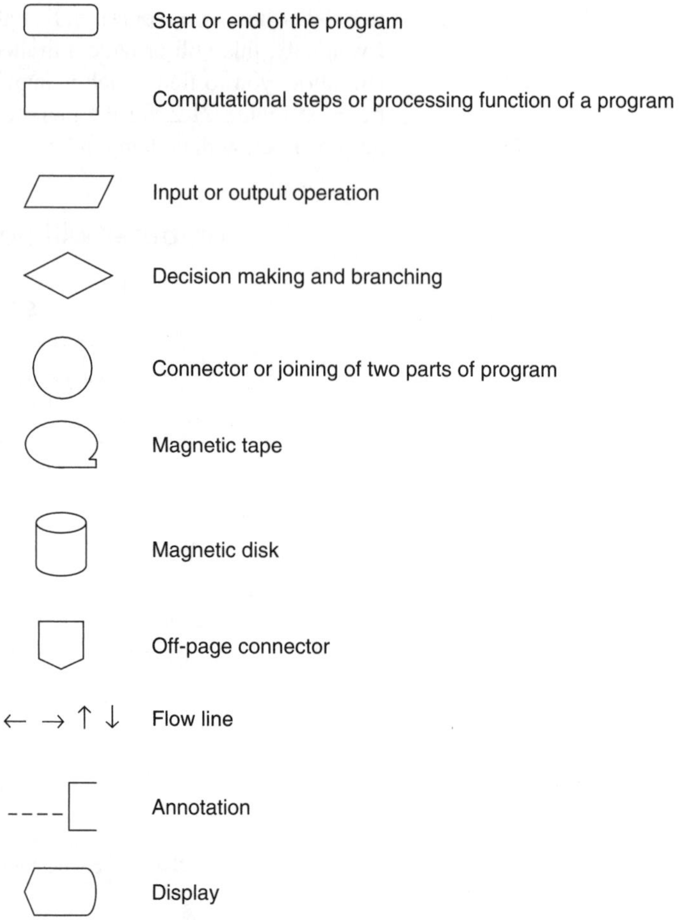

Start or end of the program

Computational steps or processing function of a program

Input or output operation

Decision making and branching

Connector or joining of two parts of program

Magnetic tape

Magnetic disk

Off-page connector

← → ↑ ↓ Flow line

Annotation

Display

**Figure 21.3** Some standard flowchart symbols

these are redundant because they are intended primarily for programming purposes.

## Servicing precautions

The most important precautions to be taken in servicing are

- electrical safety
- avoidance of damage to the equipment being serviced
- maintenance of a log of action taken and results achieved
- use of the Internet to explain and illustrate unfamiliar actions.

Electrical safety (see also Chapter 25) is less of a problem on computer equipment than on other electrical goods because of the low voltages. The only points where mains voltages are encountered are inside the sealed PSU, and this should be replaced as a unit if it is not working: fault-finding is not a rewarding exercise because the price of a new PSU is lower than the cost of service work. Care should be taken, however, to avoid short-circuits, because some of the low-voltage lines can pass very high currents. Although the PSU is protected against excessive currents, servicing can become difficult if you have just welded a screwdriver to the motherboard.

**Practical 21.3**

Carry out field service procedures for a computer mainboard unit with a hardware fault.

**Built-in diagnostic checks**

When the computer is booted, it carries out a set of tests, the POST set. Precisely what tests are carried out will vary according to the BIOS that is being used, and a full list of all the BIOS information is beyond the scope of this book. A good source of POST code and test routines is at the website: http://www.bioscentral.com/postcodes/awardbios.htm#top, which lists a wide range of BIOS used by leading manufacturers.

These diagnostic checks are particularly useful for hardware faults on the motherboard, and checks for other boards (graphics, sound, etc.) are probably best done by substitution.

Note that many Internet sources offer software for diagnosis and repair of software faults that can cause slow running, freezing, rebooting and other problems that are not hardware related. Use these if they are relevant, but do not expect them to find problems that are related to faults in hardware.

When there is a software problem (such as a corrupted driver) the machine may start Windows in *safe mode*. This uses only the drivers that are built into Windows and it allows limited use of the computer for running diagnostic checks. The machine will always boot up into safe mode until the problems are fixed, and in some cases this may require Windows to be reinstalled (particularly for Windows versions prior to XP). In general, software problems should not be the subject of an electronics servicing workshop because they are more specialized.

**Fault location and treatment**

Diagnostic checks and experience can lead to the location of a fault by the usual 'divide and conquer' methods. If the fault is in a module like the graphics card or the PSU, the best solution is the replacement of the card or unit, because this is almost certainly more cost-effective than trying to make diagnosis and repair on an unfamiliar unit. A more difficult decision arises if the fault is in the motherboard, because working on a motherboard requires considerable experience and care, and unless some obvious cause (such as a crack in a track or a dry joint) can be seen in the motherboard, replacement is often the simplest way of dealing with the problem. Once again, this has to be weighed against cost. On an old computer, it is unlikely that a replacement motherboard can be found easily, and the cost of replacing may be considerably greater than the value of the computer (and more than the price of a well-featured new machine). At a time when a new computer can cost under £300, decisions about repairs can be difficult.

## Practical 21.4

Construct a simple diagnostic flowchart for a faulty hard drive.

## Multiple-choice revision questions

21.1 A computer unit is being assembled for testing. The last connector you put in should be:
(a) the printer cable
(b) the power cable
(c) the monitor cable
(d) the keyboard cable.

21.2 Plug and play means that:
(a) you can hear music when you switch on
(b) a device will start working whenever the mains is switched on
(c) the added hardware will be automatically recognized and used
(d) games software has already been installed.

21.3 Virus checking will not be needed if:
(a) the computer has no floppy drive
(b) the computer has no CD-ROM drive
(c) the computer has no Internet connection
(d) the computer has no Internet connection and no new software has been installed.

21.4 Windows starts in safe mode. This means that:
(a) a virus cannot affect it
(b) the minimum amount of driver software has been loaded
(c) something is not connected
(d) the floppy drive is faulty.

21.5 The POST procedure means that:
(a) the computer is working perfectly
(b) new programs are needed
(c) you need to carry out checks
(d) some hardware testing has been done.

21.6 A virus is suspected. You should use virus checking software on:
(a) the contents of the hard drive
(b) all floppy discs
(c) all CD-ROM discs
(d) everything that contains software.

# 22 Keyboard, mouse and monitors

## Keyboard

The computer keyboard contains all the keys of a normal typewriter, together with a set of **function keys** (labelled F1 to F12) that can be used for controlling actions. These function keys are all programmable, meaning that their actions can be set by whatever program is running. A set of number keys is usually included at the right-hand side of the keyboard, and there are also keys marked with arrows that will move a pointer (or **cursor**) on the screen. In addition, there are keys marked Insert, Delete, Home, End, Page Up and Page Down. All these key actions can be modified to suit whatever program is running.

The layout of the three rows of main writing keys (as distinct from function keys) follows the standard typewriter QWERTY layout, named from the first six keys on the top line. This layout was originally devised more than 100 years ago to prevent fast typing, which jammed the keys of early typewriters, but it has become a standard and although alternatives (such as Dvorak and Maltron) are available for computers they are seldom seen. For laptop and other miniature PCs, the keyboard also contain a pointing device (see later for details of the mouse) which allows a pointer (usually an arrow shape) to be moved around screen by moving your fingers on a square sensing pad on the keyboard.

A keyboard is constructed from a matrix of input and output lines, with the required interconnections being provided by suitably positioned diodes and contacts made by (debounced) switches. A section of such a structure is illustrated in Figure 22.1.

The keyboard uses a set of miniature switches wired in a matrix, and pressing a key closes a switch contact and results in generating a unique code. The keyboard contains a processor that converts this code into a standard called ASCII, and communicates with the computer using a serial connection. The keyboard cable can be flat or round and conventionally uses either a PS/2 miniature connector (Figure 22.2) or a universal serial bus (USB) connector. When a keyboard uses a USB connection it is usual for the keyboard also to have a USB connector for the mouse. Adaptors for changing the connection type are available. Older computers may use a serial mouse that connects to the nine-pin serial port. The keyboard and mouse PS/2 sockets are on the rear panel of the main unit and are distinguished by symbols, illustrated in Figure 22.2. Each plug must be inserted into the correct socket.

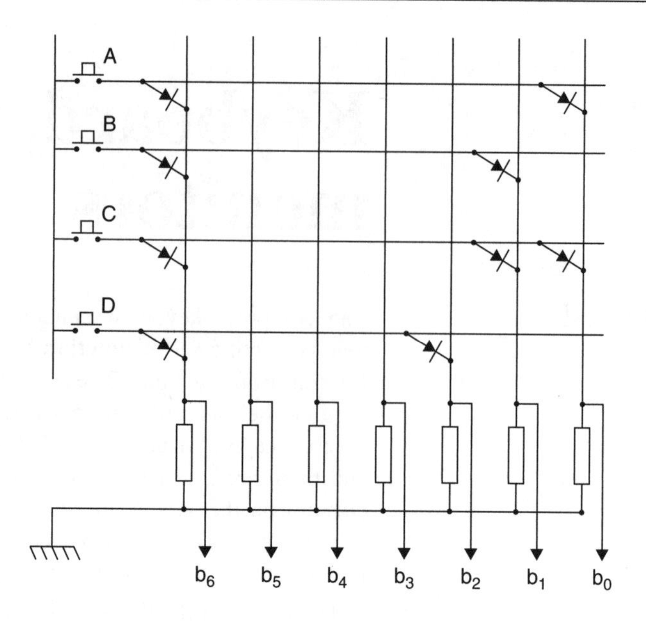


**Figure 22.1** A section of a simple four-key keyboard matrix

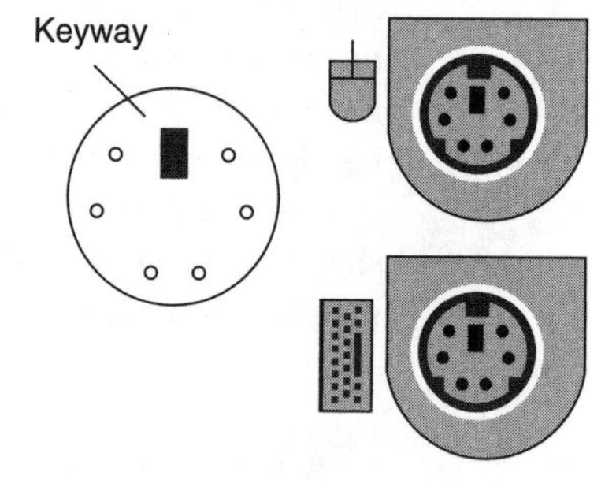


**Figure 22.2** PS/2 connector for keyboard or mouse, and the appearance of the PS/2 connectors at the rear of the computer

Keyboard problems are usually detected and notified when the computer is booted up, but the error message is often a number code that tells you nothing about the cause. Typically, PC machines use a three digit code in the range 300–399 to signify a keyboard error, and if you see this appear in an error message you need to switch off and check the keyboard connections and the keys themselves. The two most common causes of keyboard error are a loose plug and a jammed key. If in doubt, check by substituting another keyboard that is known to be perfect, because all normal keyboards use the same driver software that is built into the Windows operating system. Keyboard maintenance should be confined to cleaning at a time when the computer is switched off. The keys become coated with a grimy layer from human fingertips, and this can be cleaned using a spirit-based cleaner. You should never use spray cleaners (which can get into places where solvents can cause problems) or strong solvents.

Key action can be checked by starting a text application (such as Notepad) in Windows, and typing, using all the letter and number keys. If you have a program that makes use of the function keys, this also can be used.

## Practical 22.1

Connect up a keyboard, stating what type it is (PS/2, serial or USB). Boot the PC and check that the keyboard is working. Reboot with the A key held down to simulate the effect of a jammed key and note the error message. Carry out cleaning actions on the keyboard.

## Mouse

The **mouse** is one form of pointing device used for selection and commands by moving a pointer on the screen. This is useful for selection from a menu and also for drawing actions. The mouse consists of a casing holding a heavy steel ball covered with a rubbery coating. The movement of the ball is sensed by rollers or by optical methods so that any movement of the mouse will generate two sets of signals, one for up–down and another for side-to-side movement. In addition, two or three switches can be operated by buttons on the mouse, and the action of pressing and releasing a switch (called **clicking**) can be used to send controlling signals to the PC using a serial link. Infrared and wireless links can also be used if the computer is equipped for such links.

Some types of mouse also use a wheel that can be rotated to scroll the lines on the screen.

The mouse is an essential part of the **graphical user interface (GUI)** systems in which choices are made by moving the pointer and pressing (clicking) a button on the mouse. The best known of these GUI systems are Microsoft Windows, used on PC machines, and the (older) operating system devised for Apple computers.

Figure 22.3 shows the interior of a typical mouse. The ball is held in a hemi-spherical casing, and its movement is sensed, in this illustration, by rollers that are held against the ball. One roller detects lateral movement, and the rotation of the roller shaft is digitized by using, typically, an optical system with a set of blades and an infrared light-beam. Another roller detects movement in the direction at 90° using the same method. A third roller is a dummy, used to hold the ball against the other two. Circuits on a small printed circuit board (PCB) within the mouse shell convert the digitized signals into serial form for transmitting to the computer. Other systems are used, some not depending on the roller, that sense movement across a surface directly.

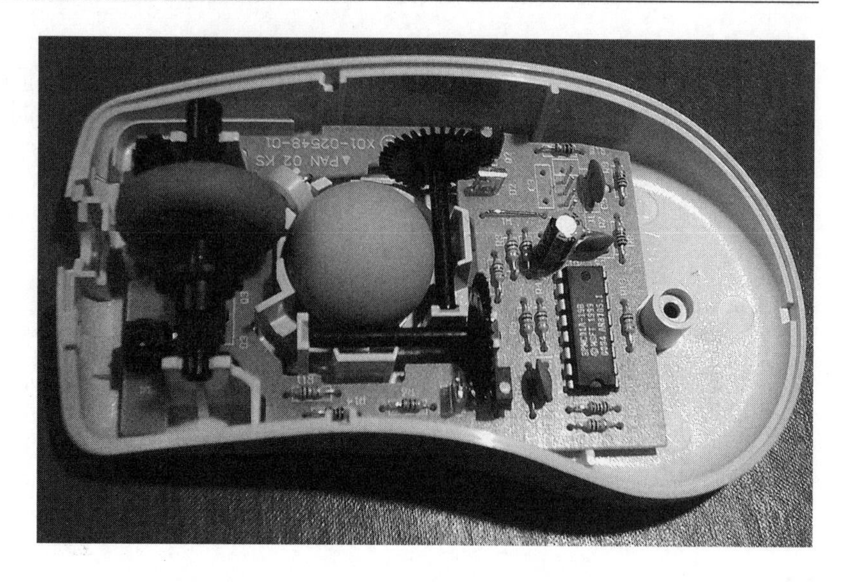

**Figure 22.3** Inside a typical mouse [photograph: John Dunton]

The mouse can normally be connected in two different ways, each of which requires a mouse with the correct cable termination, and a matching connector on the computer. Computers that use the ATX type of motherboard and casing will normally provide a mouse port with the PS/2 type of connector specifically intended for the PS/2 type mouse. An option that is starting to appear is the USB fitting, plugging into a port on the (USB) keyboard. Adapters are available so that one variety of connector can be used with the other.

For use with any computer with a serial connection, a serial mouse can be obtained that uses a nine-pin serial connector.

Connecting the mouse has no effect unless suitable mouse driver software is present and is run each time the computer is started. On modern machines that use Windows the mouse software is built in and will run automatically when Windows starts. You can use the *Control Panel* action of Windows to modify some of the mouse actions to suit your own requirements.

The main problem you are likely to encounter with a mouse is sticky or erratic pointer movement, caused by dirt either in the mechanism or, more likely, on the skids. Turn the mouse over and clean the skids (Figure 22.4a), with a moist cloth. If this removes the problems there is no need to do any more, but if the mouse still gives problems, remove the circular cover that holds the ball, revealing the rollers illustrated in Figure 22.4(b).

Dirt collects on the skids from the mouse mat, and this layer of dirt is enough to lift the mouse body slightly so that the ball can no longer roll smoothly over the mat. If the skids are dirty, you should also clean the

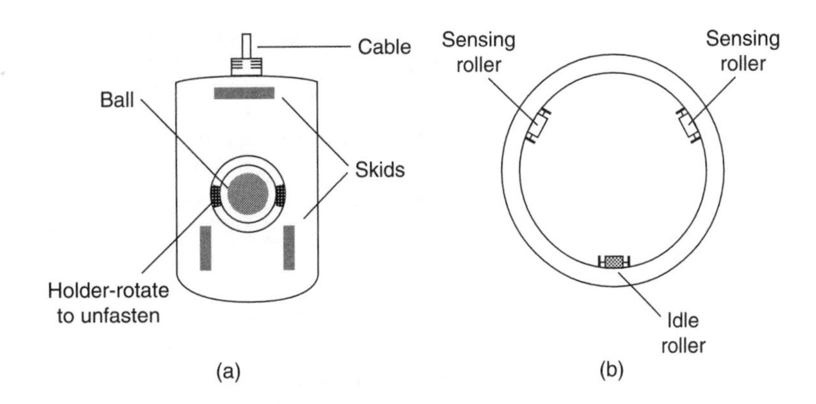


**Figure 22.4** Cleaning the mouse skids and rollers: (a) underside view, and (b) detail of rollers

mouse mat. If problems persist after this, remove the ball and clean it, using a clean moist cloth; you can use spectacle lens cleaner fluid or windscreen-cleaning fluid if you prefer. Check by rotating the rollers with a finger that these are working correctly; rotating a roller should move the screen cursor when the computer is working. If the roller action is not working, you need a new mouse. If the rollers are working correctly you should clean them. They are most easily cleaned by wrapping a piece of clean cloth onto the tips of tweezers, moistening this and wiping it across each roller, repeating several times while turning the roller slightly. Allow a few minutes for any moisture to dry and then reassemble the ball into the mouse casing and make sure that it is locked in place.

If the mouse movement is erratic from the start check that the correct driver is being used. For a machine using Windows and a standard type of mouse, it is most unlikely that the driver would be at fault, but if you have changed to another mouse or you are using a machine which has an unorthodox arrangement this is a possible cause of trouble.

Another type of problem that you may encounter is that the mouse will move the pointer easily in one direction, but needs much more mouse movement to shift the pointer in the opposite direction. This cannot be cured by cleaning the mouse rollers (although you should try it), and usually requires mouse replacement. Test first by substituting another mouse.

The standard mouse uses two buttons, but there are some three-button types, although the third button is not normally used by Windows software. A commonly used Microsoft mouse design, the Intellimouse, uses scroll wheels to permit the screen scrolling action to be carried out from the mouse. A scroll-wheel type of mouse is very useful if you work with long text documents.

### Practical 22.2

Install a mouse, stating what type of connection is used. Boot the computer, and when Windows has loaded check that the mouse is operating correctly. Clean the mouse, including ball and rollers.

There are other options for pointing devices, and one is the **trackball** (Figure 22.5). This looks like an inverted mouse, and consists of a heavy casing that remains in one place on the desk. This carries a ball, larger than the usual mouse type, that can be moved with the fingers, and a button either side of the ball. The action of ball movement and button clicking follows the same pattern as the use of the mouse, and some users prefer it, particularly for graphics applications.

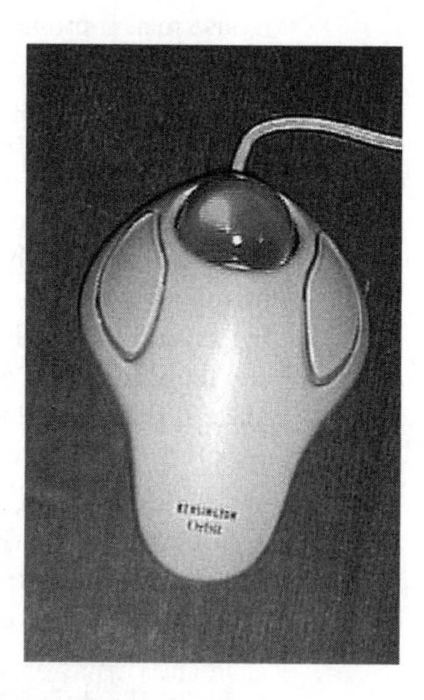

**Figure 22.5** A typical trackball

Graphics **tablets** are another form of pointer that, as the name suggests, are particularly suited to graphics work. A graphics pad or tablet looks like a rectangle of plastic with a stylus, and the movement of the pointer on the monitor screen is controlled by the movement of the stylus over the graphics tablet. This means that every point on the monitor screen corresponds to a point on the graphics tablet, so that the larger the tablet, the more precisely you can control the screen pointer. The stylus can be pressed to provide the click action of a mouse. Graphics tablets are expensive, particularly in the larger sizes.

## Monitors

The **monitor** or **VDU** (visual display unit) is a display device that is the main output for the computer, showing the effect of entering or manipulating data. At one time computers did not use monitors, and relied entirely on printed output. When monitors were first provided on large computers they were used only for checking (monitoring) that the computer was working correctly, hence the name.

Note: To avoid repeating information, refer to Chapters 18 and 19 for details of television displays. In particular, you should know the meanings for the terms **hue**, **luminance**, **saturation**, **interlacing** and **convergence**. You should be aware of the **colour triangle** for adding colours, and the structure and operation of the colour cathode-ray tube (CRT).

For the lower cost and older desktop machines, the CRT is the basis of the monitor. All colour monitors currently used for PC machines are of the RGB type, meaning that separate red, green and blue signals are sent from computer to monitor. Modern machines use the liquid crystal display (LCD) type of monitor.

The **resolution** of a monitor measures its ability to show fine detail in a picture, and is usually quoted in terms of the number of dots that can be distinguished across the screen and down the screen. The figures differ because the screen is not square, normally with a ratio of width to height of 4:3 (as for predigital television) although some, mainly laptop, machines use the wide-screen standard of 16:9 and this (wide-screen) shape is also appearing on desktop machines. These dots are referred to as **pixels** (picture elements). A figure of 640 × 480 pixels was once regarded as the standard, but modern machines are likely to use resolution figures such as 1280 × 1024, and higher.

As you might expect, the signals within the computer cannot be used directly by the CRT monitor, and the interface circuits are placed on a **graphics card**. The graphics card, usually plugged into a slot on the motherboard, also contains memory, and will convert a set of digital signals into the analogue signals or red, green and blue that are needed by the monitor. For normal text work, the graphics card can be simple (and some motherboards incorporate the circuits of such a graphics card and need no additional card), but more elaborate cards are needed for pictures and particularly to handle animated pictures, games and video. Large amounts of memory (typically 128 Mbytes) are needed on the graphics board for fast-changing picture displays for video or games. Modern monitors of the LCD type can be manufactured to use an alternative **digital video input (DVI)** in place of the 15-pin connection type used by RGB CRT monitors. A DVI connection can handle bandwidths up to 160 MHz. Older LCD monitors will use the older type of analogue connector.

Unlike television, which uses an interlaced picture format (see Chapters 18 and 19), PC monitors always use non-interlaced scanning with high field (refresh) rates. As a result, the video bandwidth can be in the order of 80 MHz, compared to the 6 MHz of a television receiver. Figure 22.6 is a simplified block diagram for a conventional CRT monitor.

Note that servicing of LCD monitors is difficult and costly, and often there is little or no help from manufacturers. A new monitor is usually the only economical solution to a faulty unit.

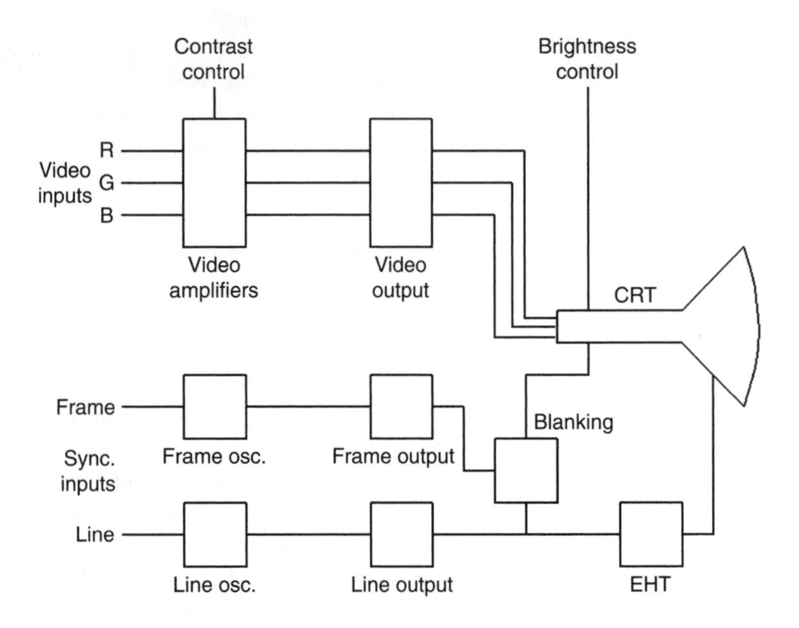


**Figure 22.6** Simplified block diagram for a computer monitor using a cathode-ray tube

Monitors using CRT displays suffer from two forms of distortion, both of which can be corrected by adjustments. **Barrel distortion** (Figure 22.7a) makes the shape of a rectangle appear to have the sides curved outwards, whereas **pincushion distortion** (Figure 22.7b) causes the opposite effect. The corrections are made by modifying the shape of the scan waveform, and these corrections are nowadays usually digitally controlled by applying pulses to correction inputs in the integrated circuit that generates the scans. A modern monitor will have controls (usually under a front panel) for applying corrections to these distortions.

The conventional monitor, using a CRT, must now conform to the EnergyStar specification for reduced power consumption. For a 15 inch monitor, this requires a maximum power consumption of around 80 W, and provision for reducing this consumption in two stages to 15 W (or less) and then to 8 W or less. The lower standby consumption figures are achieved by switching off the scanning and the video output stages when the monitor has been idle for a preset time such as 15 minutes. This action, and the time at which is operates, is controlled from the Power Options item in the Windows Control Panel. In a large office, the use of power saving can result in a considerable reduction in costs, both directly and indirectly. Using modern LCD monitors reduces power consumption appreciably.

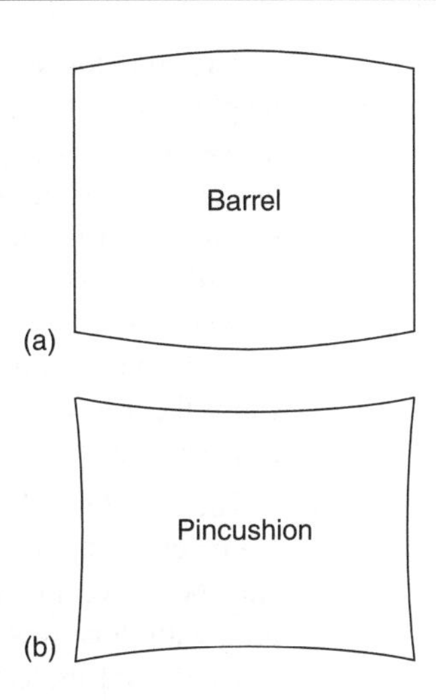


**Figure 22.7** Distortion of CRT picture: (a) barrel, and (b) pincushion

## Practical 22.3

Using the manufacturer's service manual, carry out the recommended routine maintenance actions on a CRT monitor. Using a monitor known to be faulty, carry out diagnostic and repair actions. Clean the monitor, and check the Windows settings for resolution and number of colours.

The LCD type of screen is an alternative to the CRT for monitor use, and is used universally on laptops and on most new desktop machines. Larger LCD screens (21 inch or more) are expensive for home users, but are increasingly found in business applications. Sizes of 15–19 inches are commonly found on desktop computers for home use. A drawback of some older LCD monitors was that the picture was clearly visible only if the user was directly in front of the screen. This feature is sometimes used now to provide a higher degree of privacy for the user in public places. Modern LCD screens for general use can be read easily over a large viewing angle.

Monitor faults are, for the most part, quite rare, and Chapter 18 deals with CRT construction, convergence, electromagnetic deflection and typical electrode voltages. You should not attempt to service the internal circuits of a CRT monitor unless you have experience with CRT devices and appreciation of the dangerous voltage levels. LCD monitors in general do not provide for servicing.

Installation of a monitor follows the same pattern as that of the main computer unit. Unpacking should be done carefully, and for a large CRT monitor (17 inches or more) help should be sought in lifting the monitor from its packing. Once unpacked, the monitor should be inspected for a damage that might have been incurred in transit. For modern monitors, no internal adjustments are required, and the manufacturer will actively discourage any attempt to open the monitor casing.

The monitor should be placed on the computer desk, preferably at a height that suits the user, and the unit tilted on its stand so that the user at the computer keyboard can see the display clearly without the need to hold his or her head at an awkward angle. This is important, because anyone using a VDU that is badly placed is likely to suffer neck pains. Another hazard is eye strain caused by poor monitor setup or poor positioning (such as when direct sunlight can fall on the monitor screen). Another reason for carefully positioning the monitor is that an incorrectly sited monitor can cause the user to adopt a posture that will very quickly lead to severe backache (the most common cause of absence from work).

The monitor will need connections to the mains and to the computer. The mains connection is normally made through a conventional three-pin plug, and the conventional video connector is a thick cable ending in a 15-pin plug with three rows of pins (Figure 22.8). DVI connectors are much easier to route, and the connector uses either 24 or 29 pins, depending on type. Most units use the 24-pin type. Combined digital and analogue connectors using a larger pin count are also in use.

| Pin | Function | Pin | Function |
|---|---|---|---|
| 1 | Red video | 9 | +5 V DC |
| 2 | Green video | 10 | Sync return |
| 3 | Blue video | 11 | ID bit 0 |
| 4 | Reserved | 12 | Serial data line |
| 5 | Earth | 13 | Sync (horizontal or composite) |
| 6 | Red return | | |
| 7 | Green return | 14 | Vertical sync |
| 8 | Blue return | 15 | Data clock line |

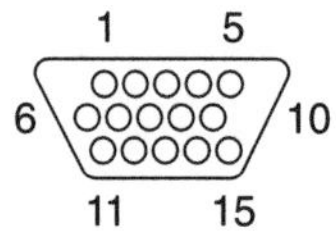

1 5
6 10
11 15

**Figure 22.8** The established 15-pin monitor connector plug

The LCD monitors used on all portable machines and increasingly on desktop machines are easily recognized by their slim profile (less than 2 inches thick) and lower weight. They are, however, still quite heavy in the larger sizes, and great care must be taken not to put any pressure on the screen, even for cleaning. The oldest types of LCD display used a purely passive form of display with signals applied to the LCD cells, but the modern type use thin-film transistors (TFTs) formed on each LCD cell, providing a colour display that is much superior to older LCD VDUs and also superior to CRT units.

## Setting up and maintenance

Normally, CRT monitors do not require any driver software to be loaded, because the drivers are built in to the Windows operating system. However, LCD monitors may need a software driver to be installed, and this should be done after the graphics card driver (see below) has been installed. In many cases, no separate LCD monitor driver is needed.

A graphics card drive is always needed, particularly if the graphics card is separate (not integrated into the motherboard). The instructions for installation of the graphics card driver should be followed carefully, particularly on modern machines, because if the plug-and-play system is allowed to install a driver it may not be the best suited to the graphics card, and it can be quite difficult to remove the default driver and install the correct one from the CD-ROM provided with the graphics card.

Once the correct drivers are installed, Windows *Control Panel* can be used to select the resolution required, the colour scheme and colour quality (in terms of number of bits used to define colour). Click *Display* from the *Control Panel* to gain access to these adjustments. The older type of monitor usually provides hardware adjustments (potentiometers or switches) for display settings of language, centring of picture, brightness, contrast, on-screen display (OSD) of settings, scan size, etc. On modern monitors these adjustments are usually made by a software menu and controlled by push-buttons on the monitor.

Cleaning of a monitor should be confined to casing and screen, using a cleaning cloth designated for the purpose. Very great care should be taken in cleaning the screen of the LCD screen, because excessive pressure can cause damage. Do not use spray cleaners intended for domestic window cleaning.

Fault-finding should start with substitution, because if the fault still persists when another monitor is used (and correctly set up) the fault is in driver settings rather then in the monitor. For the LCD monitor, replacement and return of the faulty unit to the manufacturer is the only economical course of action, but sometimes a fault in a CRT monitor can be diagnosed and repaired using normal television fault-finding techniques.

## Hazards

The long-term use of a monitor can be a hazard to health unless some care is taken. The most obvious hazard is eye strain, and you should ensure that the monitor can be clearly seen; you may need to wear spectacles to ensure easy viewing. Another hazard, very common among office workers, is back strain because of working hunched over a monitor. Servicing use of a computer is unlikely to be as intensive as office use, but you should be aware of this problem and make sure that you use the monitor while seated in a comfortable position that gives a clear view. It helps if the monitor itself is a good one. A 17 inch LCD monitor, for example, can be much clearer to read than any CRT monitor. In addition, you are much less likely to have problems with working at a monitor if you are not excessively tired from other work.

## Multiple-choice revision questions

22.1 You have plugged in the keyboard and the mouse, but neither seems to work. The most likely cause is:
(a) the plugs are not PS/2
(b) the plugs are reversed
(c) the keyboard is faulty
(d) the mouse is faulty.

22.2 The mouse action is jerky. You would first check:
(a) the driver software
(b) the connector
(c) the skids
(d) the rollers.

22.3 The monitor works, but only in 640 × 480 resolution. You need to:
(a) check with another monitor
(b) make adjustments in software
(c) check the connector
(d) change the video card.

22.4 The computer has been idle for some time. Suddenly the monitor picture disappears. This is because:
(a) the monitor is faulty
(b) the connection is faulty
(c) the computer is faulty
(d) the power-saving software has been activated.

22.5 To correct pincushion distortion on a monitor picture, you would:
(a) check the connectors
(b) adjust the monitor controls
(c) make use of *Control Panel* software
(d) run a diagnostic check.

22.6 Care must be taken with an LCD monitor because:
(a) the casing is made of plastic
(b) the unit is lighter than a CRT monitor
(c) the screen surface is sensitive to touch
(d) the screen attracts dirt more easily.

# 23 Drives

## Floppy drive

Some modern PCs use a **floppy-disk drive (FDD)**, but this is now less common, particularly on laptop machines. The floppy drive is now being replaced by the 'pen drive', a memory device that plugs into the universal serial bus (USB) socket and has a much larger capacity (typically 128 MB to 2 GB) than the old-style floppy disk. Where a floppy drive is needed for compatibility with other computers, it can be added on a modern machine by way of an external drive plugged into the USB port.

The FDD makes use of thin plastic 3.5 inch magnetic disks with a capacity of around 1.4 Mb. The use of a floppy drive is essential on older machines because it allowed a computer to be commissioned with a new unformatted hard drive. On these machines, a **system floppy**, usually included with a copy of Windows for OEM (original equipment manufacturer) use, contains the MS-DOS command files along with drivers for CD-ROM drives, allowing you to format a hard drive, copy files to it and make the CD-ROM drive available for installing Windows. On modern machines, this commissioning can all be done using a CD-ROM.

Before a new floppy disk can be used in the drive it must be **formatted**, meaning that the disk is magnetically marked into tracks and sectors with an identifying byte, and the position of the start of each sector is recorded in one of the tracks.

A **track** is a circle on the disk, and a **sector** is a portion of a track, conventionally holding 512 bytes. The identifying portion is called the **directory track** or **table of contents (TOC)**; its use allows data to be split up and recorded on different tracks and sectors, with the directory track information used to specify where each portion of data is placed. The same type of drive and sector formatting is also used on the hard drive (see later).

Unlike a hard drive, the recording system for a floppy is standardized because a floppy recorded on one PC must be readable on another. A new floppy must be formatted before use, and can be reformatted to wipe it clear of all data. The formatting can be carried out from Windows as illustrated in Figure 23.1. Click the program Windows Explorer, which will display the drives. The floppy drive (A:) is shown, and if you place the mouse cursor over this and click the right-hand button (an action called right-clicking) you will see a menu whose options include format. Now left-click (left-hand button) on this action. This is not a fast operation, and when you format a floppy you may find that the computer cannot be simultaneously used for other actions.

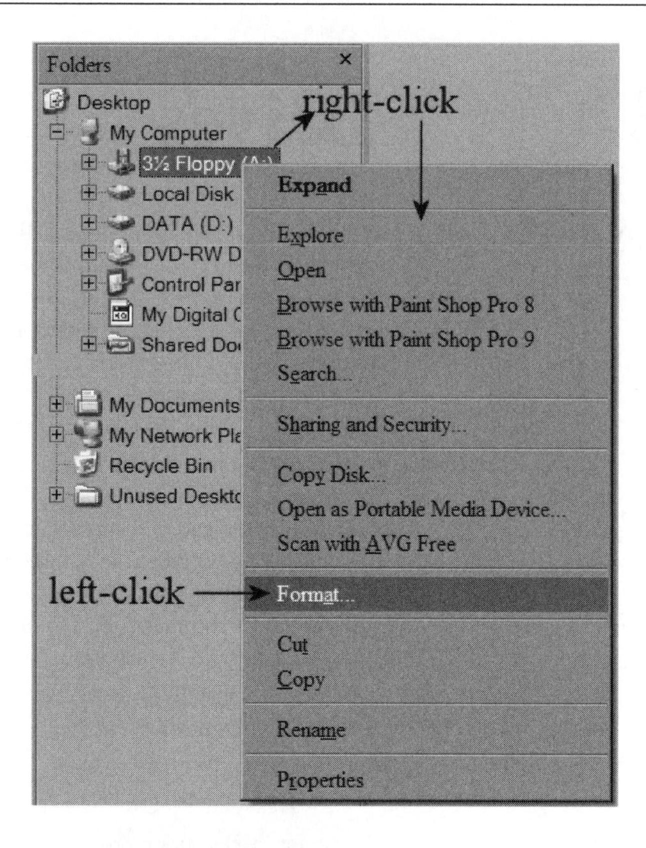


**Figure 23.1** Formatting a floppy disk

The controller for a floppy drive is on the system motherboard, so that the drive itself contains only the minimum of hardware. The connector for the floppy drive uses a 34-way flat-ribbon cable with the usual stripe to identify the pin 1 side. Cables usually allow for two floppy drives, but only one is normally used. The floppy drive also uses the same type of power connector as the hard drive, although a miniature version is sometimes used. The floppy drive connector may exist on a new motherboard even if no floppy drive is fitted to the computer.

The floppy disk itself consists of a thin plastic disk coated with magnetic material on both sides, and held in a plastic casing to protect it. When the disk is inserted into the drive, read/write heads are lightly clamped against the plastic disk on each side (opposing each other), and the disk is spun to about 300 rpm. A tab at the corner of the disk housing can be positioned so that the disk is write-protected; this is checked by the drive when a floppy disk is inserted and a signal prevents writing, although the disk can be read. Some computers will try to read from a floppy at switch-on time if one has been left inserted in the drive, and will deliver an error message because the operating system is not on the floppy. More usually, the PC can be configured to ignore a floppy when the machine is booted, so that the operating system is only ever read from the hard drive. This can cause problems if the hard drive fails, so that you need to know how to reverse this setting (in the

CMOS-RAM settings) if you are using a floppy to install MS-DOS so as to diagnose a hard-drive failure or format a new drive.

A floppy disk is much less well protected than a hard drive, and floppies need to be cared for. They should be stored in a cool, dry place, preferably in a box designed for the purpose, well away from magnetic fields (such as from loudspeakers) or strong electrostatic fields. Given such care, floppies can have a very long useful life, but carelessness can easily make the tracks on a floppy unreadable. Figure 23.2 shows a block diagram for the drive and its controller.

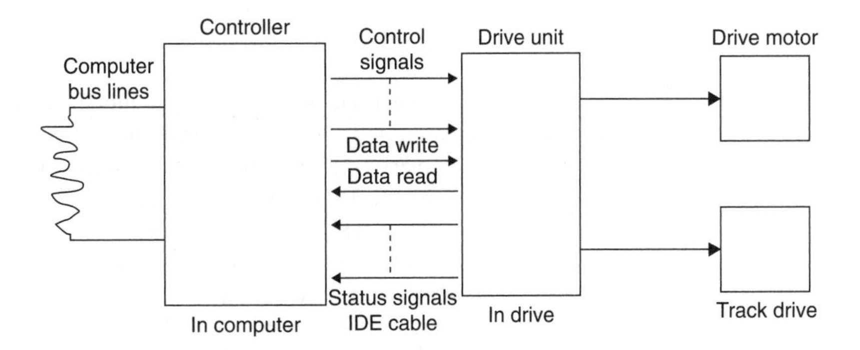


**Figure 23.2** Block diagram for floppy drive and controller

The FDD unit is now so cheap (typically £3) that any attempt to repair a faulty unit is pointless. The drive, whether internal or external, is easy to replace and this is by far the best solution to FDD problems that are not caused by poor-quality disks. For an internal FDD unit, the mounting screws are easily accessible once the cover of the computer casing has been removed, and once the power cable and data cable have been removed, the drive unit can be slid out. A replacement drive can then be installed by reversing these steps. The unit can be tested by copying a file to it and checking that the file can be used (for example, if you copy a text file, use a word processor to look at the copy on the floppy disk).

## Hard drive

The main type of disk for a modern PC type of computer is the **hard drive**. This consists of several magnetic disks (or **platters**) coated with a magnetic material, all revolving on the same shaft. Electromagnetic heads, similar in principle to tapeheads but much smaller, can be placed at any point on each platter to read or write data; all the heads are moved together, not independently. The head movement is controlled by a voice-coil mechanism, similar in construction to the movement of a moving-coil loudspeaker, controlled in a feedback system. All this is controlled by the microprocessor. The rotational speed of a typical hard drive is 7200 rpm, with some types using 10 000 rpm or more. Such speeds are possible because the record/replay heads are not in contact with the platter surfaces, but float on a thin film of air, like a hovercraft. The whole mechanical assembly is contained in an airtight casing that has been assembled in a 'clean room' to avoid the entry of dust.

A typical hard drive will have a capacity of 40 GB or more, and access times are in the order of a few milliseconds.

- Disk drive heads are manufactured using miniature thin-film coils rather than wire-wound coils. Increasingly, magnetoresistive heads are being used for reading because they provide larger signal outputs. A magnetoresistive material is one whose resistance changes when it is affected by a magnetic field, and such materials can be used in very small sensors manufactured using the same type of techniques as for integrated circuits.

At one time, the controller circuits for a hard drive were located on a card inserted into the motherboard. The action of a drive controller is to change the binary code for writing into a format that can be recorded magnetically, and perform the reverse of this set of actions for reading. Modern hard drives use **IDE** (integrated drive electronics), meaning that all the controller circuits are located on the drive itself, with a simple data interface with the motherboard. The later (faster) version of this system is known as **EIDE** (E meaning extended). An older system, SCSI (small computer systems interface), is still in use for machines that require a large number of drives.

**WARNING**: The casing of a hard drive must **never** be opened. Nothing inside the casing will be useable if the casing has been exposed to air containing dust and smoke.

Figure 23.3 shows a block diagram for a hard drive, omitting power supplies. The data cable connector supplies signals that control the drive motor and the position of the read/write heads, and the data signals are passed to and from the heads.

In addition, the higher speed of modern computers makes it undesirable for the microprocessor to control hard-drive reading or writing. This is now done using a separate chip, the **direct memory access (DMA)** chip. The main microprocessor will pass to the DMA chip the details of where the data is located, where it has to be copied and the number of bytes of data, and the DMA chip will then carry out the transfer, leaving the microprocessor free to work on another action.

Write protection is seldom used for files on a hard drive, but if required the protection can be applied by Windows software. The procedure is to right-click on the file that is to be protected, left-click on *Properties* and then left-click to place a tick in the box marked 'Read-only' (Figure 23.4).

The EIDE hard drive is connected by way of a 40-way cable (although some cables use 80 leads, with the extra 40 leads used as shields between the signal wires). At one end, a socket engages with the EIDE socket on

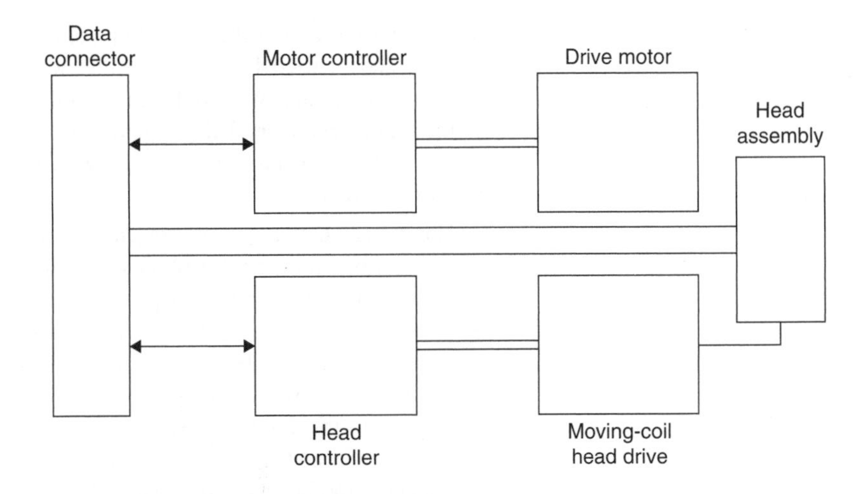


**Figure 23.3** Block diagram of hard drive, simplified

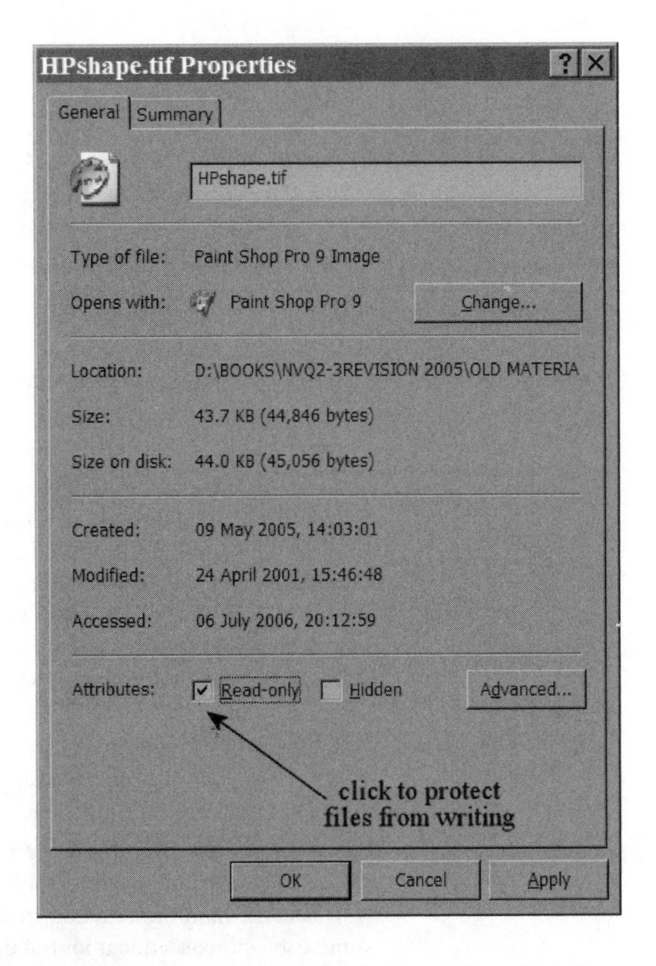


**Figure 23.4** Protecting files on the hard drive against alteration

the motherboard, and the other two sockets are used for hard drives, with the end socket used for the drive that will be lettered **C**. On modern PC machines, there are two EIDE plugs on the motherboard, referred to as the primary and secondary EIDE connectors. This allows up to four devices to be connected, typically two hard drives, a fast CD-ROM drive and a slower CD-writing drive.

The cables that are used for the parallel AT attachment (PATA) type of connections are of the ribbon type, and the connection to pin 1 of the sockets is marked by a red stripe or some similar identification on the end wire. When cables are connected up, care must be taken to ensure that this striped side of the cable is at the pin 1 end of the connector. If a connector is reversed the drive will not be correctly connected and it will not be under the control of the computer. Some new machines use a seven-wire serial AT attachment (SATA) cable replacing parallel EIDE, and requiring SATA hard drives and a motherboard that supports SATA. Figure 23.5 shows the shape of the power connector and the plugs for PATA and SATA connections.

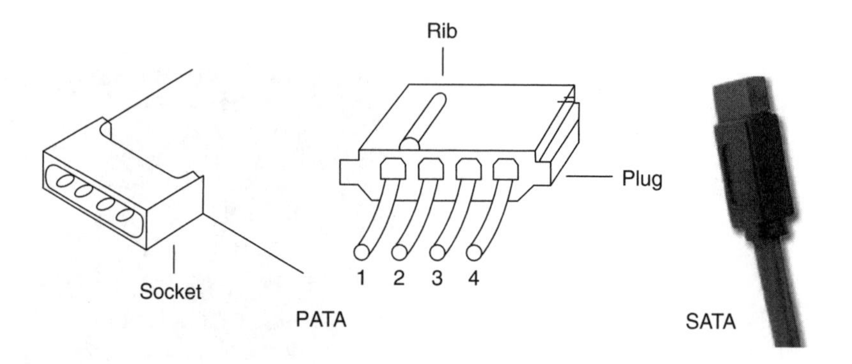


**Figure 23.5** Power plugs for PATA and SATA drive types

In addition to the data cable, each drive needs to use a power cable. These are taken from the power supply unit in the casing, and typically six will be provided to be used on hard drives, a floppy drive, a CD-ROM drive and any other drives that are needed, such as a CD writer or tape backup drives. The power plug uses a four-way cable, and the usual form of plug and socket is illustrated, left. The rib on the socket and the corresponding groove on the plug ensure that the plug cannot be inserted incorrectly. A miniature version is sometimes used for 3½ inch floppy drives.

Cable runs in a modern computer are critical, and you should take great care in servicing to ensure that any cable that has been disturbed is always replaced in the same position. An incorrect cable positioning (as for example when a cable is replaced) can cause the computer to become more sensitive to interference, resulting in rebooting during use with consequent loss of data.

Before a new hard drive can be used it must be **formatted**, meaning that a set of tracks on each platter is magnetically marked into tracks and sectors with an identifying byte, and the position of the start of each sector is recorded in one of the tracks. A **track** is a circle on a platter, and a **sector** is a portion of a track, conventionally holding 512 bytes. The identifying portion is called the **directory track** or **TOC**; its use allows data to be split up and recorded on different tracks and sectors, with the directory track information used to specify where each portion of data is placed.

Formatting removes all the data that may exist on the disk, so this should be applied only to new hard drives. The only usual exception is when a drive has to have all information removed so that it can be used on another machine, or sold as second-hand. In some circumstances, a severe problem (such as a virus that cannot be removed) may call for a drive to be reformatted and reinstalled with Windows and other software. If a machine is to be scrapped, the hard drive may contain confidential information, and simply deleting or reformatting will still leave enough traces for a determined hacker to find sensitive information. Specialist software can be obtained to remove all traces of data from the hard drive, or the drive can be physically wrecked by removing the platters and crushing them (beware of splinters).

To format a hard drive, a formatting program must be used. On older machines, the program called FORMAT.COM has to be used, and with MS-DOS floppy in the drive and in use, formatting a hard drive is carried out by typing the command:

FORMAT C:

assuming that the hard drive uses the letter C, as is normal. Formatting will take several minutes, with the exact time depending on the size of the hard drive and the speed of processing. Once the drive is formatted, the Windows operating system can be installed by using the CD-ROM drive, which can be provided with a driver from the same floppy disk that contained the MS-DOS system files. On modern machines, the formatting is carried out from the CD-ROM used for installing Windows.

The usual type of hard drive is fixed, and the main drive, lettered C, is used to hold the operating system. For a small computer, this drive may also be used for programs and data, but the drive is often partitioned, meaning that, for example, a hard drive of 120 GB can be invisibly divided by software into portions (partitions) of, say, 40 GB and 80 GB. One of these partitions can be designated as the boot drive (C:) and the other as a data drive (D:). These letter allocations are made automatically by Windows, and this can cause confusion on some machines where slots for digital camera memory cards are provided, each slot being allocated a drive letter.

A removable hard drive can also be used for backup. The hard drive fits into a casing that is connected to the computer through the parallel port or by way of USB or Firewire (IEEE1394) ports, usually with a separate power supply. Backup files can be copied to this drive, and the drive then removed from its casing and kept in a safe place. This type of backup combines fast copying with ease of use, and can be used, for example, to ensure that all the data that has been altered in a day is backed up and put in a safe each evening. The life of a drive of this type can be very long because it need not be running continually.

Built-in hard drives are mechanical components that are continually running while a PC is active, so that they have a limited life. Inevitably, a hard drive will fail at some time, although this may be after many years of use, and the computer may have been scrapped and all the data transferred before a drive reaches the stage at which failure occurs. The most serious failure is one that affects the main drive, because this makes it impossible to load the operating system. If a separate data drive (not just a partition of the main drive) fails it cannot be used, but the computer remains useable.

Typical failure problems concern either electromechanical components or disk corruption. The main drive motor can fail, making it impossible to use the drive, so that the operating system will not load. This type of failure is easy to diagnose because you will not hear the drive motor start when the computer is switched on. Another form of electromechanical failure affects the voice-coil drive for the heads, so that the tracks cannot be correctly located. In this type of failure also the operating system will not load, and the usual clicking noise of the head system will be absent.

Failure of the electromechanical system will require the drive to be replaced, but the data that was recorded on the drive will still be intact. Specialist firms can open the drive in a clean room and copy the data to a backup drive so that you can restore your data. Remember that the value of the data on a computer system used for business purposes may be very much greater than the value of the computer itself.

Drive failure that causes corruption is much more serious, particularly on a disk that contains data. A typical cause is that a motor fault slows the platters down and the 'flying' heads scrape against the platter surfaces, tearing the magnetic material. It is very difficult, or impossible, to recover data from this type of head crash. See Chapter 20 for details of backing up data.

The hard drive(s) can be checked by using the operating system. Windows provides for disk checking (Figure 23.6) and for **defragmentation**, which should be carried out at regular intervals. When a file or set of files is deleted from a hard drive, the digital bits are not, in fact, removed from the disk surfaces. All that happens is that the directory entry for the file(s) is deleted. This is why 'deleted' files can often be recovered. Because the directory entries have been deleted, however, other files can be recorded in the positions used by the 'deleted' files. This makes the deletion more permanent, but the new files are unlikely to fit the vacated space exactly, leaving the hard drive with unusable sections and splitting new files into pieces that are recorded in different parts of the disk. This is what is called **fragmentation**. Fragmentation is undesirable because it slows down disk action (too much time is spent in looking for bits of a fragmented file), so that the defragmentation action is

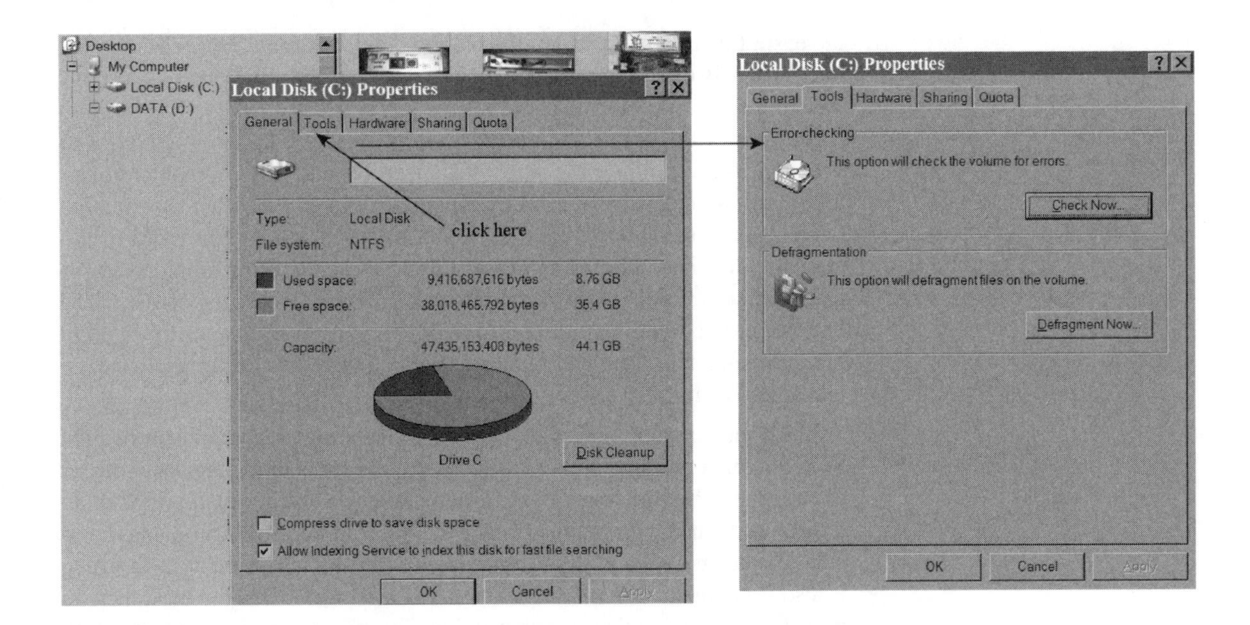


**Figure 23.6** Hard-drive maintenance, (a) error checking using Windows, and (b) defragmentation

needed in order to recover full operating speed. The defragmenting action seeks out the fragments of a file and rewrites them as a single file (a **contiguous** file), and this is continued until most of the drive is put into order. Note that some files cannot be moved because the operating system requires them to be in fixed places.

Removal and refitting of a hard drive unit is straightforward on normal PC machines (not so on laptops). The casing of the computer is removed, and the four screws that hold the hard drive in place are removed. A small piece of Blu-Tack® on the tip of the screwdriver ensures that when a screw is undone it will not drop into the casing. The data and the power cables are then disconnected, and the drive is slid forwards out of the computer. On some machines, you may need to remove a fascia to allow the drive to be removed. Reassembly or assembly of a new hard drive is done by reversing these procedures.

Never transfer an old hard drive (4 years old or more) to a new computer. A drive of that age may be reaching the end of its life if it has been intensively used, and failure of the drive means that you lose the software on it and will have to replace the drive. Note also that a really old drive will use a 40-wire connector; later types used the standard 80-wire connection.

Note that when a drive has to be replaced, all of the software that was installed on the old drive must be installed on the new drive (after it has been formatted and, if needed, partitioned). You cannot simply copy program files from the old drive to the new one. You can, however, copy data files, and the easiest method is to mount the old drive in an external drive casing, plug into the USB or Firewire port of the computer, and use Windows to copy data files.

## Optical (CD/DVD) drives

The optical drive allows the use of CD or DVDs that carry computer codes, and these are extensively used to contain programs and other information (such as multimedia text, pictures, audio or video). Older CD-ROM drives are read-only, but a later generation of drives permits both reading and writing. The conventional CD-ROM will hold about 700 Mbytes of data. The later development, the DVD read/write drive, is almost universally fitted to modern computers, and can read or write CD or DVDs, either the R (write once only) or the RW (write, read, delete, rewrite) type.

The optical system makes use of optical recording, using a beam of light from a miniature semiconductor laser. Such a beam is of low power, a matter of milliwatts, but the focus of the beam can be to a very small point, about 0.6 μm in diameter; for comparison, a human hair is around 50 μm in diameter. The beam can be used to form pits in a flat surface, using a depth which is also very small, in the order of 0.1 μm. If no beam strikes the disk, then no pit is formed, so that this system can digitally code pulses into the form of pit or no-pit. These pits on the master disk are converted to pits of the same scale on the copies. The pits/dimples are of such a small size that the tracks of the CD can be very close; about 60 CD tracks take up the same width as one vinyl LP track, and the tracks on a DVD are closer still. The pits are arranged into a spiral track starting near the hub of the disk and ending near the edge. Digital signals of 8 bits are converted into 14-bit signals to allow for error-reduction methods to be used, and also to avoid long runs of the same digit (0 or 1) occurring.

Reading a set of pits on a disk also makes use of a semiconductor laser, but of much lower power since it need not vaporize the material. The reading beam will be reflected from the disk where no pit exists, but scattered where there is a pit. The rotational speed of the disk is controlled so that the rate of reading pits is constant. This means that the disk spins more quickly at the start of the track than it does at the end. By using a lens system that allows the light to travel in both directions to and from the disk surface it is possible to focus a reflected beam on to a photodiode, and pick up a signal when the beam is reflected from the disk, with no signal when the beam falls on to a pit. The output from this diode is the digital signal that will be amplified and then processed eventually into an audio signal. Only light from a laser source can fulfil the requirements of being perfectly monochromatic (one single frequency) and coherent (no breaks in the wave-train) so as to permit focusing to such a fine spot.

Focusing and track location are automatic. The returned reading beam is split and the side beams detected so that servo circuits can locate the track and focus on it. This mechanism must not be disturbed unless you have had training in servicing optical drives.

The optical reader drive uses the laser beam reading system that was originally developed for music CDs, but the speed of rotation of the disk is usually much higher. When you buy a CD-ROM drive, you will see this speed quoted as a multiple of the normal speed of the music CD, such as 20×, 32×, and so on. This allows the disks to be read at higher speeds, so that information can be read from a CD almost as quickly as it could from old hard drives. The normal rotational speed of DVDs is much higher, so that higher speeds are more usually in the range of 2× to 10×. At the time of writing, optical drives that are being fitted to computers will take either the older single-layer writable DVD (4.7 GB) or the later double-layer type (8.5 GB). Further developments of the DVD system such as Blu-ray and HDVD can store much larger amounts of data per disk.

Note that the double-layer disks cannot (at the time of writing) be written with a continuous file of 8.5 GB; you need to record one layer to its limit and then write a separate file on the other layer.

The form of optical drives is fairly well standardized now, as a slim 5¼ inch casing. The front panel holds a drive-on light, a volume control and a headphone jack. This allows the use of the drive for playing ordinary audio CDs. The output audio signal is typically 0.6 V r.m.s. at 1 kHz.

When a button on the front panel is pressed, a tray slides out so that a disk can be placed in the tray, and pressing the button again will cause the tray to slide back in. The drive motor then spins the disk (taking a second or so to build up to full speed), and from then on the disk can be accessed, usually as drive D:. Windows allows you to specify Autostart, so that when a CD or DVD is inserted it will run its program(s) automatically.

There are three connectors at the rear of the CD-ROM drive, the power connector feeding +5 V and +12 V DC, an audio output connector for amplifiers, and the data interface. The audio interface typically uses a four-pin connector. The data interface uses a 40-pin insulation displacement connector (IDC) type.

The CD-ROM drive contains a low-power laser, but because the beam can focus to such a small point, it can cause damage to your eyes. No attempt should be made to use a CD reader with the casing removed, and if focusing problems are encountered, the unit should be returned to the manufacturer.

Figure 23.7 shows a simplified block diagram for a typical optical drive unit. The control signals set the speed of the drive motor, which is not constant; it runs faster when the inner tracks are being read and slower for the outer tracks, keeping the data speed constant. The beam position drives select the track and maintain focus, and the optical head reads the signals reflected from the track. These signals are then converted into conventional digital data in the interface.

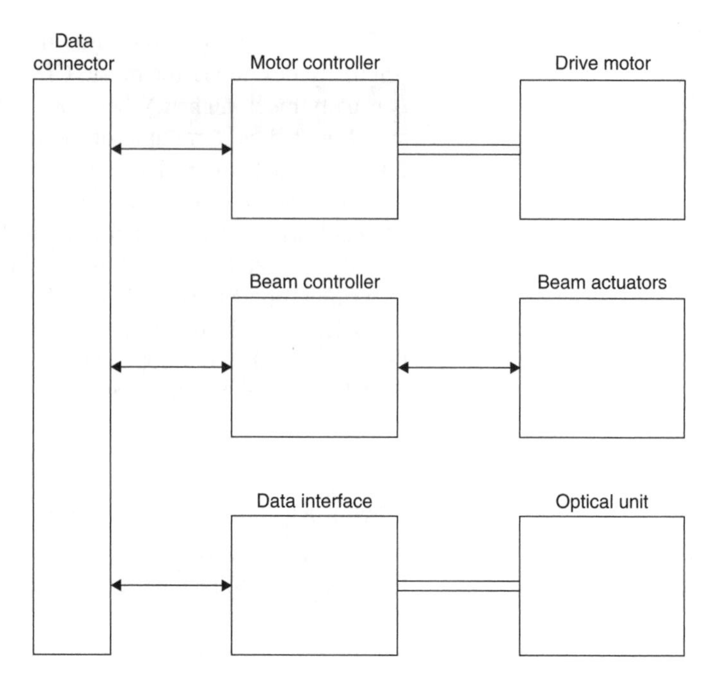


**Figure 23.7** Block diagram, simplified, for a CD-ROM drive

## Multiple-choice revision questions

23.1 A floppy is being written to and an error message appears. Which of the following **cannot** be the cause of the problem?
(a) the disk has had its write-protection set
(b) there is no floppy-disk driver software
(c) the disk is not formatted
(d) the disk is full.

23.2 Any make of hard drive intended for a PC can be used as a replacement for a faulty unit. This is because:
(a) no driver software is needed
(b) the controller on the motherboard can be used with any drive
(c) the controller is built into the drive
(d) all hard drives are of identical design.

23.3 A hard drive can have a large data capacity because it:
(a) is physically larger than a floppy
(b) runs at a faster speed than a floppy
(c) does not have to be inserted and removed
(d) uses many platters of magnetic material.

23.4 Which of these is NOT an advantage of a hard drive as compared to a floppy?
(a) it has a much greater data capacity
(b) it is faster in use
(c) it does not need to be formatted
(d) it can hold the Windows software.

23.5 A hard drive sometimes does not start until it is tapped. This is because:
(a) there is excessive friction on the drive motor spindle
(b) there is too much software on the drive
(c) the drive is not formatted
(d) air has entered the drive casing.

23.6 A CD-ROM is inserted but does not start sending data to the computer. This is because:
(a) the drive is not formatted
(b) the drive is faulty
(c) Autostart has not been activated in Windows
(d) the CD-ROM is faulty.

# 24 Printers

Many types of printer have been marketed for computers, but the dominant varieties are the impact dot-matrix, the laser printer and the inkjet types. Virtually all of them use the standard Centronics connector or the later universal serial bus (USB) connector so that you can use any printer with any computer. On modern home computers, the inkjet type is predominant, and the impact dot-matrix type is now used only in offices where the ability to produce several carbon copies is required. The laser printer is used mainly where high-speed black-and-white output is needed, although colour laser printers are now available at reasonable prices and can be a better option when colour is needed in text documents.

## Inkjet printer

The inkjet printer uses the principle of a set of tiny jets used to squirt a fast-drying ink at the paper. The two main systems currently are the bubblejet™ and the piezoelectric jet. The bubblejet system, devised by Canon, uses a tiny heating element in each jet tube, and an electric current through this element will momentarily vaporize the ink, creating a bubble that expels a drop of ink through the jet. The piezoelectric system, devised by Epson for its Stylus™ models, uses for part of the tube a piezoelectric crystal material which contracts when an electrical voltage is applied, so expelling a drop of ink from the jet. Both methods can provide high resolution and are adaptable to colour printing.

The ink is contained in one or more cartridges (Figure 24.1). The cartridges are contained in line in a **cradle** that can be moved from side to side in the course of printing. This cradle must **never** be moved by hand; always use the control buttons when the cradle is to be moved for replacement of a cartridge. Nearly all inkjet printers manufactured now are colour types, and the usual cartridge scheme for general purposes is to use one black-ink cartridge and one three-colour cartridge.

Some models, designed for photographic reproduction, use a separate cartridge for each colour, and some have as many as six cartridges, each operating its own set of jets. Each cartridge is clipped securely into the cradle so as to ensure that the electrical contacts are firmly pressed together and that the ink nozzles are correctly lined up. Printer software contains programs for lining up the colour cartridge(s) so that coloured lines match the positioning of black lines. See later in this chapter for details of how to replace or install a cartridge. The ink is designed to dry quickly so that sheets do not smear when they are ejected from the printer. Spilt ink must be wetted and mopped up immediately because it can be almost impossible to remove from fabrics, particularly carpets, once dry.

The two other main variations in design concern ink tanks and jetheads. Some designs, notably Hewlett-Packard (H-P), use a combined inktank and jethead assembly, so that when the ink runs out, the whole assembly

**Figure 24.1** Typical inkjet cartridges: (a) with tank and printhead, and (b) tank only

is changed. This is expensive, but it ensures that a new set of jets is bought with each ink change. On other machines, the ink tank can be renewed separately, giving lower running costs, but with the risk that if an ink jet becomes blocked (usually because of ink drying out when the printer is not used daily) the cost of replacing the jets will be high (often as much as a completely new printer).

The cartridge cradle can move from side to side on rails, controlled by a toothed belt that in turn is driven by a motor controlled by software. The electrical connections are through a ribbon cable that is long enough to cope with the full range of movement of the cradle. The head drive motor is controlled by the printer software. The combination of head movement and paper movement allows software to control where each dot of ink lands, so building up an image that can be text, a photograph or both.

The paper is usually in single-sheet form fed from a hopper. Some printers place the hopper at the rear of the printer, feeding sheets down into the printing plane and then out at the front. This system, used by Epson and Canon among others, relies partly on gravity for feeding, but the paper path is fairly straight, allowing thick sheets to be used. Another method, favoured by H-P, is to place the paper in a horizontal holder that is at the front of the printer, and pull each sheet through, bending the paper through 180° to the printing plane, then ejecting the paper on to a tray above the feed tray. This makes the printer more compact, but it cannot handle thicker paper without seriously bending them. Figure 24.2 shows examples of both styles of paper feed.

The differences are important if you use coated inkjet paper that is coated on one side, or if you want to print both sides of a sheet. The rear-hopper system requires the coated side of the paper to face towards the printer, whereas the front-holder method requires you to load the paper with the coated surface downwards.

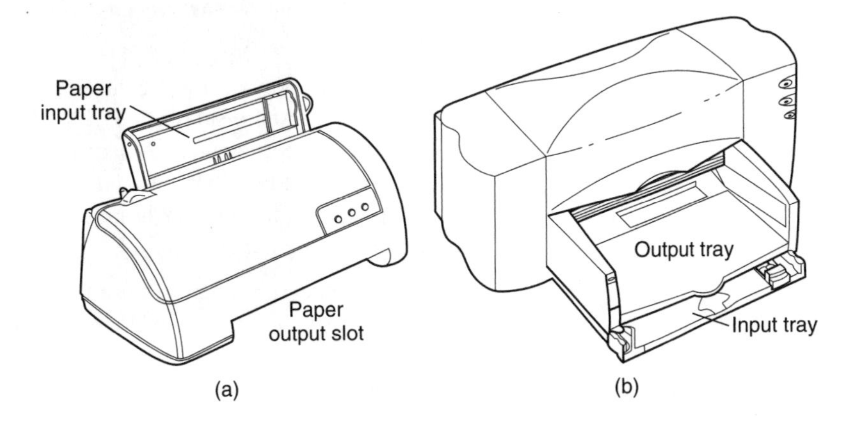


**Figure 24.2** Paper feed methods: (a) rear hopper, and (b) underside hopper

Printing speed was at one time quoted in terms of text characters per second, and the figure referred to rather artificial conditions, so that it served only as a guide to the speed of printing that you might observe. Speeds are now more often quoted in pages per minute, and most printers can use more than one printing speed, depending on print quality required. Typically, you may have the choice of printing in *Draft* mode at 10 pages per minute, in *Normal* mode at 5 pages per minute, or in *Best* mode at 3 pages per minute. These are full-page black text printing speeds, and for most purposes the draft mode is perfectly acceptable on most machines. On a few types of machine, draft mode is too faint for purposes other than speed of printing. Printing speeds will be much slower if a document contains graphics or elaborate text fonts, compared to a document that uses a simple font throughout. Colour printing is always considerably slower than monochrome, and an A4 colour page may need several minutes to print in 'best' mode.

The **font**, or fount, means the design of the printed characters.

The selections of print quality are made by controlling the nozzle movement so as to alter the number of ink-dots per inch. In draft mode, modern inkjet printers use a resolution of 300 × 300 dots per inch (the same resolution both horizontally and vertically). In the higher quality modes, the resolution is normally 600 × 600 dots per inch or more. For graphics, the printer head drive and paper feed can be controlled so that dots overlap, so that resolutions that appear to be much higher can be obtained, typically up to 1700 dots per inch. The true resolution, however, is that set by the number of nozzles on the jethead. Attempts to obtain very high apparent resolution require a combination of ink and paper that avoids one dot of ink spreading too much on the paper, and high-resolution pictures often come off the printer quite wet and need to be carefully dried to avoid smudging.

At one time, printers used control panels with a set of switches and, often, a liquid crystal display (LCD). Modern printers are totally controlled by software in the computer, so that the controls on the printer itself nowadays amount to a few switches and light-emitting diode (LED) indicators. Some types even dispense with an on/off switch and are set to a standby mode by the computer.

A typical modern printer will use two buttons and three LEDs. One button is used for power on/off, and the other to resume normal printing after an interruption (such as replacing paper or a cartridge). The power button may act only on a low-voltage d.c. supply if the printer takes its power from a mains adapter, and you may want to ensure that the power supply unit can be switched off at the mains when all printing is finished. One LED will be used to indicate power on, one to show that the *Resume* button needs to be pressed, and one to indicate that a cartridge needs to be replaced.

The cables that need to be connected to the inkjet printer are the power (mains) cable and the printer cable. Some printers contain a power supply internally and are connected directly to a.c. mains; more commonly now, the printer uses a low-voltage d.c. supply, and is powered through a transformer/rectifier supply that plugs directly into a power socket, or has an a.c. power cable attached. Typical power requirements are in the order of 5 W in standby and 30 W when printing.

The printer data cable needs more consideration. Older printers could use an ordinary parallel (Centronics) type of cable, but later printers needed the bidirectional type termed IEEE 1284. This type of cable allows two-way communication between the printer and the computer, so that the printer can signal to the computer when it is out of paper or ink. The IEEE 1284 cable also requires the computer to be set up so that its parallel port is of the extended compatability port (ECP) type; this usually has to be done using the CMOS-RAM setup action. At the time of writing, all printers can use a USB connection, and this is the preferred method of connection. Currently, only a few models now provide a Centronics socket. A few machines, such as the H-P Color Laserjet, allow for both USB and Ethernet connections. The Ethernet connection allows the printer to be used by any computer on the same network.

Paper for text work on inkjet machines can be any normal office paper marked as suitable for photocopying, laser or inkjet printers. The specification of paper thickness is in terms of the weight of a square metre of paper, and typical weights for inkjet printing are 75–90 g/m$^2$. The use of expensive coated papers is justified only if colour photographs are being reproduced. The best quality thick photographic paper can cost 50 pence or more per sheet. If you need to reproduce photographs on thick paper, make sure that the printer can handle such paper.

Paper needs to be stored in a cool, dry place. Before loading paper into an input tray, the sheets should be shuffled to ensure that they are not sticking together: if more than one sheet enters the printer it is likely to cause a jam. Note that glossy paper suitable for inkjet printers may not be suitable for laser printers.

**Commissioning an inkjet printer**

A new inkjet printer will be delivered in a cardboard case, and this should be opened carefully to avoid damaging the contents or spilling packing materials. A typical pack will contain the printer, its power supply/mains cable, ink cartridges, software on a CD-ROM and an instruction manual. These should be checked to ensure that you have all that you need. The printer cable is not usually included with the printer.

You should always keep all packing material and cases in the event of having to return or move the printer.

When the unpacking is complete and the items have been checked, the printer can be set up. Internal packing is normally used to prevent the printheads from moving while the printer is being transported, and all of this packing must be removed before any attempt is made to connect up the printer. Other moving parts may be taped to prevent movement, and the tape must also be removed. The packing and taping will be described either in the printer manual or in a separate document. Keep the packing and the information in the event of having to repack the printer. Check carefully for this packing and taping, because if it is not removed you will not be able to print and may cause damage to the printer.

Once the packing has been removed, check that the printer is in good condition, with no apparent damage, and place it on a flat surface. This need not necessarily be where it will finally be used, but it is often more convenient to install the ink cartridge(s) when the printer is easily accessible. When you have to replace a cartridge, it will not matter if the printer is less accessible because you will already have gone through the procedure. The printer can now be connected to the power supply for testing.

Note that many printers require you to install software before you attach the data cable between the computer and the printer. This is to prevent Windows from recognizing the connection of the printer and automatically loading a driver that, although permitting printing of text, will not allow you to make full use of the facilities of the printer.

First, the ink cartridge(s) need to be inserted. Each cartridge comes in a pack and must be carefully removed. The ink nozzles on the cartridge are usually covered by a strip of transparent tape which has to be removed before the cartridge is inserted. Before removing the tape, check how the cartridge fits into the cradle, and where the retaining clip engages. When you are certain that you know how this is done, remove the tape, press the cartridge into place and clip it securely. A few machines use only one (large) cartridge for both black and coloured inks. More usually, four cartridges are used, with separate black, yellow, cyan and magenta inks. Photographic quality printers may use six or more ink cartridges.

Different machines use different methods of retaining cartridges, so that you will have to check with the supplied manual.

Note that a colour printer will usually need to have its colour cartridge installed even if you never intend to print in colour. In general, machines that use more than one cartridge will not work if a cartridge is missing.

Some printers can be self-tested at this point. If so, load in some paper, and press the self-test button to see a sheet printer. It is likely that a modern machine will make no provision for this action.

The next step is to install the Windows driver. This software converts the computer data into signals that the printer can use to produce text and graphics on the paper, and it is important to use a driver that is correct for your printer. The importance of carrying out this step before the printer is connected is that Windows may install a default driver if the printer is connected and switched on, and this driver may conflict with the one provided on the software that comes with the printer.

Driver installation starts with establishing the correct printer description. You need to know the name of the manufacturer and the precise model number of the printer. For example, if you are installing an Epson Stylus Photo R400, it is not good enough to use the driver for any other Stylus model; you must look for the precise model.

With the computer switched on but not connected to the printer, run Windows and insert the CD-ROM that came with the printer. This should start automatically and will install the driver software. If the installation does not start automatically (because the default settings of Windows have been altered), use Windows Explorer to find the optical drive, and look for a file called, typically, SETUP.EXE. Click this file to start the installation action.

Note that an inkjet printer can take some time, after switching on, to print the first page of a document. You will hear a considerable amount of clicking and whining as the motors are activated, although there is no paper movement. This is a routine of activating the jets and moving the cradle, and in the course of this ink is squirted into pads over which the cradle rests. At some (late) stage in the life of the printer these pads will become saturated with ink, mostly dry, and will have to be replaced before there is a risk of ink escaping from the printer body and causing stains on the desk.

## Practical 24.1

Install an inkjet printer and print a sample page using both monochrome and colour.

Modern inkjet printers require little maintenance other than cleaning and cartridge replacement. The exterior of the printer should be wiped at intervals with a soft, moist cloth. You must not use any form of solvents or cleaning fluids, and you should not normally attempt to clean the interior of the printer unless a spillage of ink has occurred. You should not lubricate the rail(s) on which the cartridge cradle slides. The cartridge cradle may need to be cleaned (check with the manual), using a clean, preferably synthetic, cloth. Manufacturers strongly recommend that you do not make any attempt to clean the printheads because it is all too easy to block the tiny jets and so make the heads unusable.

When an ink tank is nearly empty it will have to be replaced. Many printers require a colour tank to be replaced when it is empty, even if all of your printing is in black. Others allow empty running on unused tanks, but it is not good economics to allow a tank to become almost empty, because this usually has the effect of allowing the jets to dry out, so that replacing the tank will not restore correct colour printing even if you can regain use of the jets. Always replace a tank by the time it has only 5% of its ink supply remaining.

A typical inkjet printer fault is the appearance of fine white lines on a printed page. This is a certain indicator of a blocked nozzle, possibly more than one. In severe cases, colour pictures may be unusable because all the jets of one colour are blocked. A less common fault is the appearance of black (or coloured) lines. These are caused by unsuitable paper which deposits fibres on the jets so that the wet fibres leave smears on the paper.

> Print nozzles on modern inkjet printers are smaller than they were on older models and they block much more easily. Blockage is unusual if the printer is used each day, printing both in black and in colour. If you require colour only at intervals (perhaps once a week) then you should print a colour test-page document daily just to keep the nozzles clear. This may seem costly in terms of ink usage, but it is much cheaper than the process of cleaning a set of jets.

If nozzles become blocked and cannot be unblocked, then the printheads should be replaced. On H-P machines, each cartridge is combined with a printhead, so that it is very unusual to have a problem with blocked nozzles. On other machines that use a separate ink tank and printhead, the printhead usually may to be replaced after a stated number of tank replacements (often five). If the area around the nozzle block is dirty, it can be cleaned with a synthetic cloth, taking great care *not* to wipe the nozzle area itself. On some types of printer, replacement heads may cost more than the price of a new printer.

In an emergency, a printhead that is giving problems may have to be cleaned. The first area to clean is the set of electrical contacts, gently wiping these with a moist cloth that will not deposit strands on the contacts. Try out the head after cleaning the contacts. If this has no effect, you can sometimes use the head-cleaning utility program that is part of the software installed with the printer driver. This will deal with minor blockages, but has no effect if a complete set of jets has dried out. Another tactic, which is often successful, is to use a cleaning cartridge. This replaces the ink tank that feeds the blocked jets, and after printing several cycles with this tank, the ink tank can be replaced. Check the website http://www.cartridgesave.co.uk/100.html for information on cleaning blocked jets. If this fails, printhead replacement or printer replacement is the only option. The cost of this in terms of ink and cleaning tanks is quite considerable, and this is why some users find that a colour laser printer can be more economical to use despite the high cost of toner cartridges (because each toner cartridge can have a very long, trouble-free life).

> Note that you will get little or no sympathy from a printer manufacturer if you have bought cartridges from any other supplier or have filled cartridges yourself.

Cartridge replacement needs to be done when a cartridge is empty, usually indicated by warnings on the monitor when you try to print, or by faint or streaky printing on older models that do not use ink-monitoring software. The printer manual will indicate the precise procedure for your printer, but the general method is as follows:

1. With the printer switched on, use the control buttons to move the cradle to the centre of its rail(s). On some machines, this is automatically done when the top cover is opened.
2. Switch the printer off and unplug it.
3. Unclip the cartridge and lift it out. Lay the cartridge on a scrap piece of paper.
4. Unpack the new cartridge.
5. Remove the tape covering the nozzles of the new cartridge.
6. Insert the new cartridge and clip it into place.

Once a cartridge has been replaced, restore power to the printer and use the controls to return the cradle to its rest position. The empty cartridge should then be returned for recycling in the packing provided with the new cartridge.

Software diagnostics can be used for printers. For an inkjet printer, the main error messages are likely to concern out of paper, paper jam and empty cartridge situations. The utility programs typically allow for aligning cartridges where more than one cartridge is used, for cleaning nozzles, printing a test page, and checking that there is communication in each direction between the printer and the computer.

The nozzle-cleaning action is a utility that is used on several types of inkjet printer. It works by applying pulses of ink to each nozzle in turn, aiming to clear any blockages by pressure of ink. The routine should *not* be used if the ink supply is low, because it uses considerably more ink than normal printing, and the printhead will be damaged if it is operated with no ink in the tank.

Note that some Epson printers use ink tanks with an embedded chip that transmits ink content information to the computer. When a cartridge is replaced you must ensure that the correct type is used.

**Practical 24.2**

Carry out routine maintenance and field service procedures on an inkjet printer.

## Laser printers

The laser printer has become the standard office printer on the grounds of high-speed printing and quiet operation. The principle is totally different from that of inkjet printers, and is much closer to the photocopier principle, based on xerography. The heart of a laser printer is a drum made from light-sensitive material. This drum is an insulator, so that it can be electrically charged, but the electric charge will leak away in places where the drum has been struck by light. The principle of the laser printer (and the photocopier) is to charge the drum completely and then make the drum conductive in selected parts by being struck by a laser beam. The beam is switched on or off and scanned across the drum as the drum rotates, all controlled by the pattern of signals held in the memory of the printer, and enough memory must be present to store information for a complete page. A schematic for a typical monochrome laser printer is shown in Figure 24.3.

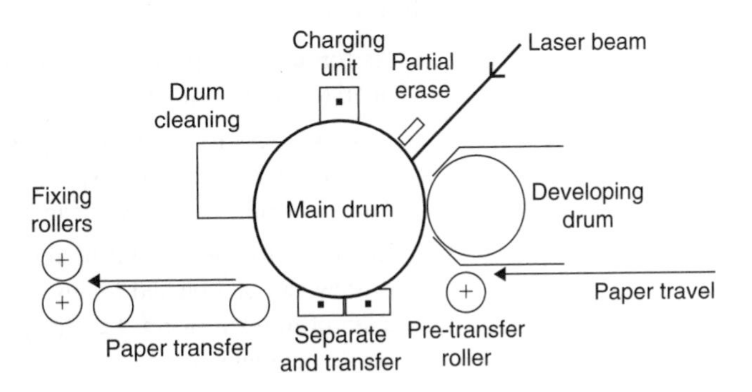


**Figure 24.3** Outline of a monochrome laser printer mechanism

This system requires built-in memory of about 0.5 Mbyte as a minimum to store data for text work, and 2 Mbytes or more if elaborate high-resolution graphics patterns have to be printed. Once the scanning process is complete, the drum will contain on its surface an electrical voltage 'image' corresponding exactly to the pattern that exists in the memory, which in turn corresponds to the pattern of black dots that will make up the image. Finely powdered resin, the **toner**, will now be coated over the drum and will stick to it only where the electric charge is large: at each black dot of the original image.

The coating process is done by using another roller, the **developing cylinder**, which is in contact with the toner powder, a form of dry ink. The toner is a light, dry powder which is a non-conductor, and a scraper blade ensures that the coating is even. As the developing cylinder rolls close to the main drum, toner will be attracted across where the drum is electrically charged.

Rolling a sheet of paper over the drum will now pass the toner to the paper, using a corona discharge to attract the toner particles to the paper by placing a positive charge on to the paper. After the toner has been transferred, the charge on the paper has to be neutralized to prevent the paper from remaining wrapped round the drum, and this is done by the static-eliminator blade. This leaves the toner only very loosely adhering to the paper, and it needs to be fixed permanently into place by passing the paper between hot rollers which melt the toner into the paper, giving the glossy appearance that is the mark of a good laser printer. The drum is then cleared of any residual toner by a sweeping blade, recharged and made ready for the next page.

The main consumables of this process are the toner and the drum. The toner for most modern copiers is contained in a replaceable cartridge, avoiding the need to decant this very fine powder from one container to another. The resin is comparatively harmless, but all fine powders are a risk to the lungs and also carry a risk of explosion. Drum replacement will, on average, be needed after every 80 000 copies, and less major maintenance after every 20 000 copies. Paper costs can be low because any paper that is suitable for copier use can be used; there is little advantage in using expensive paper, and some heavy-grade paper may cause problems of sticking in the rollers.

- As for an inkjet printer, the correct Windows driver must be installed to operate a laser printer satisfactorily.
- Laser printers for office use are generally black-ink printers, and colour laser printers are more expensive and need more expensive cartridges. These cartridges, however, can have a very long life in normal use.
- Empty cartridges should be returned to the manufacturer for recycling, or they can be donated to any charity shop that will accept them. Do no attempt to open or refill cartridges for yourself, as toner dust is dangerous to the lungs if inhaled.

Colour laserjets use four toner cartridges: black, yellow, cyan and magenta, all with a long life. Colour printing can be done on plain paper because the toner looks glossy, and there is a further advantage that accidentally wetting the paper does not cause smearing. Machines such as the H-P 2600N can match ink jets for running costs and provide very pleasing results.

The cheapest monochrome laser printers for home use can be sold at much the same price as a colour inkjet model, but if we compare like for like, laser printers are more expensive to buy. For black-ink uses, the laser printer provides better looking text and line drawings, but the difference between a laser print and a print from a good inkjet machine can be seen only with the aid of a magnifying glass. The laser lines look smoother, with no trace of separate dots or of smears. For colour reproduction, the laser copy is sharper and brighter, but if high-quality paper is used along with a photo-quality ink jet, the differences are small.

Printers of any type that are specified for office use are built to standards that assume almost continuous use at fast printing speeds, and are priced accordingly.

Inkjet printers can cost more in running costs, per printed page, due mainly to the cost of cartridges. The life of a laser cartridge (costing typically £40) is much longer than that of a typical inkjet cartridge at £25. If expensive inkjet paper is used, the difference in running costs becomes greater.

In general, laser printers are faster, although the claimed speeds can be very misleading. The claimed speed for a laser printer is usually for making a large number of copies of a single page, and printing speed for different pages is lower, because of the time needed to create a different pattern on the drum for each page. Claimed speeds range from 6 pages per minute to more than 20 pages per minute in monochrome. The fastest laser printers can turn out monochrome documents at around twice the rate of the fastest inkjet machines.

Replacement of a toner cartridge is very much faster and simpler than replacing inkjet tanks or cartridges. Typically, the cover is opened and the empty toner cartridge taken out and replaced by a new cartridge (after removing the seal of the new cartridge, usually by pulling on a tape at the side of the cartridge). The old cartridge should be sent for recycling and not disposed of in a dustbin.

## Multiple-choice revision questions

24.1 The two different inkjet technologies are:
(a) laser and bubble
(b) bubblejet and piezo
(c) piezo and laser
(d) impact and laser.

24.2 Fine parallel lines appear in a printed photograph. These are due to:
(a) inferior quality ink
(b) poor quality paper
(c) unsuitable software
(d) one or more blocked jets.

24.3 An inkjet printer has been connected to a computer, but only text printing is satisfactory. The most likely reason for this is that:
(a) some ink jets are blocked
(b) the correct printer driver has not been installed
(c) the paper is unsuitable
(d) the computer is faulty.

24.4 The cleaning routine for jets has been used on a printer that uses fixed heads and removable tanks but has no effect. The next step is to:
(a) reinstall the driver
(b) change the heads
(c) use a cleaning cartridge
(d) buy a new printer.

24.5 The first page out of an inkjet printer always takes longer. This is because:
(a) it takes some time to pass the data to the printer
(b) the printer runs a routine to clear jets before starting printing
(c) the paper has to be rolled into the correct position
(d) the cartridges have to be correctly lined up.

24.6 An inkjet printer is to be used for colour photographs and a laser printer for text. How would you connect and use the printers?
(a) connect only the printer you want to use
(b) connect both printers, but install only the driver for the one you want to use
(c) connect both printers, install both drivers, and select from Windows
(d) connect one printer using USB and the other using the Centronics port.

# 25 Health and safety

The Health and Safety at Work Act of 1974 became fully operative in October 1978, and since then the regulatory conditions have been further extended. This affects all aspects of servicing and other workshop activities. The legislation ensures that any work must be carried out in a way that ensures maximum safety for the employee and for anyone else present. Employers must provide a safe working environment and see that safety rules are obeyed. All employees (including students or apprentices) have a duty to observe the safety precautions that have been laid down by their employers, and to carry out all work in a safe way.

The 1974 Act can be summarized as follows:

- All employers, including teachers, tutors and instructors, must ensure that any equipment to be used or serviced by employees, apprentices or students is as safe to operate as it can reasonably be made. Equipment which cannot be made safe should be used only under close supervision.
- Employers, tutors and instructors must make sure that employees and students at all times make use of the required safety equipment such as protective clothing, goggles, ear protectors, etc., which the employer must provide for them.
- Employees and students must ensure that they carry out all their work in such a way that it does not endanger them, people working around them, or the subsequent users of equipment that they repair.
- Any accidents must be logged and reported, and measures taken to ensure that such incidents cannot happen again.
- Failure to observe the provisions of the Act is reasonable grounds for dismissal.

**Accident prevention** depends on:

- recognition of hazards
- elimination, if possible, of hazards
- replacement of hazards
- guarding and/or marking hazards
- a sense of personal protection
- continuing education in safety.

## Hazards

We shall concentrate on electrical hazards because these are the main problems for a servicing workshop, but others (notably fire) are also important. One particular difficulty that faces anyone working on servicing domestic electronics equipment is the wide variety of working places, which can vary from a well-designed workshop, as safe as possible in the light of current experience, to a home in which a television receiver has failed, and in which almost every possible hazard, from bad electrical wiring to the presence of children around the television receiver, exists. This huge difference in working conditions makes it vital for anyone working on electronics servicing to be aware of safe working methods and to practise them at all times, whether supervised or not.

The types of hazards that can lead to accidents, as far as electronic servicing is concerned, are:

- **electrical**, particularly where industrial equipment is serviced
- **fire**, particularly where flammable materials are used
- **asphyxia**, particularly because of degreasing solvents
- **mechanical**, particularly in cramped conditions.

Current health and safety legislation emphasizes accident **prevention** as well as dealing with accidents. Dealing with accidents is a matter of provision of facilities for, mainly, first aid and fire-fighting, and is very much in the hands of the employer. Prevention of accidents depends on attitudes of mind and is the responsibility of everyone concerned. The person who recognizes a hazard is the person who can imagine what might happen if an emergency arose, or who knows how to look for danger, and the Act implies that this person must not be treated as a trouble-maker or whistle-blower, but as a valuable contributor to safety who can save money in the long run.

The legislation now extends to the Control of Substances Harmful to Health (**COSHH**). In particular, this relates to cleaning solvents, paint, varnish, adhesives, sealants and rosin-cored solder. Under these regulations a responsible person must be nominated to ensure that all related materials are stored and used safely. Such a person should also be familiar with the basic principles of chemical first-aid procedures. With particular respect to the fumes emitted from rosin-cored solder, it is now well established that this can be the cause of asthma. **Soldering** should therefore ideally only be carried out within the confines of a fume extractor system. Recently developed rosin-free fluxes are now available, but even these should be used within a fume extractor environment. The cleaning solvents used for the removal of flux residues from printed circuit boards should **not** be chlorofluorocarbon (CFC) based, as these create environmental problems. The toxicity of lead is also recognized and as a result of this, tin/lead alloy solders are progressively being replaced by such as silver/copper alloy solders. The use of lead-containing solder is now confined to equipment (such as military electronics) in which the need for reliability overrides the concern about the use of lead.

The regulations regarding the reporting of injuries and accidents at work have also been significantly reinforced. The Reporting of Injuries, Diseases and Dangerous Occurrences Regulations (**RIDDOR**) first appeared in 1986, but this has now been updated by **RIDDOR95**, which requires that employers and the self-employed must notify the local environmental health department of any such occurrences. See the website http://www.riddor.gov.uk/# for further information, including reporting hazards.

## Responsibility

The safety legislation requires both employer and employee to observe, maintain and improve safety standards and the adoption of safe working practices. The responsibilities of an employer are:

- To ensure that the workplace is **structurally** secure. Typical hazards are insecure floors, leaking roofs, blocked windows, restricted doorways and so on.
- To provide **safe plant and equipment**. Work benches and equipment must be adequate for the job. All tools must be fitted with safety guards. Larger power tools should be screened or fenced. If necessary, floors should be marked with safe walking areas.
- To lay down **safe working systems**. Employees do not have to use makeshift equipment or methods. Protective clothing should be provided if needed.
- To ensure that **environmental controls** are used. The workshop must be kept at a reasonable working temperature. Humidity levels should be controlled if needed, and ventilation must be adequate. Employees should not be subjected to dust and fumes, and there must be washing facilities, sanitation for both men and women, and provision for first aid in the event of an accident.
- To ensure **safety in the handling, storing and movement of goods** (see below). Employees should not be required to lift heavy loads (a useful guide, enforced by law in some EU countries, is that an employee cannot be expected to lift without assistance a load of more than 15 kg). Note that most widescreen television receivers using cathode-ray tubes (CRTs) display weight more than this, some models more than three times this limit. Mechanical handling must be provided where heavy loads are commonplace. Dangerous materials must be identified and stored where they do not cause any hazard to anyone on the site.
- To provide a system for **logging and reporting accidents**. Such a log is not to be simply a list of happenings, but a guide to better work practices.
- To provide **information, training and updating of training** in safety precautions, along with supervision that will ensure safety. This includes signs to indicate hazards.
- To devise and administer a **safety policy** that can be reviewed by representatives of both employer and employees.

## Employees' responsibilities

The law recognizes that an employer can provide only a framework for safe working, and that it can be difficult to force an employee to work in a safe way. The Act therefore emphasizes that the employee also has responsibilities to ensure safe working. This is a matter not only of personal safety, but also of the safety of fellow workers and members of the public. Most accidents are caused by human carelessness in one form or another and although standards can be drawn up for safe methods of working, it is impossible to ensure that everyone will abide by these standards at all times.

The employees' responsibilities, which apply also to the self-employed, include:

- **Health-care**: this means, for example, avoiding the use of alcohol or drugs if their effects would still be present in working hours. Remember that the term 'drugs' can include such items as painkillers, antihistamines for hay fever, antidepressants, and 'alternative medicines' such as herbal treatments. In some countries, blood alcohol level can be checked following industrial accidents as well as following a road accident. Working while excessively tired can also be a cause of accidents owing to relaxed vigilance.
- **Personal tidiness**: clothing should provide reasonable protection, with no loose materials that can be caught in machinery or even (as has been reported) melted and set alight by a soldering iron. Long hair is even more dangerous in this respect, and must be fastened so that it cannot be caught.
- **Behaviour**: carelessness and recklessness cannot be tolerated, and practical jokes, no matter how traditional, do not belong in any modern workplace. Legislation in several countries now treats this type of behaviour in the workplace as seriously as it is treated on the roads.
- **Competence**: an employee must know, from experience, discussion with a colleague or from reference to manuals, what has to be done and the safe way of carrying out the work. In several EU countries, it is an offence to carry out work for which you are not qualified, and if you are working abroad you will have to find out to what extent UK qualifications apply in other countries.
- **Deliberately avoiding hazards**: self-employed service personnel should insure against third party claims, and employees must remember that they can be sued if their careless or reckless behaviour leads to injury. Misusing safety equipment is a criminal offence, quite apart from endangering the lives of others. It is also an offence to fail to report a hazard which you have discovered.

## Electric shock

The main electrical hazard in servicing operations is working on live equipment, because of the ever-present risk of shock. Electric shock is caused by current flowing through the body, and it is the amount of current and where it flows that is important. The resistance of the body is not constant, so that the amount of current that can flow for a given voltage will vary according to the moistness of the skin. The most important hazard is of electric current flowing through the heart. Ways of avoiding this include the following:

- Ensure that only **low voltages** (lower than 50 V) are present in a circuit.
- Ensure that no **currents** exceeding 1 mA can flow through the body in **any** circumstances.
- Keep the hands dry at all times, because moist hands conduct much better than dry hands.
- Keep workshop floors dry, and wear rubber-soled shoes or boots.
- Avoid two-handed actions, particularly if one hand can touch a circuit and the other hand is touching a metal chassis, metal bench or any other metal object.

The greatest hazard in most electronic servicing is the **mains supply**, which in the UK and in most of Europe is at 240 V a.c. The correct use of three-pin plugs and sockets, correctly wired and fused, is essential. As a further precaution, earth leakage contact breakers (**ELCBs**) can be used to ensure that any small current through the earth line will operate a relay that will open the live connection, cutting off the supply.

If live working is unavoidable, try to avoid the possibility of current passing through the heart in the following ways:

- Work with the right hand only, making a longer electrical path to the heart in the event of touching live connections.
- Wear insulating gloves.
- Cover any metalwork which is earthed.

Note that the outer conductor of coaxial (coax) aerial cables is usually electrically isolated.

This allows a considerable a.c. voltage to be induced in the outer conductor from the house mains, often more than 50 V. Because this is an induced voltage the impedance is high, but it is often possible to experience a tingle if one hand is touching the coax outer and the other brushed against an earthed object such as a radiator. If the household includes a baby, it is wise to ensure that all coax connectors are of the modern type using a plastic cap rather than the metal one formerly used. Another possibility is to earth the outer of the coax at some convenient point.

## Power tools

Power tools present a hazard in most workshops, and must be electrically safe. Some useful points are:

- Always disconnect when changing speed or drill bits.
- Never use a power tool unless the correct guards and other protective devices have been correctly fitted.
- Ensure that the mains lead is in perfect condition, with no fraying or kinking, and renew if necessary.
- Metal power tools should be earthed, and in some conditions can be powered from isolating transformers. Another option is to use only battery-powered tools.
- Most domestic power tools that employ plastic linings or inner casings to provide double insulation are not necessarily suitable for industrial applications.
- Test equipment must itself be electrically safe, using earthing or double insulation as appropriate.
- Test equipment can often be bought in battery-operated form.
- Mains-powered test equipment must use the correct mains voltage setting, and have power leads in good condition and the correct fusing for the load (usually 3 A).

- All test equipment, particularly if used on servicing live equipment, should be subject to safety checks at regular intervals.
- Users should maintain such equipment carefully, avoiding mechanical or electrical damage.

Low-power industrial equipment often makes use of standard domestic plugs, but where higher power electronic equipment is in use the plugs and socket will generally be of types designed for higher voltages and current, often for three-phase 440 V a.c. In some countries, flat two-pin plugs and sockets are in use, with no earth provision except for cookers and washing machines, although eventually uniform standards should prevail in Europe, certainly for new buildings. Uniform standards will almost certainly match those used in Germany.

## Mains power

Any mains electrical circuit should include:

- A **fuse** or **contact-breaker** whose rating matches the maximum consumption of the equipment. This fuse may have blowing characteristics that differ from those of the fuse in the plug. It may, for example, be a fast-blowing type which will blow when submitted to a brief overload, or it may be of the slow-blow type that will withstand a mild overload for a period of several minutes.
- A **double-pole switch** that breaks both live and neutral lines. The earth line must never be broken by a switch.
- A **mains warning light** or indicator which is connected between the live and neutral lines.

All these items should be checked as part of any servicing operation, on a routine basis. In addition:

- As far as possible, all testing should be done on equipment that is disconnected and switched off. Use a *don't switch on* sign if needed.
- The absence of a pilot light or the fact that a switch is in the OFF position should **never** be relied on as a sign that it is safe to touch conductors.
- Mains-powered equipment being repaired should be completely isolated by unplugging from the mains.
- If the equipment is permanently wired then the fuses in the supply line must be removed before the covers are taken from the equipment.

Many pieces of industrial electronics equipment have safety switches built into the covers so that the mains supply is switched off at more than one point when the covers are removed.

The ideal to be aimed at is that only low-voltage, battery-operated equipment should be operated on when live. When mains-powered equipment must unavoidably be tested live, the meter, oscilloscope or other instrument(s) should not be connected until the equipment is switched off and isolated, and the live terminals should be covered before the supply voltage is restored. The supply line should be isolated again before the meter or other instrument clips are removed.

The main dangers of **working on live circuits** are:

- the risk of fatal shock through touching high-voltage exposed terminals, which can pass large currents through the body
- the risk of fatal shock from the discharge of capacitors that were previously charged to a high voltage
- the risk of damage to instruments or to the operator when a mild electric shock is experienced: the uncontrollable muscular jerking that is caused by an electric shock of any kind can cause the operator to drop meters, and to lose his or her balance and fall on to other, possibly more dangerous equipment.

Even low-voltage circuits can be dangerous because of the high temperatures that can be momentarily generated when a short-circuit occurs. These temperatures can cause burning, sometimes severe. As a precaution, chains, watchstraps and rings made of metal should not be worn while servicing is being carried out.

Older types of television produced in the UK used 'live chassis' circuits in which the neutral lead of the mains supply was connected to the metal chassis of the equipment. Incorrect wiring of the plug or a disconnected neutral lead would cause the chassis to be live at full supply voltage. Later models of television receiver used considerably lower internal voltages, but some still used a live chassis approach, and a few were wired so that the chassis was at about half of supply voltage, irrespective of how the plug was connected. As a rule, equipment that uses SCART sockets will usually have an isolated chassis.

When carrying out service work either in the workshop or at a customer's premises, the equipment should only be powered via a **mains isolation transformer**. This device has a 1:1 turns ratio with its magnetic core and metallic casing earthed to the mains supply input. The secondary winding is completely floating (isolated) so that there is no direct connection between the equipment being serviced and the mains power supply. For portable working, a transformer rated at about 150–250 W is light enough to be readily portable, while a unit rated at about 500 W is suitable for the workshop bench.

Other electrical safety points are:

- Cables should not be frayed, split, or be sharply bent, too tightly clamped or cut.
- Damaged cables must be renewed at once.
- Hot soldering irons must be kept well away from cable insulation.
- Cables must be securely fastened both into plugs and into powered equipment. The supply cable to a heavy piece of equipment should be connected by way of a plug and socket which will part if the cable is pulled.
- The live end of the connector must have no pins that can be touched.
- Every electrical joint should be mechanically as well as electrically sound. It must be well secured with no danger of working loose and with no stray pieces of wire.
- The old motto of 'volts jolt but mills kills' implies that only a few milliamps of through-body current can be fatal.

The old-fashioned type of lead/acid accumulator releases an explosive mixture of hydrogen and oxygen while it is on charge, and this can be ignited by a spark. No attempt should ever be made to connect or disconnect such accumulators from a charger or from a load while current is flowing. An explosion is serious enough, but the explosive spray of acid along with sharp glass or plastic fragments is even worse.

## Portable appliance testing

It is a Health and Safety Executive regulation requirement that the flexible leads/cables, plugs and sockets of all electrical equipment should be examined and tested at regular intervals, typically at least annually and always after repairs. This operation should be carried out by a technically competent person. Once tested, each appliance should be suitably labelled and a recording of the results kept in a readily accessible database.

The cables and connectors should be physically examined for serviceability and defects, and to ensure that they meet the current technical standards. The plugs should have shrouded live and neutral pins, the cables must be correctly colour coded and no component should be damaged in any way. The cord grip screws should prevent the cable being twisted from beneath its clamp. If a line connector and socket are in use, then the socket must be connected nearest to the mains supply point to ensure that contact with the live mains is prevented. Similarly, the connections at the appliance end should be sound and securely clamped (see also Chapter 2).

If there are externally accessible fuse holders, these should be undamaged and loaded with fuses of the correct rating. The mains supply on/off switch should be undamaged and work correctly. If the output is greater than 50 V, then the short-circuit current must be limited to less than 5 mA.

Both single and double insulation types of equipment have to be considered and in the latter case, the appliance is not directly earthed and the case should carry the double square symbol (one within the other). The earth bonding test, which applies only to the former types of appliance, consists of measuring the earth wire resistance while passing a current of up to 25 A for up to 5 seconds. This ensures that any poor earthing connections will become open-circuit under test. The total resistance including the earth line and the resistance to the casing should not exceed 0.75 Ω.

**Insulation testing** is carried out at a potential of at least 500 V d.c. For directly earthed appliances the line and neutral pins are shorted together and the resistance is measured between them and the earth pin. A value of at least 2 Ω is considered to be satisfactory. For double-insulated appliances, the resistance between the shorted pins and any exposed metalwork should be measured where a value of more than 7 MΩ is acceptable.

Portable appliance testing kits range from relatively low-cost, handheld systems to larger, less portable devices that carry liquid crystal display (LCD) screens and contain interfaces to a PC/printer for data logging.

## Fires

Workshop fires present several types of hazard. The obvious one is flesh burns, but injuries from other causes are common, such as falling when running from a burning workshop. Clear escape routes should be marked and these must be kept clear at all times. If there is further risk of suffocation by smoke, low-level emergency lighting should be present so that a route to a safe exit is marked.

A fire can start wherever there is material that can burn (**combustible material**), air or oxygen-rich chemicals that can supply oxygen for burning, and any hot object that can raise the temperature of materials to the burning point. A fire can be extinguished by removing all combustible material, by removing the supply of air (oxygen), by smothering with non-combustible gas or foam, or by cooling the material to below the burning temperature. The most dangerous fires are those in which the burning material can supply its own oxygen (such materials are classed as **explosives**) and those in which the burning material is a liquid that can flow about the workplace, taking the fire with it.

Another major danger is of **asphyxiation** from the fumes produced by the fire. In electronics workshops the materials that are used as switch cleaners, the wax in capacitors, the plastics casings, the insulation of transformers and the selenium that can still be found in some old-fashioned metal rectifiers will all produce dangerous fumes, either choking or toxic, when burning. Good ventilation can reduce much of this particular hazard.

The key points for fire safety are:

- good maintenance
- no naked flames and no smoking permitted
- tidy working, with no accumulation of rubbish
- clearly marked escape paths in case of fire
- good ventilation to reduce the buildup of dangerous fumes
- suitable fire extinguishers in clearly marked positions
- knowledge by all staff of how to deal with fire/explosion
- regular fire drills and inspection of equipment.

**Figure 25.1** Standard pictograms for use on fire extinguishers illustrating the types of fire they can be used on.

Because of the variety of materials that can cause a fire, more than one method of extinguishing a fire may have to be used, and use of the wrong type of extinguisher can sometimes make a fire worse. Until January 1997, when a new harmonized European standard – BS EN3 – came into effect, providing a single standard for fire extinguishers across Europe, the UK used a variety of colours to distinguish the different type of extinguisher and its uses. While fire extinguishers in European countries are now coloured red (with standard pictograms to illustrate the types of fire that the extinguisher can be used on), a concession has been given for the UK to display a small colour zone on the top half of the front of the extinguisher body using the following colour codes:

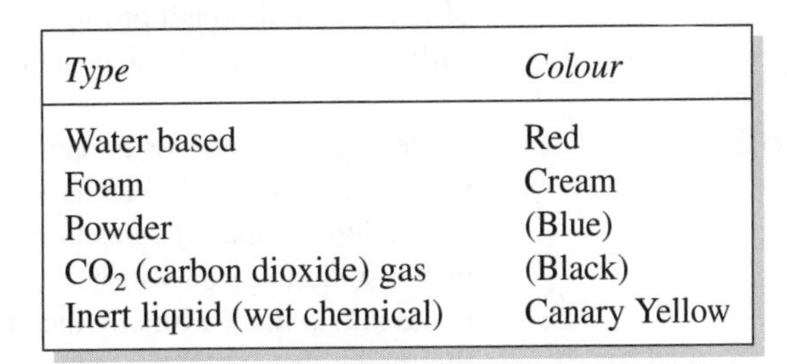

| *Type* | *Colour* |
|---|---|
| Water based | Red |
| Foam | Cream |
| Powder | (Blue) |
| $CO_2$ (carbon dioxide) gas | (Black) |
| Inert liquid (wet chemical) | Canary Yellow |

The water-based type of extinguisher (for a class A fire) is most effective on materials such as paper, wood or cloth, using the extinguisher on the base of the fire to soak materials that are not yet burning and to cool materials that have caught light. Water-based extinguishers must **never** be used on electrical fires or on fires that occur near to electrical equipment.

To help users select the correct type of extinguisher fires have been categorised into classes, these are:

Class A fires involving organic solids, e.g. wood, paper
Class B fires involving flammable liquids
Class C fires involving flammable gases
Class F fires involving cooking oil and fat

Fire extinguishers may be suitable for one or more of these classes.

Class B fires involve burning liquids or materials that will melt to liquids when hot. The main hazards here are fierce flames as the heat vaporizes the liquid, and the ease with which the fire can spread so as to affect other materials. The most effective treatment is to remove the air supply by smothering the fire, and the foam, dry powder or $CO_2$ gas extinguishers can all be useful, although such extinguishers should be used so that the foam, powder or gas falls down on to the fire, because if you direct the extinguisher at the base of this type of fire the liquid will often simply float away, still burning. Fire blankets can be effective on small fires of this type, but on larger fires there is a risk that the blanket will simply act as a wick, encouraging the fire by allowing the burning liquid to spread.

For dealing with electrical fires, the water- or foam-based types of extinguisher should be avoided because the risk of electric shock caused by the conduction of the water or foam is often more serious than the effect of a small fire. Inert liquid extinguishers are effective on electrical fires, but the liquids can generate toxic fumes and can also dissolve some insulating materials. **Powder** extinguishers and **$CO_2$** types are very effective on small fires of the type that are likely to develop in electronics workshops. They must be inspected and check-weighed regularly to ensure that internal pressure is being maintained.

A few **sand buckets** are also desirable. They must be kept full of clean sand and never used as ashtrays or for waste materials. A **firemat** is also an important accessory in the event of setting fire to the clothing of anyone in the workshop. The firemat should be kept in a prominent place and everyone in the workshop should know how to use it.

In the event of a fire, your order of priorities should be:

- to raise the alarm and call the fire brigade even if you think you can tackle the fire
- to try to ensure evacuation of the workshop
- to make use of the appropriate fire extinguishers.

You must **never** try to fight a fire alone, or put others at risk by failing to sound the alarm. The most frightening aspect of fire is the way that a small flame can in a few seconds become a massive conflagration, completely out of hand, and although prompt use of an extinguisher may stop the fire at an early stage, raising the alarm and calling for assistance is more important.

## Soldering

Soldering should never be carried out on any equipment, whether the chassis is live or not, until the chassis has been completely disconnected from the mains supply and all capacitors have been safely discharged. The metal tip of the soldering iron will normally be earthed, and should not be allowed to come into contact with any metalwork that is connected to the neutral line of the mains, because large currents can flow between neutral and earth. Any contact between the soldering iron and the live supply should also be avoided.

Soldering and desoldering present hazards that are peculiar to the electronics workshop. The soldering iron should always be kept in a covered holder to prevent accidental contact with hands or cables. Many electrical fires are started by a soldering iron falling on to its own cable or the cable of another power tool or instrument. Although it can be very convenient to hang an iron up on a piece of metal, the use of a proper stand with a substantial heat sink is the safe method that ought to be used.

In use, excessive solder should be wiped from the iron with a damp cloth rather than by being flicked off and scattered around the workshop. During desoldering, drops of solder should not be allowed to drip from a joint; they should be gathered up on the iron and then wiped off it, or sucked up by a desoldering gun which is equipped with a solder pump.

Soldering advice, including many excellent photographs, is available from Alan Winstanley's website (http://www.epemag.wimborne.co.uk/solderfaq.htm). A guide to the use of lead-free solders is available at http://www.leadfreesoldering.com/

Care should be taken to avoid breathing in the fumes from hot flux. Where soldering is carried out on a routine basis, an extractor hood should be used to ensure that fumes are efficiently removed. Materials other than soldering flux can cause fumes, and some types of plastics, particularly the PTFE types of materials and the vinyls, can give off very toxic fumes if they are heated to high temperatures. Extraction can help here, but it is preferable to use careful methods of working which ensure that these materials are not heated.

## Toxic materials

Virtually every workplace contains toxic materials, or materials that can become toxic in some circumstances, such as a fire. **Industrial solvents**, such as are used for cleaning electronics subassemblies, switch contacts, etc., are often capable of causing asphyxiation and even in small concentrations can cause drowsiness and stupor. Many common insulating materials are safe at normal temperatures, but can give off toxic fumes when hot.

In addition, some very poisonous materials are used within electronics components. Such materials can be found in CRTs, fluorescent tubes, valves, magnetrons (used in microwave ovens), metal rectifiers, power transistors, electrolytic capacitors and other items. The local hospital should be informed of the toxic materials that are present and, if possible, should advise on any first aid that may be effective in the event of these materials being released. In particular, CRTs and fluorescent tubes must be handled with care because they present the hazard of flying glass as well as of toxic materials if they are shattered.

Transistors and integrated circuits (ICs) should **never** in any circumstances be cut open. No servicing operation would ever call for this to be done, but if a faulty component has to be cut away there might be a danger of puncturing it. Some power transistors, particularly those used on transmitters, contain the solid material beryllium oxide whose dust is extremely poisonous if inhaled, even in very small quantities. All such transistors that have to be replaced should be removed very carefully from their boards and returned to the manufacturers for safe disposal.

## Hand-tools

The incorrect use of small hand-tools is a very common cause of accidents, and safe methods of working must always be used, especially when you are in a hurry.

**Never** point a screwdriver at your hand. This seems obvious, yet a common cause of stabbing accidents is using a screwdriver to unscrew a wire from a plug that is held in the palm of the hand rather than between fingers or, better still, in a vice.

- Small tools have to be suited to the job and used correctly. Never use a blade screwdriver when a Philips or Pozidrive type is needed, and remember that screwdriver size must be matched to the work.
- Files should be fitted with handles and used with care because, being brittle, they can snap. Never use a file as a tommy bar, for instance.
- Box spanners and socket sets should be used in preference to open spanners where possible.
- Snipping wire with side cutters can be a hazard to the eyes, either of the user of the cutters or of anyone standing close. Ideally, the wire that is being snipped should be secured at both ends so that no loose bits can fly around. One safe method is to hold the main part of the wire in a vice and the other part in a Mole wrench.
- Protective clothing should be worn wherever the regulations or plain common sense demand it. The workshop is no place for loose ties or cravats, for long untied hair or strings of beads. A workshop coat or boiler suit should be worn whenever workshop tools are being used, and eyeshields or goggles are nearly always necessary. Goggles and gloves must always be worn when CRTs are being handled and safety boots should be worn when heavy objects have to be moved. Goggles are also a useful protection when wire is being snipped, particularly for hard wire like Nichrome.

## Back injury

Back injury is the single most common cause of absence from work in the UK. In the course of electronics servicing work many large and heavy items have to be shifted. Injury is most commonly caused by incorrect methods of lifting heavy objects or holding such objects before putting them into place.

- The correct method of lifting is to keep a straight back throughout the whole of the lifting operation, bending your knees as necessary but never your back.
- Never attempt to shift anything heavy by yourself.
- Even if you bear all the weight of carrying a load, help in lifting and steering the load can make the difference between safe and unsafe work.
- In several countries, there is a statutory upper limit of load that one operator is allowed to lift without help. This can be as low as 15 kg (33 lbs), which is less than the usual 20 kg (44 lbs) weight allowance for luggage on flights.

- Workshop benches and stools should be constructed so that excessive lifting is not required.
- If a large number of heavy items have to be moved to and from benches, mechanical handling equipment should be used.
- Stool heights should be adjustable so that no user needs to stoop for long periods to work on equipment.

## Reporting hazards and injuries

An employer must ensure that a workspace is kept free of dangerous and badly sited materials. Someone in the workshop, however, may have put the hazard into place, and quite certainly someone who works near the hazard will be the first to notice it. The responsibility cannot be shrugged off as being entirely that of the employer.

Two **logbooks** should be maintained for any workshop. One is used for reporting potential hazards so that action can be taken (or if the worst happens, blame apportioned). The other logbook is used to record actual accidents, so that a written account is available of every incident. Referring to these logs at regular intervals, perhaps at the start of each month, is a valuable way of revising safety policies.

In every workshop at least one person, and preferably two, should be trained in **first-aid** procedures. Particular attention should be given to the procedures for resuscitating victims of electric shock, but you should remember that most first-aid cases will not involve elaborate treatment, and is concerned with minor cuts and burns. A fairly basic first-aid chest, along with good facilities for washing, is adequate for at least 90% of incidents.

Treatments for exposure to toxic materials or fumes call for much more specialized equipment and knowledge. It is particularly important never to work alone with potentially toxic materials.

Training to deal with **electric shock** should be a first priority of first-aid instruction for servicing workshops. The power supply must be cut off before the victim is touched, otherwise there may be two victims instead of one. Mouth-to-mouth resuscitation, preferably by way of a plastic disposable mouthpiece, should then be applied as soon as the power is off, and continued until the victim is breathing or until an ambulance arrives.

- If it would take too long to find and operate a mains switch, push the victim away from live wires using any insulated materials, such as a dry broom handle, or pull away by gripping (dry) clothing.
- Do not touch the skin of the victim. This endangers the rescuer and also the victim (because of the additional current that will flow, perhaps through the heart).
- Speed is important, because, even if breathing can be restored, the brain can be irreparably damaged after about 4 minutes.
- Always leave an unconscious victim on his or her side, never on the back or on the face.
- Never try to administer brandy or any other liquid to anyone who is not fully conscious because the risk of choking to death is at least as great as that of electric shock itself.

Once the victim has been removed from the danger of continued shock, a 999 call can be made and mouth-to-mouth resuscitation can be given. Practical experience in this work is essential, and should form an important part of any first-aid training. A summary will remind you of the steps, and posters showing the method in use should be displayed in the workshop.

- Place the victim on his or her back, loosen any clothing around the neck, and remove any items from the mouth, such as false teeth or chewing gum.
- Tip the head back by putting one of your hands under the neck and the other on the forehead. This opens the breathing passages.
- Pinch the nose to avoid air leakage, breathe in deeply, and blow the air out into the victim's lungs. If possible, use an approved mouth-to-mouth adapter to avoid any risk of transferring disease, but never waste time looking for one.
- Release your mouth and watch the victim breathe out. You may have to assist by pressing on the chest.
- Repeat at a slow breathing rate until help arrives or the victim can breathe unattended. Do not give up just because the victim is not breathing after a few minutes, because these efforts can sustain life and avoid brain damage even if the victim is unconscious for hours.

**Electric shock** is often accompanied by the symptoms of burning, which will also have to be treated, although not so urgently. The workshop telephone should have permanently and prominently placed next to it a list of the numbers of emergency services such as doctors on call, ambulance, fire, chemists, hospital casualty units, police, and any specialized services such as burns and shock units. This list should be typed or printed legibly, maintained up to date and stuck securely to a piece of plywood or hardboard. Part of any safety inspection should deal with checking that this list is updated, well placed and legible.

- Severe burning must be treated quickly at a hospital. First aid can concentrate on cooling the burns and treating the patient for shock.
- Apply cold water to the region of the burn, and when the skin has had time to cool, cover with a clean bandage or cloth.
- Never burst blisters or apply ointments, and do not attempt to remove burned clothing because this will often remove skin as well.
- The reaction to burning is often as important as the burn itself, and any rings, bracelets, tight belts and other tight items of wear should be removed in case of swelling.
- Try to keep the patient conscious, giving small drinks of cold water, until specialist help arrives.

The treatment of **minor wounds** is a frequent cause of a call on first aid. The most important first step is to ensure that a wound is clean, washing in water if there is any dirt around or in the wound. Minor bleeding will often stop of its own accord, or a styptic pencil (alum stick) can be used to make the blood clot.

More extensive bleeding must be treated by applying pressure and putting on a fairly tight sterile dressing. One dressing can be put over the top of an older one if necessary, rather than disturbing a wound. Medical help should always be summoned for severe wounding or loss of blood, because an anti-tetanus injection may be required even if the effects of the wound are not serious.

The effects of chemicals require specialized treatment, but first aid can assist considerably by reducing the exposure time. In electronics servicing work, the risk of swallowing poisonous substances is fairly small, and the main risks are of skin contact with corrosive or poisonous materials and the inhalation of poisonous fumes.

If a corrosive chemical has been swallowed or spilled on the skin, large amounts of water should be used (swallowed or used for washing) to ensure that the material is diluted to an extent that makes it less dangerous.

Common solvents like trichloroethylene and carbon tetrachloride degreasing liquids give off toxic fumes. These liquids should be used (other than in very small quantities) only under extractor hoods or in other well-ventilated situations.

Never try to identify a solvent by sniffing at a bottle. Bottles should be correctly labelled, preferably with a hazard notice.

Some solvents, like acetone and amyl acetate, are a serious fire hazard in addition to giving off toxic fumes.

## Damage to equipment

While equipment is being serviced, it is the responsibility of the servicing engineers to ensure that the equipment is not damaged. Damage in this respect means mechanical or electrical damage, and although such hazards should be covered by insurance, the customer is unlikely to feel well disposed to any servicing establishment that cannot take good care of equipment that is being serviced.

Mechanical damage covers such items as:

- damage to either the cabinet or the functioning of the circuitry, or both, caused by dropping the equipment
- tool damage, such as burn marks on a cabinet caused by a soldering iron
- marks or stains, such as can be caused by a hot coffee cup or by carelessness with solvents.

Such mechanical damage might be due to an unavoidable accident, but only too often it is the result of carelessness, and the ultimate loser, regardless of insurance, will be the staff of the workshop. Never assume that an old piece of equipment is of low value to the customer, or that a customer will not worry about damage. Since mechanical damage to the cabinet is the form of damage that is most obvious to the customer, all cabinets should be wrapped both when being handled and in the course of servicing. Bubble-pack is particularly useful for avoiding damage caused by knocks, and this can also protect against marking or staining. Damage from soldering irons can be minimized by using holders for irons, and ensuring that cabinets are kept well away from tools.

One other hazard of this type is ingrained swarf, including fragments of solder caused by shaking a soldering iron. If the workshop is not kept clean, metal swarf can become embedded in the underside of a cabinet. Although this will not be visible to the customer, it can cause severe scratching on the table that the equipment stands on, and the customer is not likely to forget or forgive the damage.

Electrical damage in servicing is typically caused by:

- incorrect use of servicing equipment such as signal generators
- carelessness in removing ICs
- ignorance of correct operating conditions
- replacement of defective components by unsuitable substitutes
- operating equipment with incorrect loads (including o/c or s/c load)
- failure to check power supply components when the initial fault has been caused by a power-supply fault.

One feature that is common to many of these items is incorrect or unavailable information. Equipment should not be serviced if no service sheet is available. Granted that there are some items which are so standardized that no service sheet is necessary, but these items are the ones, such as radios and personal stereo players, that are uneconomical to service in any case. Equipment such as television receivers, hi-fi, personal computers and video recorders will require information to be available, and that information should be as complete as possible. Service engineers must be aware which components must be replaced only by spares approved by the manufacturers, and why. Substitution should be considered only if the manufacturer no longer supplies spares; remember that some manufacturers will no longer supply spares after a comparatively short period, as short as 5 years.

## Installation

Outside installation work, as distinct from outside servicing, is unavoidable, and for consumer electronics often refers to aerials (FM, DAB, television or satellite), downleads and interconnections. The installation of personal computers, however, also comes into this class, and although aerial work is specialized and will usually be contracted out (and will not be further considered here), the installation of hi-fi and computers will often have to be done by the supplier. As always, this should be realistically priced. If the customer suggests that he or she can buy the boxes elsewhere for a lower price, point out by all means that you will supply an installation service at a price that reflects a fair return, and that such installation is an integral part of the price that you quote for equipment.

Outside installation follows the same pattern as was mentioned earlier for outside repairs, but with no option for taking the equipment to the workshop. The customer needs to be consulted about positioning of equipment – the loudspeaker may, after all, clash with the colour of the curtains – but the installer should be able to point out tactfully that a full stereo effect will not be experienced if the loudspeakers are placed close to each other, facing in different directions, or even in different rooms or behind curtains. For a multispeaker installation there is considerable latitude on the position of the subwoofer (because the listener is not aware of the position of any source of very deep notes), but the other speakers should be set up as close as possible to the ideal of a rectangle with the speakers at the corners and the listener at the centre of the rectangle.

Similarly, it is up to the installer to show that the monitor of a computer should be placed where the user of the keyboard can see it without needing to lean forward or sideways, that the mouse should come conveniently to hand, and that the flow of air through the main casing is unobstructed.

Although the interconnection of equipment is often so standardized that you can work without manuals, you should be aware of any peculiarities of equipment, limitations to cable lengths (particularly for parallel printers for computers), and so on. A useful hint for computers that are used with a large number of powered peripherals (scanner, modem, printer, etc.) is to connect the power take-off socket on the main computer case to a set of distribution sockets, which can be used to supply to peripherals. This allows the user to switch everything on and off together using only the main switch on the computer. A similar scheme can be used for the more elaborate type of hi-fi setup.

Never leave an installation without testing it adequately: do not assume that because it sounds good playing a cassette the CD player will be as good. In some cases, you may need, for example, to alter sensitivity levels at inputs. In addition, never leave an installation without making sure that the customer knows how to operate the equipment and what peculiarities of installation are present. Few users, for example, know how to deal with the connections between a modern DVD recorder and a television receiver, and many assume that video replay from a VCR still needs the use of a spare channel rather than the AV option. Always check what happens when equipment is used, because, for example, some DVD recorders do not release the SCART cable, so that the user must return from AV to television manually.

## Maintenance

System servicing and maintenance normally operates within the bounds of two extremes: 'If it ain't broke, don't fix it' and 'If anything can go wrong it will and always at the most inconvenient time'. This section deals with the type of maintenance needed for equipment that is used commercially. If a television receiver or CD player fails, it is nothing more than an inconvenience, but when the monitor or hard drive of a computer control system goes down, the effect can become catastrophic.

Unexpected failures within a production or control environment can create loss of production or service, which leads to loss of customer confidence. This is most important in **'just in time' (JIT)** applications where there is virtually no buffer stock of components to allow production to continue. Out-of-hours repairs by highly qualified service technicians/engineers, not only add to system/production costs, but also increase the health and safety hazards. Responding to unexpected failures is often described as 'fire-fighting'. It is better not to have the fire.

The basis for a **planned maintenance** scheme is built on the system manufacturer's recommendations and requires a significant constant retraining element to ensure that all service personnel can respond quickly and accurately to any indication that a failure is imminent. An early response and a temporary fix can often allow proper maintenance to be initiated at a more convenient slack period.

Service documentation for a system needs to be constantly updated in line with operation experience. Data logging is therefore essential and many computer software packages are available to support this effort. Over a period of time it becomes possible to predict areas of the system that need more regular service attention, so historical records become very important. System downtime costs money and loses orders.

For the larger systems that operate on a 24-hours-a-day basis, the annual holiday period with its complete shutdown provides something of a respite for service personnel. Maintenance is then not just a case of fire-fighting. Smaller systems may have to be serviced overnight. Maintenance of a major nature thus has to be fitted in with predicted non-operation times.

When an unplanned halt occurs, it may provide a useful period in which to carry out service on other parts of the system, particularly those areas next due for planned maintenance, thus making use of such downtime to minimize the duration of planned breaks at a later date.

Planned maintenance normally operates on a time basis and many machines actually carry a run time meter to help. Some areas need daily attention for, say, lubrication and running adjustments, while other areas need daily, weekly or even annual attention. With such a variability in a system maintenance scheme, it is important that a readily available and up-to-date database is maintained. The recording of failures should lead to the pinpointing of stock faults which may then be minimized by more regular servicing.

The most important features of a planned maintenance scheme can now be restated and include:

- obtaining a thorough understanding of the system operation
- continuous retraining for personnel
- accurate maintenance of historical records
- management plan modified in the light of experience
- ensuring that maintenance routines are written in a clear and understandable manner.

## Multiple-choice revision questions

25.1 If you cannot eliminate or replace a reported hazard you should:
(a) avoid the hazard
(b) guard or mark the hazard
(c) tell your friends
(d) ignore the hazard.

25.2 If you have to work on live equipment without gloves, you should:
(a) make sure you are earthed
(b) use battery-powered test equipment
(c) remove fuses
(d) work with one hand only.

25.3 A cable has been damaged by contact with a soldering iron. You should:
(a) wrap insulating tape round the damaged part
(b) put silicone varnish over the damaged part
(c) replace the cable immediately
(d) put a warning label on the cable.

25.4 A fire extinguisher is coloured red. It must NOT be used on:
(a) a paper fire
(b) an electrical fire
(c) a wood fire
(d) a cloth fire.

25.5 The most common injury causing absence from work is:
(a) electrocution
(b) poisoning
(c) gassing
(d) back injury.

25.6 A colleague has touched a live wire and is unconscious. You must **first**:
(a) turn off the power supply
(b) apply mouth-to-mouth resuscitation
(c) pull the victim away from the wire
(d) administer brandy.

# Answers to multiple-choice questions

| | |
|---|---|
| **Chapter 1** | 1.1(c); 1.2(b); 1.3(d); 1.4(a); 1.5(d); 1.6(a) |
| **Chapter 2** | 2.1(b); 2.2(a); 2.3(b); 2.4(b); 2.5(d); 2.6(a) |
| **Chapter 3** | 3.1(d); 3.2(b); 3.3(a); 3.4(d); 3.5(c); 3.6(c) |
| **Chapter 4** | 4.1(c); 4.2(b); 4.3(c); 4.4(a); 4.5(a); 4.6(c) |
| **Chapter 5** | 5.1(a); 5.2(c); 5.3(b); 5.4(d); 5.5(d); 5.6(c) |
| **Chapter 6** | 6.1(b); 6.2(d); 6.3(d); 6.4(b); 6.5(c); 6.6(b) |
| **Chapter 7** | 7.1(b); 7.2(a); 7.3(d); 7.4(d); 7.5(b); 7.6(a) |
| **Chapter 8** | 8.1(d); 8.2(b); 8.3(a); 8.4(c); 8.5(b); 8.6(d) |
| **Chapter 9** | 9.1(c); 9.2(b); 9.3(d); 9.4(a); 9.5(a); 9.6(c) |
| **Chapter 10** | 10.1(c); 10.2(d); 10.3(b); 10.4(a); 10.5(c); 10.6(d) |
| **Chapter 11** | 11.1(c); 11.2(d); 11.3(d); 11.4(b); 11.5(c); 11.6(a) |
| **Chapter 12** | 12.1(b); 12.2(d); 12.3(c); 12.4(c) |
| **Chapter 13** | 13.1(a); 13.2(c); 13.3(d); 13.4(a); 13.5(d); 13.6(c) |
| **Chapter 14** | 14.1(b); 14.2(a); 14.3(c); 14.4(d) |
| **Chapter 15** | 15.1(c); 15.2(b); 15.3(a); 15.4(b) |
| **Chapter 16** | 16.1(a); 16.2(c); 16.3(d); 16.4(d); 16.5(b); 16.6(c); 16.7(c) |
| **Chapter 17** | 17.1(b); 17.2(c); 17.3(c); 17.4(b) |
| **Chapter 18** | 18.1(c); 18.2(a); 18.3(c); 18.4(c); 18.5(b) |
| **Chapter 19** | 19.1(b); 19.2(b); 19.3(d); 19.4(c); 19.5(a); 19.6(c); 19.7(a) |
| **Chapter 20** | 20.1(d); 20.2(b); 20.3(a); 20.4(c); 20.5(d); 20.6(b) |
| **Chapter 21** | 21.1(b); 21.2(c); 21.3(d); 21.4(b); 21.5(d); 21.6(a) |
| **Chapter 22** | 22.1(b); 22.2(c); 22.3(b); 22.4(d); 22.5(b); 22.6(c) |
| **Chapter 23** | 23.1(b); 23.2(c); 23.3(d); 23.4(c); 23.5(a); 23.6(c) |
| **Chapter 24** | 24.1(b); 24.2(d); 24.3(b); 24.4(c); 24.5(b); 24.6(c) |
| **Chapter 25** | 25.1(b); 25.2(d); 25.3(c); 25.4(b); 25.5(d); 25.6(a) |

# Index